# SOLID-STATE DEVICES AND CIRCUITS

**S.P. BALI**
Former Faculty Member
Military College of Electronics
& Mechanical Engineering
Secunderabad, (A.P.)

PUBLISHING FOR ONE WORLD
**NEW AGE INTERNATIONAL (P) LIMITED, PUBLISHERS**
New Delhi · Bangalore · Chennai · Cochin · Guwahati · Hyderabad
Jalandhar · Kolkata · Lucknow · Mumbai · Ranchi
Visit us at **www.newagepublishers.com**

Published by New Age International (P) Ltd., Publishers
First Edition: 1995
Reprint: 2009

---



***Branches:***

- 36, Malikarjuna Temple Street, Opp. ICWA, Basavanagudi, **Bangalore**. ✆ (080) 26677815
- 26, Damodaran Street, T. Nagar, **Chennai**. ✆ (044) 24353401
- Hemsen Complex, Mohd. Shah Road, Paltan Bazar, Near Starline Hotel, **Guwahati**. ✆ (0361) 2543669
- No. 105, 1st Floor, Madhiray Kaveri Tower, 3-2-19, Azam Jahi Road, Nimboliadda, **Hyderabad**. ✆ (040) 24652456
- RDB Chambers (Formerly Lotus Cinema ) 106A, Ist Floor, S.N. Banerjee Road, **Kolkata**. ✆ (033) 22275247
- 18, Madan Mohan Malviya Marg, **Lucknow**. ✆ (0522) 2209578
- 142C, Victor House, Ground Floor, N.M. Joshi Marg, Lower Parel, **Mumbai**. ✆ (022) 24927869
- 22, Golden House, Daryaganj, **New Delhi**. ✆ (011) 23262370, 23262368

---

**ISBN (10): 81-224-0694-7**

**ISBN (13): 978-81-224-0694-8**

**Rs. 195.00**

**C-09-01-3302**

Printed in India at A.H. Print Pack, Delhi.

**PUBLISHING FOR ONE WORLD**
**NEW AGE INTERNATIONAL (P) LIMITED, PUBLISHERS**
4835/24, Ansari Road, Daryaganj, New Delhi-110002
Visit us at **www.newagepublishers.com**

*This book is dedicated to*

*All those scientists, technocrats and engineers who have endeavoured in the field of electronics to bring it to the present state of the art and to make the short span of human life all the more enjoyable.*

# PREFACE

This book is the outcome of over three decades of association with the field of electrical communication. Having been associated with the training of electronics (Radio and Line) technicians at apprentice and three year diploma level (Armament Artificer) in Military College of Electronics and Mechanical Engineering, Secunderabad, and also with vocational training of new entrants in the field of electronics, I have strongly felt the need of a 'state of the art' text that bridges the long maintained traditional gap between theory and practice. It is with this thought in mind that this book was written.

Intended primarily for use at technical college level, as an introduction to the theory and practice of semiconductor (solid-state) devices and circuits, and also suitable for secondary school (10+2) use, as a reference for students of advanced technology, and for adult education, the relevant topics are explained in words and descriptions free of cumbersome mathematics. The only prerequisite is a basic and workable knowledge of ac and dc theory.

The book is divided into three parts covered in fifteen chapters. Part 1 of the book deals with *devices*. Starting with a straight forward and typical explanation of atomic structure and its closely associated electrical concepts, intrinsic and extrinsic semiconductors and the versatile diode action in semiconductor devices, its relevance in practical rectifying and amplifying circuits is explained in Chapter 1. The student is then introduced to special semiconductor devices (tunnel diode, field-effect transistor and thyristor). The transition from transistor to integrated circuits (ICs) and the principles and construction of ICs are then explained in Chapter 2. A brief discussion of functional devices, which employ the physical effects on the volume of semiconductors, is given in Chapter 3. Charge-coupled devices, which form the basis of solid-state photoelectric imaging devices, are discussed in Chapter 4. Optoelectronic devices, which include emitters, sensors and couplers are dealt with in detail in Chapter 5 of the book. Chapter 6, covers sensors, information measurement elements that have functions with information conversion as their object. An elementary coverage of microprocessors and microcomputors is given in Chapter 7.

Part 2 of the book covers a wide range of *circuits* in which the devices covered in Part 1 find practical application. Different types of power supplies, both regulated and unregulated, power supply filters and switching regulators are given adequate coverage in Chapter 1. A very wide range of analog circuits is discussed in Chapter 2 with special emphasis on differential and operational amplifiers, relaxation oscillators, amplitude and frequency modulators and detectors, and frequency changers and translators. The modern trend is towards an ever increasing use of digital circuits. These are covered in Chapter 3 and include logic gates and logic circuits, binary, decade and ring counters, encoders and decoders, frequency dividers and digital clocks. Electronic control exploits the switching capabilities of solid-state devices. The practical application of these devices in control circuits is discussed in Chapter 4.

Part 3 of the book, *Solid-State Clinic*, deals entirely with the practical aspect of devices and circuits — the so called bridge between theory and practice. During the last decade, there has been an increasing trend towards replacement, rather than repair, of functional parts in electronic equipment. Any one can change parts. The forte of a (technically) mature technician is his ability to rapidly track down a defect. This is where most of the technicians falter — 'start on the wrong foot'. The basic philosophy of trouble shooting is discussed in Chapter 1. Service Aids necessary to perform alignment and repair of 'state of the art' solid-state equipment are given a wide coverage in Chapter 2. A simple and logical explanation of 'how-to-service' solid-state devices and circuits is given in Chapter 3. Technical tips— 'do's and don'ts, or tricks of the trade, save a lot of valuable time and avoidable headache for the technician. These are detailed in Chapter 4.

One illustration tells more than ten thousand words. Keeping this in mind, the text is supplemented by a copious amount of illustrations dispersed throughout the book, each illustration intended to bring home an idea and to build understanding step-by-step.

No author writes a book without help, especially a book like this. I would like to thank the various establishments for being kind enough to grant permission, where applicable, to reproduce extracts from copyright material.

Suggestions for improvement of the text will be thankfully acknowledged.

S.P. Bali

# CONTENTS

# PART 1
# SOLID-STATE DEVICES

1. SOLID-STATE DEVICES
2. INTEGRATED CIRCUITS
3. FUNCTIONAL DEVICES
4. CHARGE-COUPLED DEVICES
5. OPTOELECTRONIC DEVICES
6. SENSORS
7. MICROPROCESSORS

# CHAPTER 1

# SOLID-STATE DEVICES

*The worldwide changeover from valves to solid-state devices has marked a turning point in the trend of electronic design. The latter have played and still continue to play, a crucial part in the modern development of ultra miniature and integrated circuitry. Valve technique never really lent itself to this trend and would have tended to retard development along these lines, rather than to facilitate it. Transistors, diodes and similar devices are by their nature very suitable for this technique. Transistor and its various 'cousins' in the semiconductor 'family' form the subject of study in this chapter.*

## 1.1 Atomic Structure

In a much simplified picture, the atom consists of a central nucleus containing positively charged protons, around which rotate a number of negatively charged electrons. The negative charge of the electron is exactly equal and opposite to the positive charge of the proton. In a neutral atom, the number of electrons orbiting the nucleus is equal to the number of positively charged protons in the nucleus.

Different elements have different number of electrons and those electrons do not all rotate at the same distance from the nucleus, but in several separate rings, as can be seen in Fig. 1.1. The electrons located in the rings close to the nucleus are *tightly bound*, where as those in the outer-most ring are comparatively *loosely bound* and can, under suitable circumstances, be dislodged from the atom. The properties of an element are largely determined by the conditions prevailing in it's outer ring of electrons. The outer ring is called the *valence ring* and the electrons in it the *valence electrons*. These are normally the only electrons entering into chemical reactions and so it is convenient to imagine a simplified picture of the atom as a central nucleus with the valence electrons in one ring around it.

## 1.2 Conductors and Insulators

Materials having the valence electrons *very loosely bound* to the atom are good *conductors* of electricity. Obvious examples are silver, copper, gold and aluminium, in decreasing order of electrical conductivity. This is shown in Fig. 1.1.

At room temperature the thermal energy present in conductors (copper for example) is sufficient to cause approximately one electron to become detached from each atom. In an isolated bar of copper, these free electrons will move in a random haphazard fashion. If a voltage is applied across the ends of the bar, the electrons will be attracted towards the positive end and a current will flow. The electrons

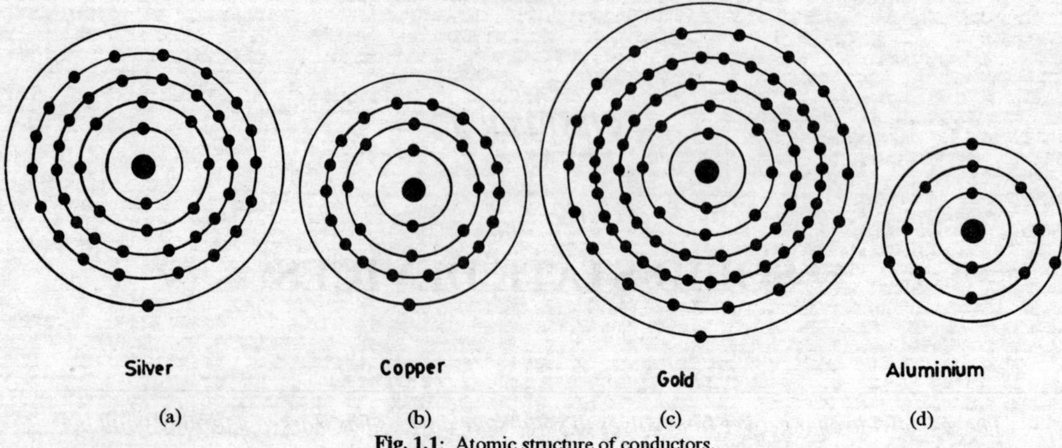


**Fig. 1.1:** Atomic structure of conductors.

will flow out of the bar to the battery at the positive end and an equal number will flow from the battery into the negative end of the bar so that the number of electrons in the bar is always the same. When an electron becomes detached from an atom, the atom is left with a net positive charge, and even though there are many free negative electrons in the bar, *it does not become electrically charged* as there are an equal number of positively charged atoms.

Thus we see that in metals which are good conductors of electricity, there normally exist a number of electrons which are free to move when an electrical potential is applied.

*Insulators* are materials in which the valence electrons are *tightly bound* to the nucleus, and need a large amount of energy to release them. There will normally be only a very small number of free electrons in the material and so practically no current will flow when a voltage is applied.

## 1.3 Semiconductors

Between the two major categories of conductors and insulators is the class of materials called *semiconductors.* As the name implies, a semiconductor is a material with resistivity roughly midway between that of conductors and insulators. However, a semiconductor is not just a poor conductor; it has two other important properties. First, *its resistance normally decreases with increase of temperature*. Second, *flow of current in a semiconductor can be by two mechanisms*; by a flow of negative electrons as in the case of conductors, or by a process known as a flow of positive *holes* in the opposite direction. This concept of hole flow is most important in the understanding of semiconductors.

In many solid materials the atoms form themselves into a regular shape called a *crystal*. The action of the valence electrons holds the crystal together. The valence electrons are to a certain extent the glue that holds the atoms together in the crystal. The inner electrons on the other hand remain bound in orbits to their respective atomic nucleus. The glue of the valence electrons remains to a certain extent fluid. This means that the individual valence electrons in a metal can move about freely in the crystal although at the same time they also bind the crystal together. Thus the valence electrons in metals are also *free electrons*.

The valence electrons in semiconductors and insulators are not normally free to move about; these valence electrons are *bound electrons*.

In a particularly important type of crystal structure, the *valence crystal*, a bond between two neighboring atoms is always produced by a pair of valence electrons, one valence electron in each bond being supplied by each atom. The number of double (*covalent*) bonds between an atom and its immediate neighbors is the same as the number of valence electrons the atom originally had.

Semiconductor materials which are currently of interest, silicon and germanium, form covalent bonds (see Fig. 1.2). In both cases there are four valence electrons per atom which means four bonds per atom in the crystal.

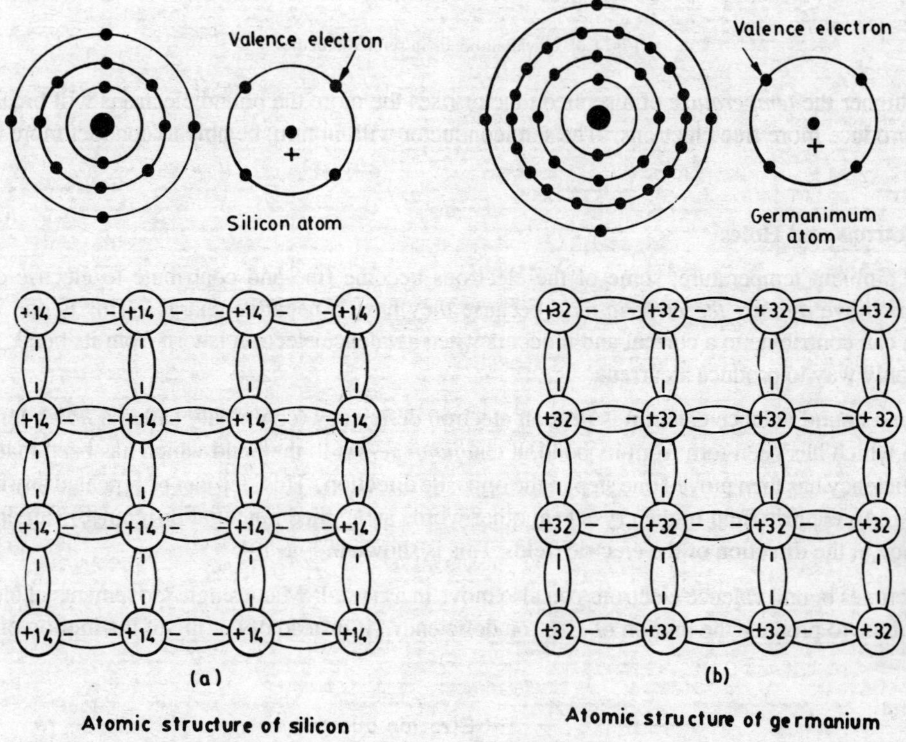


**Fig. 1.2:** Atomic structure oi semiconductors.

The valence electrons in semiconductor and insulating materials cannot normally carry current. However, *a valence electron can be turned into a free electron by an external influence.* For this to occur, the bound electron must be torn from its bond by a powerful force.

The simplest method, however, of turning bonded electrons in a semiconductor into free electrons, is to heat the crystal. The atoms will then go into increasingly violent oscillations, straining thereby at the atomic bonds. At some point the bond will be torn, which means that the valence electron concerned will become free. In this case the energy required to tear the bond is not supplied by an *external particle*, but by the *crystal itself*.

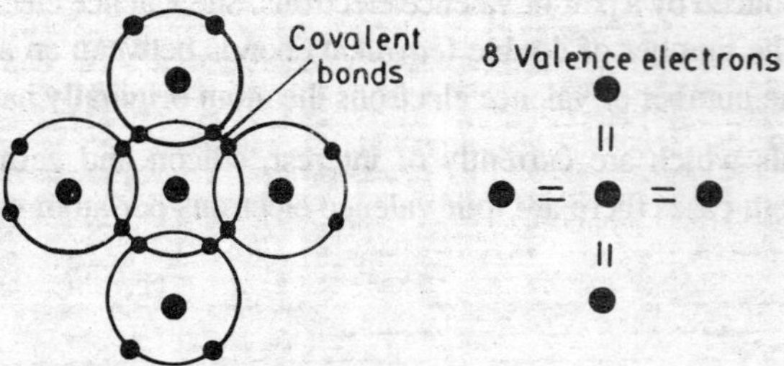


**Fig. 1.3:** Covalent bonds in semiconductors.

The higher the *temperature* of a semiconductor rises the more the bound electrons will break loose and so produce more free electrons. The semiconductor will, in turn, be able to conduct more electric current.

## 1.4 Electrons and Holes

Even at ambient temperature, some of the electrons become free and contribute to electric current. Electrons move *against the electric field* because they have a negative charge. This is one way an electron can contribute to a current and it occurs when a valence electron is torn from its bond. This is not the only way to produce a current.

When a bound valence electron is near an electron deficiency (commonly called a *hole*), left by an electron which has been torn from its bond, it can *jump across* to the bond which has been made free. This deficiency has then moved one step in the opposite direction. This step can be repeated any number of times. An electric field in the crystal (in other words an applied potential difference) can drive the deficiency in the direction of the electric field. This is shown in Fig. 1.4.

Sometimes bound valence electrons can also move in a crystal. Many single movements of electrons can combine to produce the motion of electron deficiency. If instead of talking of the motion of bound

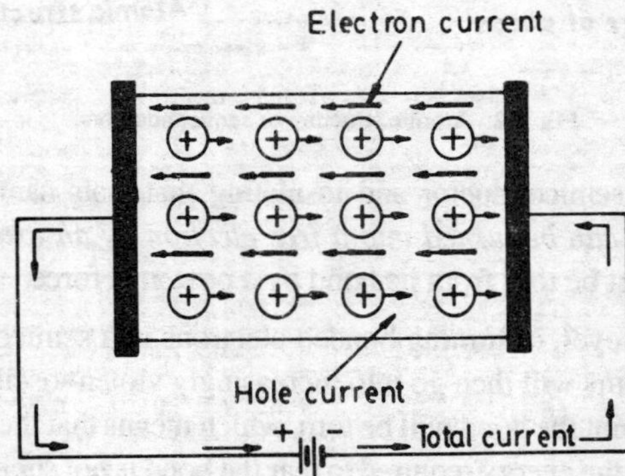


**Fig. 1.4:** Electron and hole currents in semiconductors.

valence electrons, we talk in terms of the motion of electron deficiency, the description of this state of affairs becomes easier to understand and we consider the deficiency as a particle in its own right and call it a *missing electron* or simply a *hole*. The hole is a positively charged particle, since it moves *in the direction of the electric field.*

## 1.5 Intrinsic Conduction

Electric conduction in semiconductors can be attributed to the production of *pairs of free electrons and holes* as the result of heating. This type of conduction is called *intrinsic conduction* and it *rises sharply with temperature*. The magnitude of this conduction at a specific temperature is an important characteristic of semiconductor materials.

There are types of conduction which are not dependent upon the semiconductor itself but upon foreign atoms (called *dopants*) which have been introduced into the crystal.

If all the electrons in a semiconductor crystal, which were torn free by thermal agitation, were to remain free, after a certain time all the valence electrons will be turned into free electrons.

In addition to the tearing of electrons from their bonds, there is also a reverse process. It is not only possible to produce a free electron and a hole from a valence electron but also a valence electron from a free electron and a hole. There always exists a certain probability that, when a free electron passes a hole, it will *fall* in after which it is once more bound and can no longer contribute to an electric current. This 'falling in' is called *recombination.*

At a specific temperature, in a given period of time, the number of recombinations is the same as the number of new electron-hole pairs created. In this way the number of electrons and holes remains constant.

## 1.6 n-Type Semiconductors

There is a completely different method of producing conduction from that discussed so far. This can be achieved by placing a suitable foreign or impurity atom into the crystal, for example a phosphorus atom in a germanium crystal. Phosphorus does not have four valence electrons like germanium, but five (see Fig. 1.5). This would correspond to a free electron in the crystal which had not been torn from a bond and therefore had not produced a hole.

The fifth valence electron is not needed for bonding and can tear free from the phosphorus atom very easily. It then becomes a free electron. Even at room temperature there are as many free electrons in the germanium crystal as there are phosphorus atoms.

An impurity atom, like the one above, is also called fault point. Fault points can also be produced using agents other than impurity atoms. A fault point that generates one or more free electrons is called a *donor*. Donors produce in a semiconductor crystal extra electrons, or in other words *n-type conduction* (from the negative charge on the electron).

Introducing an impurity atom into a crystal is called doping, which in the case of donors is *n-doping*. Doping is technically the most important method of increasing conduction in semiconductors, as conduction produced by doping, unlike intrinsic conduction, is *only marginally dependent upon temperature*. At room temperature, the intrinsic conduction is appreciably less than that produced by doping.

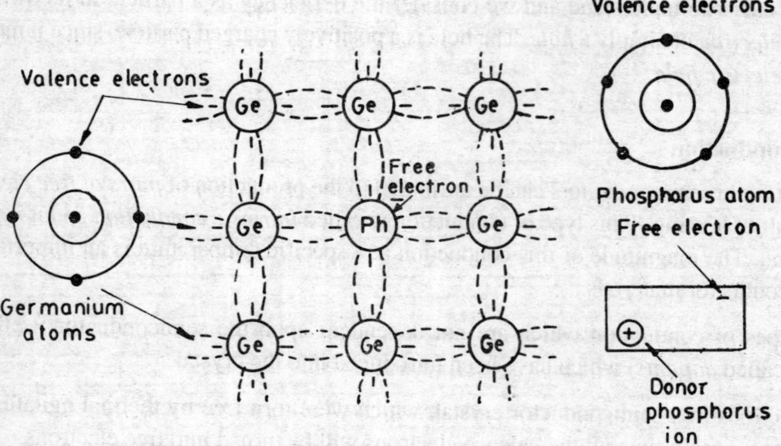


Fig. 1.5: n-Type semiconductors.

The more the impurity atoms in a semiconductor, the higher the temperature must be before intrinsic conduction becomes the major influence.

## 1.7 p-Type Semiconductors

There is yet another important method of producing conduction by doping. This is achieved by placing a suitable foreign or impurity atom into the crystal, for example an indium atom into a germanium crystal. Indium has only three valence electrons, which means that one of the bonds to the neighboring atoms is missing. In other words, there is a hole which, just like the extra electron of the phosphorus atom, moves freely even at room temperature. There are then, in the germanium crystal, as many holes for current conduction as there are indium atoms.

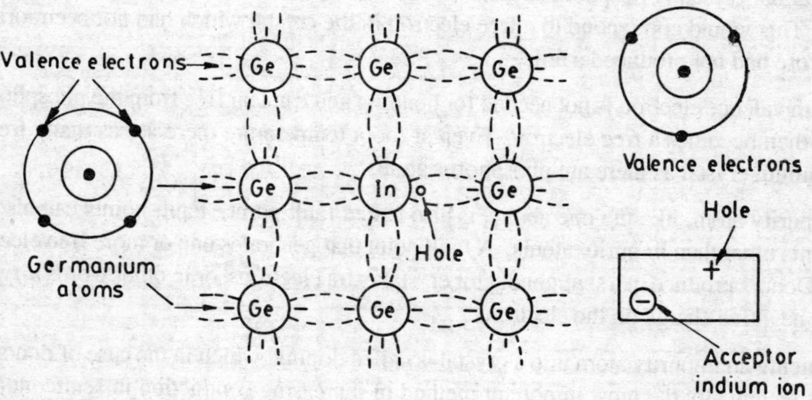


Fig. 1.6: p-Type semiconductors.

The indium atom is also a fault point in the germanium crystal. A fault point which generates one or more holes, or looked another way, absorbs one or more bound electrons, is called an *acceptor*. Acceptors produce deficiency conduction in semiconductor crystals (because of the deficiency of electrons), in other words *p-type conduction* (because of the positive charge of a hole). The crystal is thus *p-doped.*

In the case of conduction produced by doping, either p- or n- type, the number of holes or electrons is the same as the number of impurity atoms, at least when only one type of atom is involved. It follows that conduction can be made very small when the crystal is very pure, assuming that temperature does not produce too high an intrinsic conduction. On the other hand, *the introduction of a proportionately small impurity increases conduction significantly.*

This is completely opposite to the behavior of a metal. In a metal there are at least as many free electrons as there are metal atoms and this produces a high conduction which cannot be noticeably influenced.

### 1.8 Extrinsic Conduction

We cannot get enough useful current from an intrinsic semiconductor. One way to increase conductivity is by doping. Doped semiconductors are known as extrinsic semiconductors. The phenomenon of conduction of current through extrinsic semiconductors is referred to as *extrinsic conduction.*

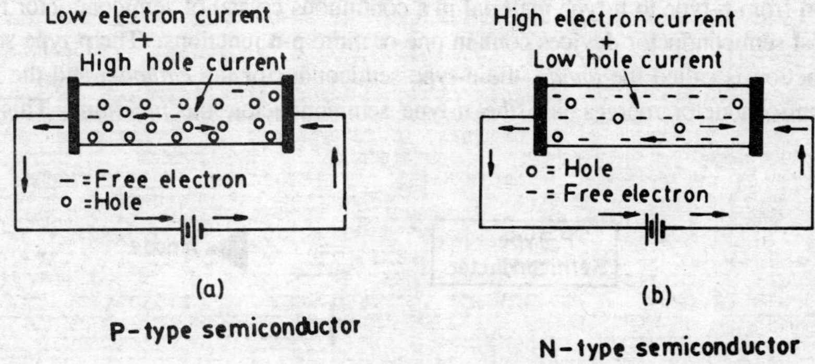


**Fig. 1.7:** Electric current in extrinsic semiconductors.

By doping, an excess of electrons over holes, or vice versa, can be created. In the case of p-type material doped with an acceptor impurity, there is an excess of holes, which are therefore called *majority carriers* in such material. With n-type material, doped with donor impurity, there is an excess of electrons, and in this case the electrons are called majority carriers.

When a p-n junction is formed, some of the majority carriers in each section diffuse across the junction, thereby becoming *minority carriers*, and eventually neutralise each other. The action of semiconductor devices is largely based on the effects of biasing p-n junctions so as to *control* the currents flowing across them.

## 1.9 p-n Junction

If a source of EMF is connected across a doped semiconductor, the following sequence of events takes place (refer to Fig. 1.7).

(1) Let us suppose that under the influence of temperature;

(1) one covalent bond is broken;

(2) two charge carriers are generated;

(3) the free electron moves towards the positive electrode of the battery; and

(4) an electron from the negative electrode of the battery moves towards the right side of the extrinsic semiconductor to make good the loss of the electron and combines with a hole generated by the breaking of the covalent bond.

The *electron current* is confined to the connecting wires and the source of EMF, whereas both the electron and hole currents flow within the extrinsic semiconductor. The *hole current* is thus confined to the boundary of the extrinsic semiconductor.

By itself, a piece of n-type semiconductor is purely resistive, that is, reversing the battery connected across it will merely reverse the direction of current flow without affecting its amplitude. In contrast, a piece of semiconductor with a p-n junction in it has *rectifying properties* — when a voltage is applied in one direction current flows, but if the voltage is reversed no current flows.

A transition from p-type to n-type material in a continuous crystal of semiconductor is called a p-n junction. Most semiconductor devices contain one or more p-n junctions. The p-type semiconductor of the p-n junction is called the *anode*, the n-type semiconductor the *cathode*, and the region where the p-type semiconductor merges into the n-type semiconductor, the *junction*. This is shown in Fig. 1.8.

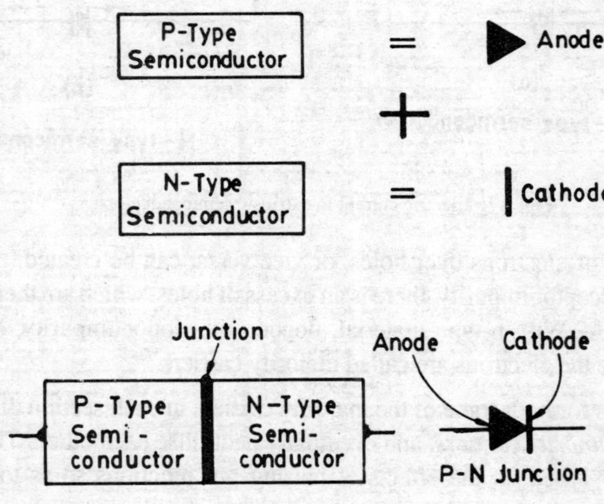


**Fig. 1.8**: p-n Junction.

## 1.10 Diffusion Current

In any crystal the atoms are oscillating, and these oscillations increase with increase in temperature. This not only tears valence electrons loose, but also makes free electrons (and holes) to wander around in a random manner, particularly in the absence of an external electric field.

These free electrons, like the molecules in a gas, try to spread themselves out uniformly. This movement of charge is an electric current, which is called a *diffusion current* and differs from that caused by an electric field, which is called a *conduction current*.

The following events take place during the formation of a p-n junction. Holes predominate in the p-type semiconductor whereas electrons are few in number. *Diffusion hole current* is set flowing from the p-semiconductor towards the n-semiconductor to bring about an equitable distribution of holes in the bulk of p-n semiconductor. As more and more holes diffuse from the p-semiconductor, it starts acquiring an increasing negative charge. This is shown in Fig. 1.9(a).

By the same token, electrons predominate in the n-type semiconductor. *Diffusion electron current* is set flowing from the n-type semiconductor to the p-semiconductor to bring about an equitable distribution of electrons in the bulk of p-n semiconductor. As more and more electrons diffuse from the n-semiconductor, it starts acquiring an increasing positive charge. This is shown in Fig. 1.9(b).

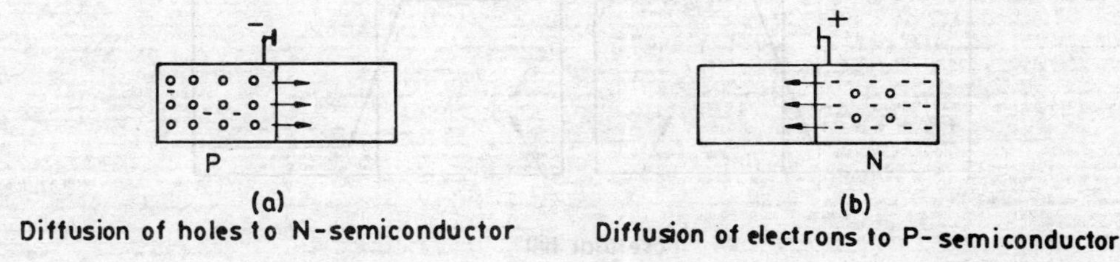


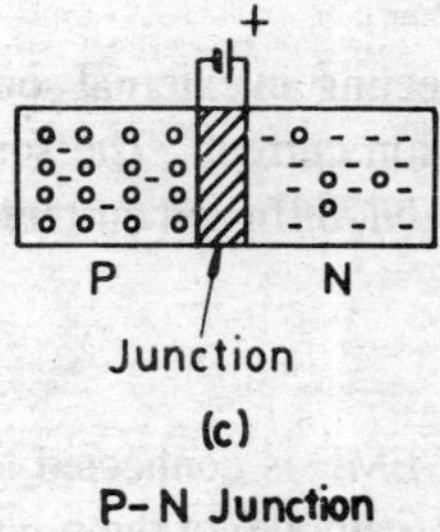


**Fig. 1.9:** Diffusion current in p-n- junction.

Electrons and holes combine at the boundary of the p-n junction. The area in the immediate vicinity of the p-n junction is thus depleted of charge carriers. This area is called *carrier depletion layer*, illustrated in Fig. 1.9(c). Carrier depletion layer acts as a barrier for transport of electric charge across the p-n junction and hence it can also be considered as a *potential hill* which has to be overcome by the charge carriers.

## 1.11 Potential Hill

Diffusion of holes and electrons across the p-n-junction continues for sometime till a state of equilibrium is reached. Diffusion of holes and electrons across the p-n- junction is then inhibited by the potential of the barrier or *potential hill*.

If the kinetic energy of the charge carriers is capable of overcoming the force of repulsion due to the barrier, diffusion current once again starts flowing. This continues for sometime and again a state of equilibrium is reached. This is shown in Fig. 1.10.

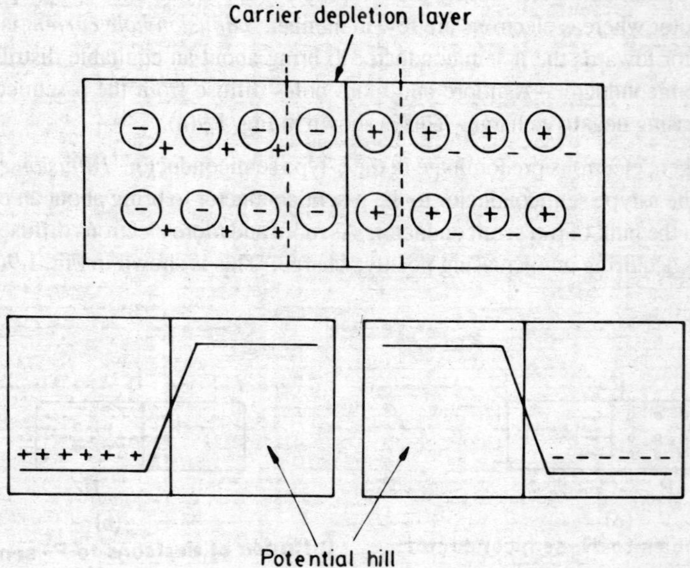


**Fig. 1.10:** Potential hill.

Potential hill can be overcome by connecting an external source of EMF across the junction in such a way as to induce further flow of diffusion current. The source of EMF can be connected in two different ways so as to *either assist, or inhibit*, diffusion current.

## 1.12 Forward Biasing a p-n Junction

If the positive electrode of the source of EMF is connected to the p-semiconductor(anode) and the negative electrode to the n-semiconductor (cathode) of the p-n junction, the *width* of the carrier depletion layer *decreases* and hence the potential barrier is *reduced* or overcome. Transport of electric charge can now take place across the junction. This is shown in Fig. 1.11.(a).

The p-n junction is now said to be *forward biased.* Maximum current flows in the forward direction. This is referred to as *forward current* and the p-n junction offers minimum resistance, referred to as *forward resistance*. When forward biased, a p-n junction acts as a *closed switch.*

A switch, when closed, has zero contact resistance. A forward biased p-n- junction does have *some resistance* even though it is minute.

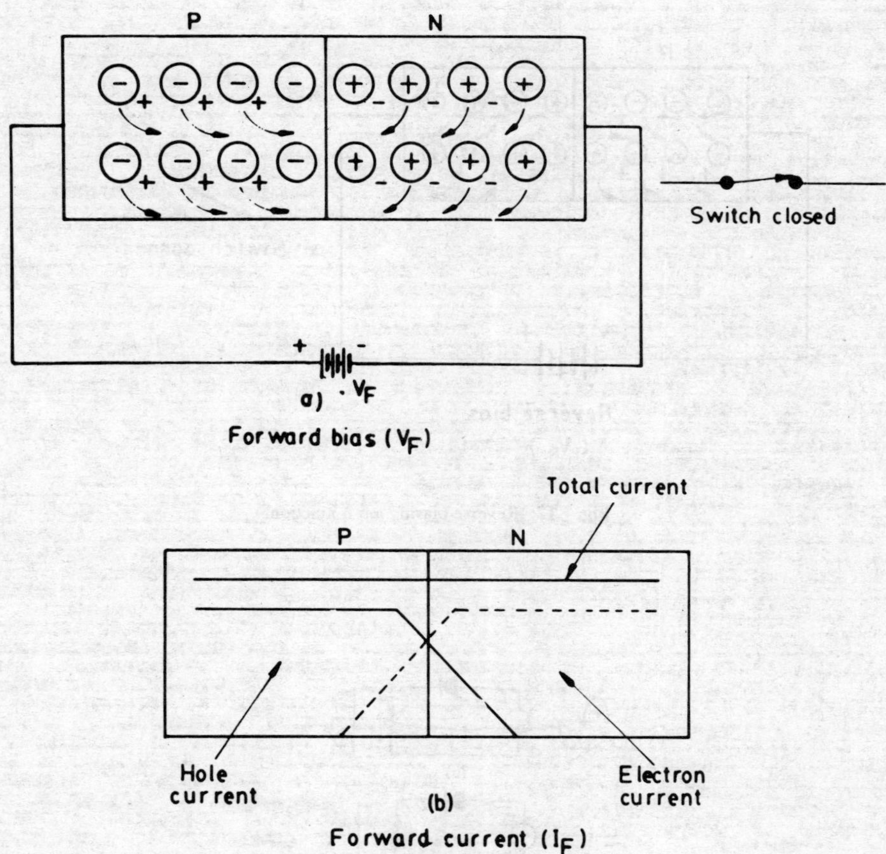


**Fig. 1.11:** Forward biasing a p-n junction.

## 1.13 Reverse Biasing a p-n Junction

If the external source of EMF is now inverted, the positive electrode connected to the n-semiconductor (cathode) and the negative electrode to the p-semiconductor(anode), the *width* of the carrier depletion layer *increases* and hence the potential barrier is still *further increased.* Transport of electric charge cannot take place across the junction. This is shown in Fig. 1.12.

The p-n junction is now said to be *reverse biased.* A very minute current leaks across the junction. This is referred to as *leakage current.* The p-n junction now offers maximum resistance, referred to as *reverse resistance.* When reverse biased, the p-n junction acts as an *open switch.*

A switch, when open, has infinite resistance. A reverse biased p-n junction has a *finite resistance*, even though it is extremely high.

## 1.14 Rectification

If an alternating voltage is applied across a p-n junction, the barrier presented by the junction to the flow of charge carriers will be *alternately* strengthened and weakened. In other words, the resistance of the

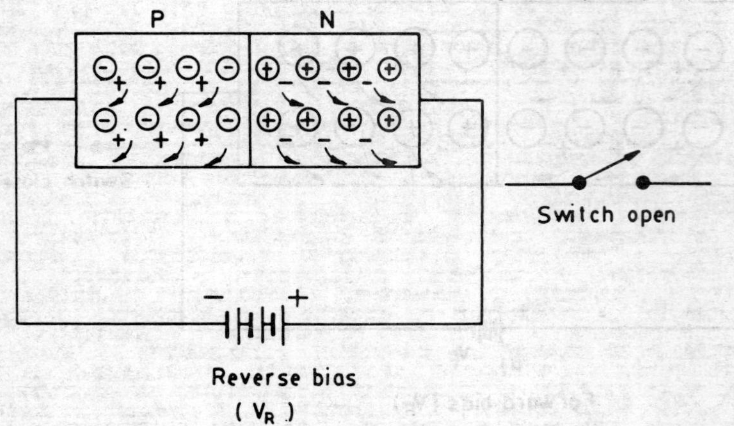


Fig. 1.12: Reverse biasing a p-n junction.

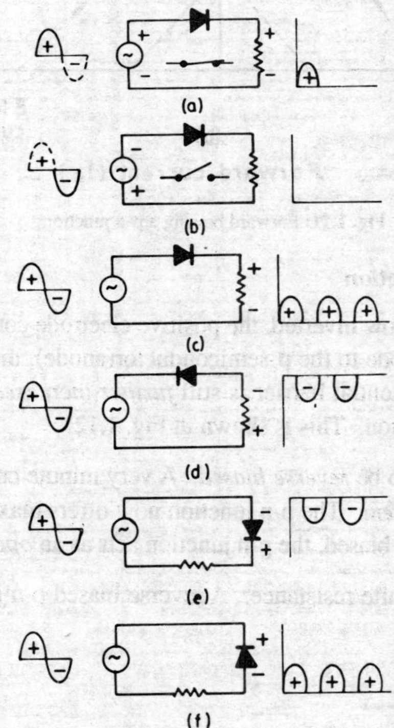


Fig. 1.13 Half-wave rectifying action of a junction diode.

junction to current flow in either direction depends on the *polarity* and *amplitude* of the voltage applied across it.

A rectifying action is thus obtained and the p-n- junction is simply referred to as a *junction diode* or *rectifier*. Note that the junction is a transition from p- to n-type semiconductor material within a continuous crystal structure.

The junction diode now behaves as a switch which operates at the *frequency* of the applied voltage. If a suitable load is connected across the diode, half-wave rectified pulses will develop across the load when the diode is *forward biased* and no output will develop across the load when the diode is *reverse biased.* This is shown in Fig. 1.13.

In any rectifying action positive-going output is obtained from the end of the load connected to the cathode. The output of the rectifier can be taken from across the load, or the rectifier itself, depending on requirement. *Output polarity* of a rectifier diode is further elaborated in Fig. 1.13.

## 1.15 Electrical Characteristics of a p-n Junction

*Silicon* p-n junctions used for power rectification, are bigger in size, and can withstand temperatures upto a maximum of 200°C. *Germanium* p-n junctions used for signal detection, are comparatively smaller in size, and can withstand temperatures upto a maximum of 110°C.

When the forward voltage across a p-n junction is zero, almost no current flows through the junction. With a slight increase in forward voltage, the forward current also increases, because the potential barrier is *partially overcome*. This change in current is not linearly related to the change in voltage. This is represented by the *lower nonlinear portion* of the v-i characteristic in Fig. 1.14(b).

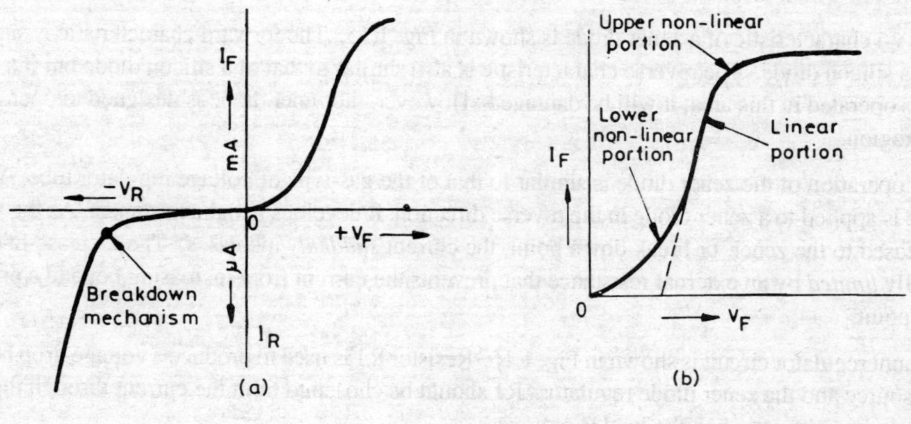


**Fig. 1.14:** v-i-characteristics of a p-n junction.

The forward voltage required to overcome the potential barrier is 0.1 to 0.3 volt in the case of germanium diodes, and from 0.3 to 1.0 volt in the case of silicon diodes.

If the forward voltage is increased still further, an *abrupt change* in forward current is obtained. This corresponds to the *linear portion* of the v-i characteristic.

Further increase in forward voltage brings about a state of *saturation* in forward current, wherein an increase in forward voltage hardly brings about a noticeable change in forward current. This is represented by the *upper nonlinear portion* of the v-i characteristic. Increasing forward voltage beyond this point will make itself manifest in the form of *increased junction temperature* without hardly any change in current.

When the reverse voltage across a p-n junction is initially increased, there is *hardly* any increase in reverse current. It is of the order of a few micro-($10^{-6}$) amperes. Minority carriers contribute to this current.

With further increase in reverse voltage, the minority carriers acquire sufficient kinetic energy, *collide* with the electrons in the covalent bonds of the adjacent atoms and make them free by breaking down the covalent bonds. This phenomenon is called *breakdown mechanism.* An appreciable increase in current is then brought about. If it happens at reverse voltages less than 6 volts, it is referred to as *zener breakdown,* whereas if it takes place at reverse voltages in excess of 6 volts, it is referred to as *avalanche breakdown.* This is shown in Fig. 1.14(c).

## 1.16. Zener Diode

The breakdown mechanism in reverse biased p-n junctions is exploited in the manufacture of *zener* diodes which have wide applications in electronics industry.

The zener diode has a normal forward conduction characteristic, a high resistance blocking state and a peak inverse voltage (PIV) rating. In conventional diodes, the PIV rating is never exceeded, but this is the *area* in which the zener diode is designed to operate.

The v-i characteristic of a zener diode is shown in Fig. 1.15. The forward characteristic is similar to that of a silicon diode. The reverse characteristic is also similar to that of a silicon diode but if a silicon diode is operated in this area, it will be damaged. However, the zener diode is designed to operate *only* in this region.

The operation of the zener diode is similar to that of the gas-type of voltage regulator tube. When a voltage is applied to a zener diode in the reverse direction, it develops a high resistance. As the voltage is increased to the zener, or break down point the current *suddenly* increases. The increase in current is usually *limited* by an external resistance that prevents the current from increasing beyond a predetermined point.

A shunt regulator circuit is shown in Fig. 1.16. Resistor R1 is used to produce a voltage drop between the dc source and the zener diode regulator. R1 should be chosen to limit the current through the diode to the rated maximum when the load is removed.

When the load is connected across the diode, the current will be divided between the diode and the load. The voltage across the diode and load will remain relatively constant over a specified range of *changing load conditions.*

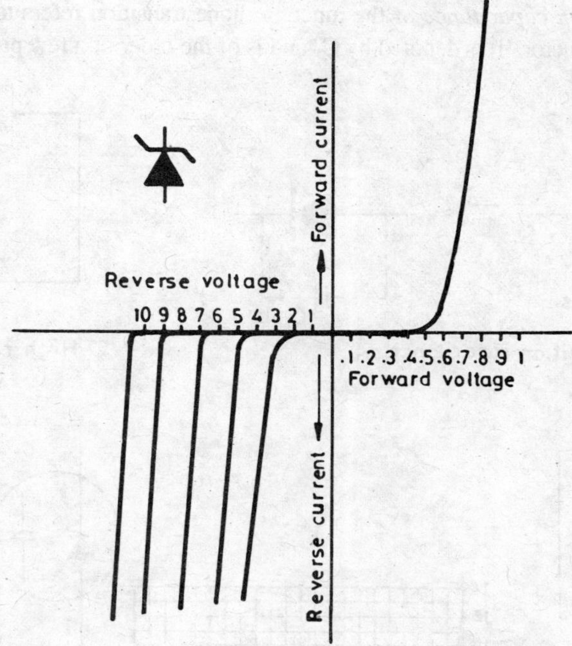


**Fig. 1.15:** v-i Characteristic for silicon zener diodes.

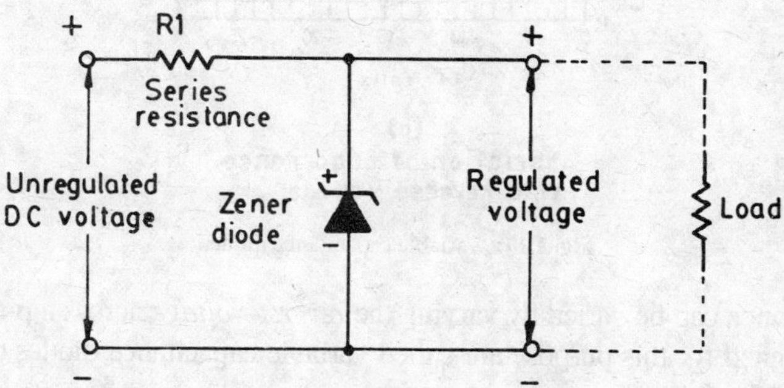


**Fig. 1.16:** Zener diode shunt regulator circuit.

The applications of the zener diode far exceed those for the gas-type voltage regulator because of the reduced size and the almost *limitless range of zener voltages* that are available.

## 1.17 Varactor Diode

The carrier depletion layer is, in fact, an insulator because it is depleted of charge carriers. It is comparable to the *dielectric medium* of a capacitor. The p-and n-type semiconductors on either side of the carrier depletion layer are conductors of electricity because they conduct current. These are comparable to the *plates* of a capacitor. Thus the p-n junction exhibits a certain amount of capacitance. This is referred to as the *transition capacitance* of the junction diode–transition refers to the transition from p-type to n-type semiconductor. It is denoted by $C_T$ and is of the order of a few pico ($10^{-12}$) Farads.

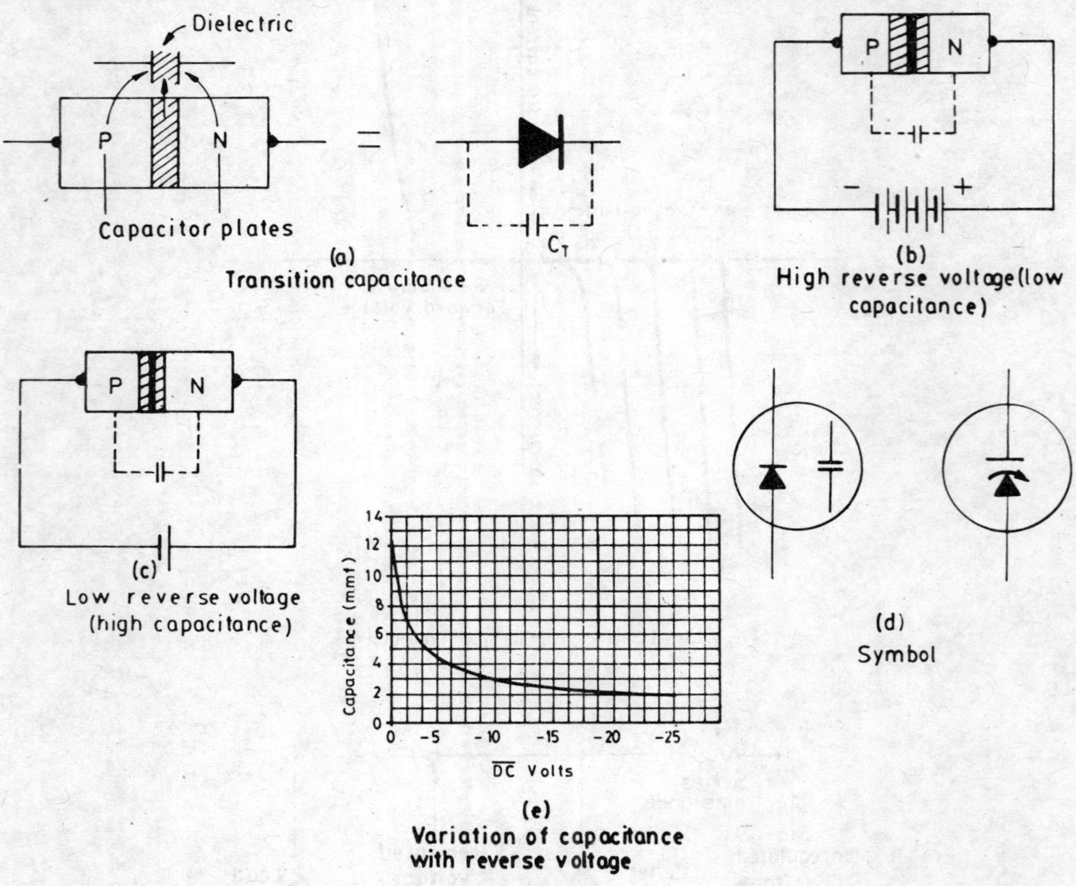


**Fig. 1.17:** Variable capacitance diode.

Transition capacitance can be varied by varying the *reverse voltage* across a p-n junction. Junction diodes specially designed for this purpose are called variable capacitance diodes or *varicap diodes* or *varactors* (see Fig. 1.17).

With an increase in reverse voltage the carrier depletion layer becomes *wider* and transition capacitance *decreases*. The converse is also true. This capacitance is *voltage controlled.*

Data sheets for varactors usually list a reference value of capacitance measured at a specified reverse voltage, typically –4 volt. For instance, the data sheet of a 1N5142 lists a reference capacitance of 15 pF at –4V. Data sheets also include a *tuning range* defined as:

$$TR = C_{max}/C_{min} \quad 1.1$$

over the specified voltage range of the varactor. For 1N5142 it is 3 : 1 for a voltage range of –4 to –60 V. This signifies that the capacitance decreases from 15 to 5 pF when the voltage varies from –4 to –60 V. *The tuning range of a varactor diode depends on the concentration of doping.*

### 1.18 Switching Diode

The switching characteristic of a p-n junction is shown in Fig. 1.18. Suppose that the forward conduction current has been allowed to flow through the diode junction from time $t_1$ to time $t_2$. At time $t_2$ the applied voltage is abruptly reversed in polarity resulting, what might be expected, in a relatively tiny reverse current. *The anticipated transition does occur, but not instantaneously.* Instead, the flow of a high reverse current occurs and persists until time $t_3$. From time $t_3$ to $t_4$, this current decays to its ultimate low value.

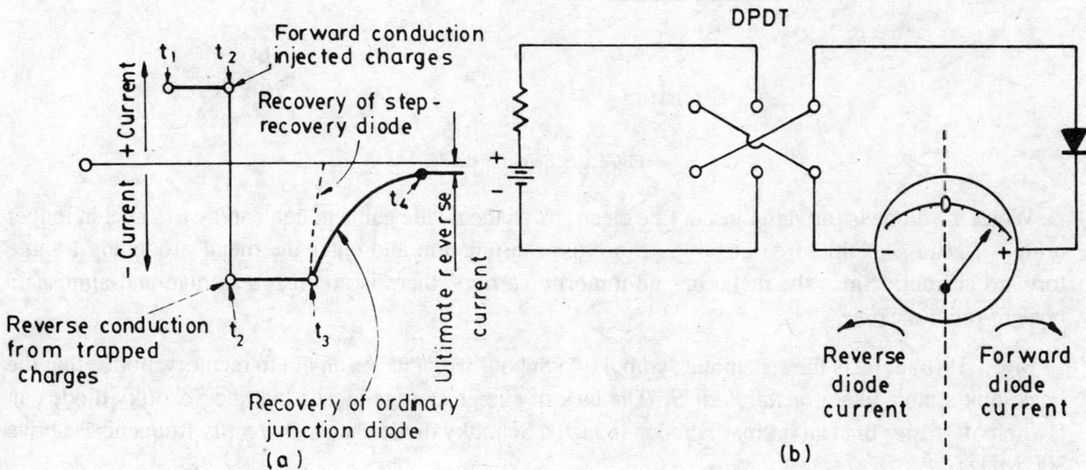


**Fig. 1.18:** Typical switching characteristics of junction diodes.

The flow of forward conduction current is attributed to the movement of minority charge carriers through the semiconductor material. These charge carriers are *trapped* in transit when the voltage impressed across the diode is reversed in polarity. They are, therefore, available to produce the temporary high reverse current. During the time these excess charges remain in the diode, *the junction is effectively forward biased despite the polarity of the applied terminal voltage.* These charges are, after a time, depleted due to collection and recombination. Then the junction becomes internally biased, and the reverse current falls to the small value due to thermal generation of charge carriers in the depletion region of the junction. The applied reverse voltage is not high enough to cause breakdown by zener or avalanche action.

### 1.19 Schottky Diode

At lower frequencies rectifier diodes can easily turn off when the operation changes from forward bias to reverse bias. As the frequency increases, a point is reached where the ordinary diode cannot turn *fast enough* to prevent significant reverse current. This means that *an ordinary diode no longer is an efficient rectifier at high frequencies* because it remains 'closed' during the initial part of the reverse cycle.

A Schottky diode is a special purpose diode that uses a metal like gold, silver, or platinum on one side of the junction and doped silicon(usually n-type) on the other side as shown in Fig. 1.19(a). When a Schottky diode is unbiased, free electrons on the n-side are in smaller orbits than free electrons on the metal side. This difference in orbit size or energy level is called the *Schottky barrier*.

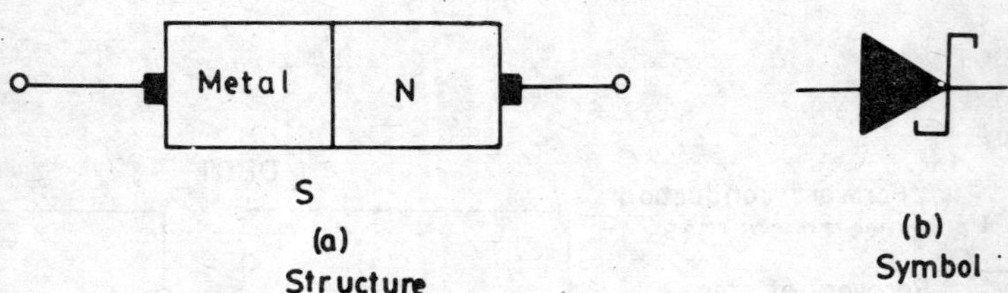


**Fig. 1.19:** Schottky diode.

When the diode is forward biased, free electrons on the n-side gain enough energy to travel in longer orbits. Because of this, free electrons can cross the junction and enter the metal producing a large forward current. Since the metal has no minority carriers, there is no charge storage and almost no *reverse-recovery time.*

Fig. 1.19(b) depicts the schematic symbol of a Schottky diode. As an aid to memory, notice that the lines look almost like a *rectangular S.* The lack of charge storage means that the Schottky diode can 'switch off' *faster* than an ordinary diode. In fact, a Schottky diode can easily rectify frequencies above 300 MHz.

An important application of Schottky diodes is in digital computers. The speed of computers depends on how fast their diodes and other devices can turn 'on' and 'off'. This is where the Schottky diode is useful.

## 1.20 Tunnel Diode:

If the concentration of impurity atoms is greatly increased (by about 1000 times) in a p-n- junction, the characteristics are drastically changed, yielding a tunnel diode. Initially called the *Esaki diode* after its inventor, who developed it in 1958, this diode's characteristics are shown in Fig. 1.20. *The explanation of the tunnel diode volt-ampere characteristic is found in the zener effect.*

A zener diode breaks down at a very small voltage (from 2 to 6 V) due to heavy doping, causing a very narrow depletion region, in which the zener effect is responsible for reverse current. *If the concentration of doping is increased still further, there is no reverse blocking, and the diode appears to be broken down for any reverse voltage.* This increased doping obviously affects the forward current also.

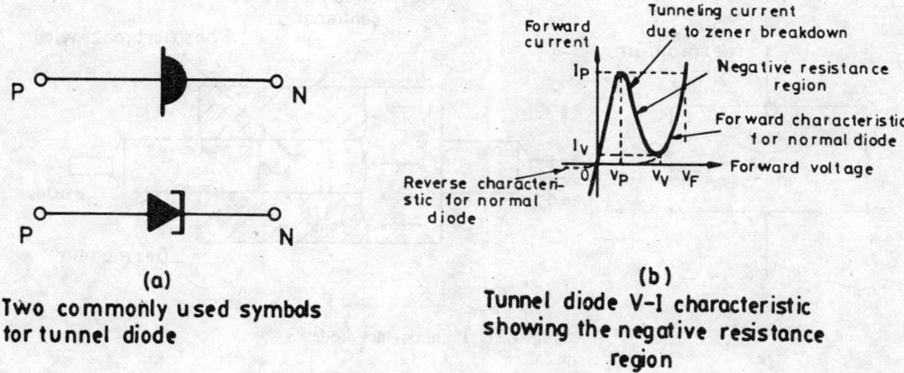


**Fig. 1.20:** Tunnel diode.

Forward current is the sum of two components due to two different mechanisms. One of these components arises from an external voltage reducing the potential barrier across the depletion region and allowing current to flow due to majority carriers, shown dotted in Fig. 1.20(b). The other component of forward current is due to the zener effect. The extremely *narrow* depletion region results in a very high electric field which permits carriers on one side of the junction to *tunnel* through the junction without having to overcome the potential hill. This current increases to some peak value $I_P$ at a forward voltage $V_P$, which is typically 65 mV in a germanium tunnel diode and 169 mV in a galliumarsenide tunnel diode.

*The tunneling mechanism takes place at the speed of light. As a consequence of this, a tunnel diode is an ultra-high frequency switching device.* Tunnel diodes are also used in microwave oscillators at frequencies of the order of 10 GHz ($10^{10}$ Hz). The *negative resistance* of the tunnel diode allows its use in relaxation oscillators.

## 1.21 Point Contact Diode

The point contact diode is the modern form of the crystal detector (cat's whisker) used in early radio receivers, and was developed originally for operation at radar frequencies where the *low inter-electrode capacitance and short transit time* provided a great improvement over the thermionic diode.

Structurally, the point contact diode consists of a springy, tungsten wire which presses against a crystal of germanium. During manufacture, a current is passed between the whisker and the crystal to *form* the diode. A small region of p-material is formed beneath the point of the whisker. Rectification takes place between this region and the remaining material which is n-type (see Fig. 1.21).

Although the point contact diode is effectively a junction diode of small interface area, there are considerable differences. At low reverse voltage, the current in the point contact diode approaches the theoretical value, but the shape of the characteristic, thereafter, rises upto the turnover voltage. Generally speaking, the forward voltage across the point contact diode is comparatively high. On the other hand, the reverse capacitance and usually the carrier storage are lower than for the junction diode. The

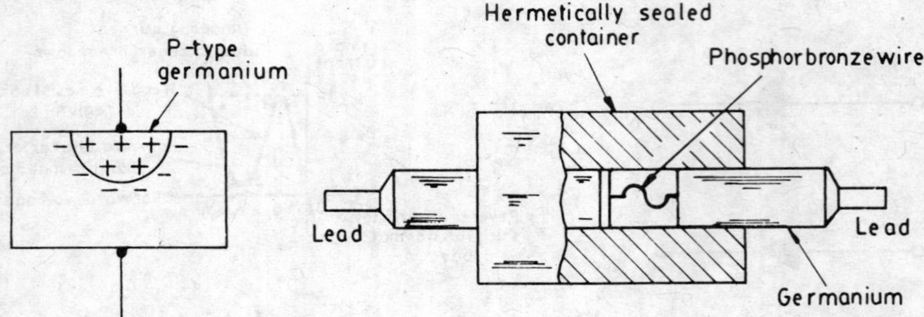


**Fig. 1.21:** Point contact diode.

temperature sensitivity of the two types is similar. The point contact diode is generally the one most suitable for RF detectors and mixers

## 1.22 Semiconductor Diode Polarity

The semiconductor diode is represented by the symbol shown in Fig. 1.22(a). The *bar* is connected with the polarity of a cathode and the *arrow head* with that of an anode. The thermionic diode on the left shows this polarity.

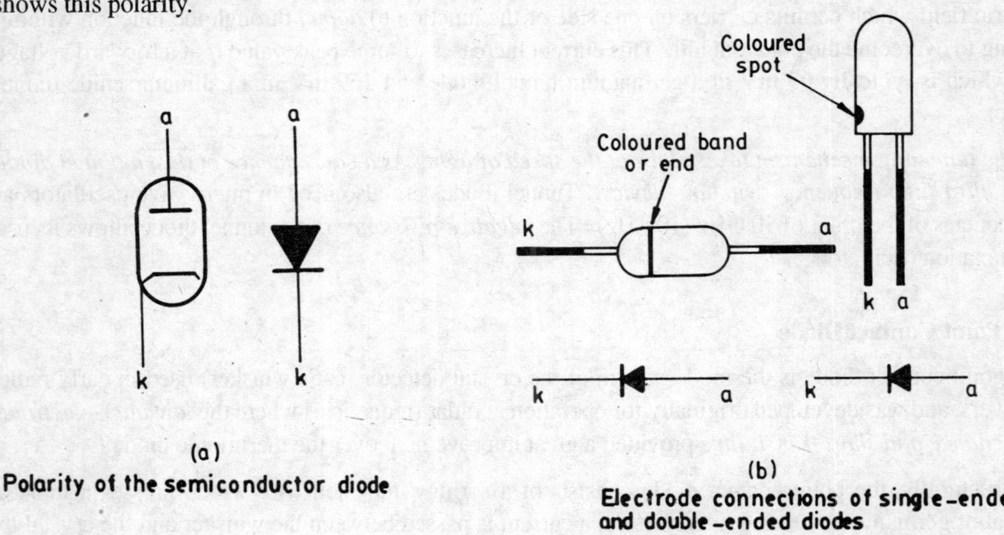


**Fig. 1.22:** Semiconductor diode polarity.

The smaller semiconductor diodes are made in either a *single-ended* or a *double-ended construction*, as shown in Fig. 1.22(b). In the single-ended construction, a coloured *spot or dot* is placed on the same side of the body as the cathode. In the double-ended construction, a *coloured band* is placed on the end of the body nearer to the cathode; alternatively the whole of the tip may be coloured.

## 1.23 Testing Semiconductor Diodes

A *resistance check* may be used as a rough test of semiconductor diode's operation. The polarity of the terminals of the battery contained within an ohmmeter appears at the leads of the ohmmeter. First, *the polarity of the ohmmeter leads is determined* by connecting a voltmeter across the ohmmeter terminals.

An ohmmeter test of a diode, which is operating normally, will reveal that the diode has a *low forward resistance* and a *high reverse resistance*. On the other hand, if a diode measures low forward and low reverse resistance, or high reverse and high forward resistance, it is defective. In the first case it is either *shorted* or it has *abnormal (excess) leakage* and in the second case it is *open*.

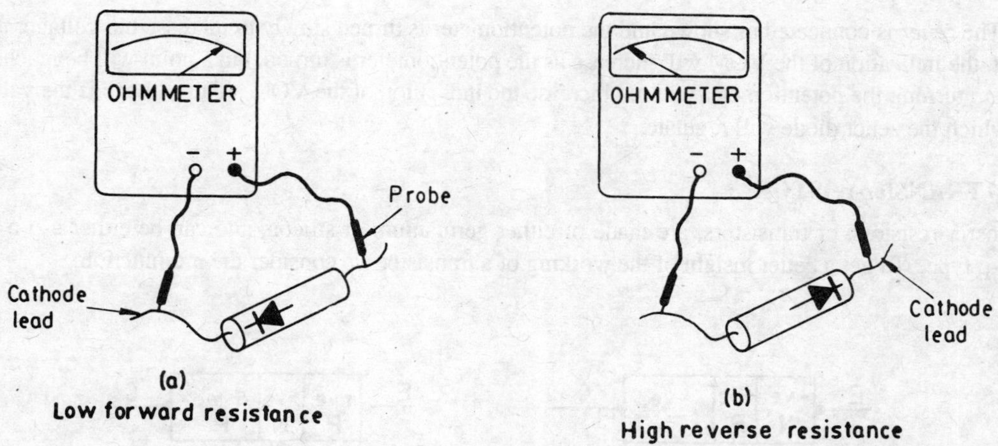


**Fig. 1.23:** Testing semiconductor diodes.

If the diode is *unmarked*, it is simple to determine by a resistance check which is the *anode*, which the *cathode*. Connect the ohmmeter leads in position which measures the forward resistance of the diode. In this position, the positive ohmmeter lead is connected to the anode, and the negative lead to the cathode.

The battery in volt-ohm-milliammeters (VOM) is 1.5V or higher. Hence it can forward bias a germanium diode beyond 0.3V. Similarly it can forward bias a silicon diode beyond 0.7V. In trouble shooting some semiconductor circuits, however, low power (LP) electronic ohmmeters are used whose lead voltage is lower than 0.3V. The *low-power ohms* (LPΩ) function of this type of ohmmeter can not be used to measure the forward resistance of a diode, nor can it be used to identify the polarity of a diode. Fortunately, manufacturers provide, in addition to low-power function, a *normal-ohms* function. Resistance tests of semiconductor diodes may be made using the normal-ohms function of the meter.

The VOM is not of much use in checking zener diodes. If a source of EMF, a potentiometer, and a resistor are available, the circuit of Fig. 1.24 can be used.

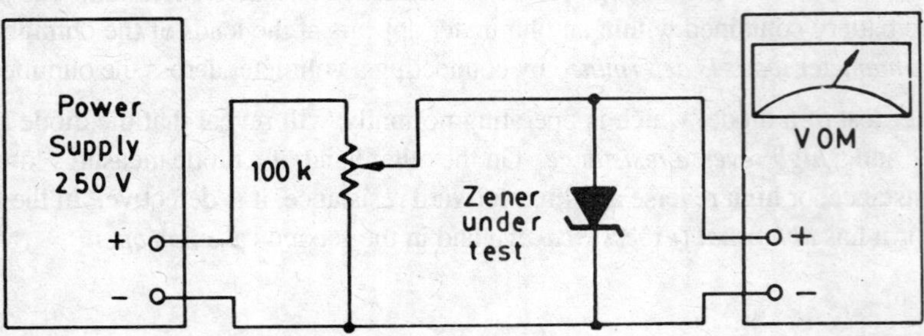


**Fig. 1.24:** Testing zener diodes.

The *zener* is connected as shown and the potentiometer is turned slowly to increase the voltage. At first, the indication of the VOM will increase as the potentiometer is turned, but a point will be reached where turning the potentiometer will not increase the indication of the VOM. This voltage is the value at which the zener diode will regulate.

## 1.24 TRANSfer-reISTOR

Transfer-resistors, or transistors, are made of either germanium or silicon, and can be either n-p-n or p-n-p type. To get a better insight of the working of a transistor we consider the p-n junction.

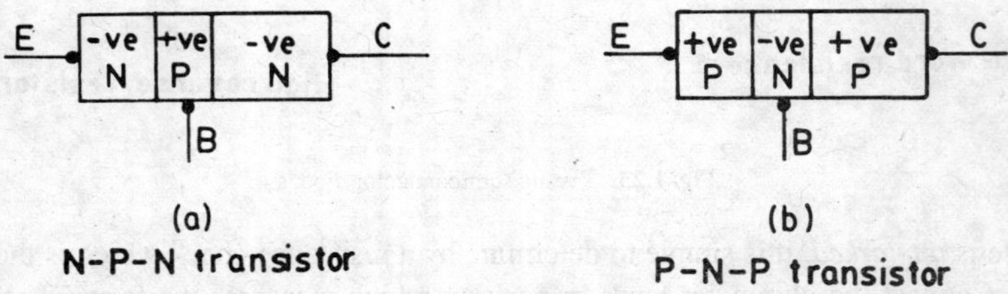


**Fig. 1.25:** Transfer-resistor.

If the transistor is divided into two portions from the centre of its base, we obtain two p-n junctions so connected that either the anodes or the cathodes of both of these p-n junctions face outwards, as can be seen in Fig. 1.26. This amounts to an *inverted connection* of the two diodes. One of these diodes is called the *input diode*. Herein we feed the signal. The other diode is called the *output diode*. Here from, we derive the output signal.

The central region, p-type in n-p-n and n-type in p-n-p transistors, is called the *base* of the transistor. The regions on either side of the base are referred to as the emitter and collector of the transistor. Thus the base of a transistor is common to the input as well as the output of the transistor.

A transistor can possibly be biased in any one of the following three ways i.e. forward-forward (F-F) bias, reverse-reverse (R-R) bias and forward-reverse (F-R) bias. For normal operation of the transistor,

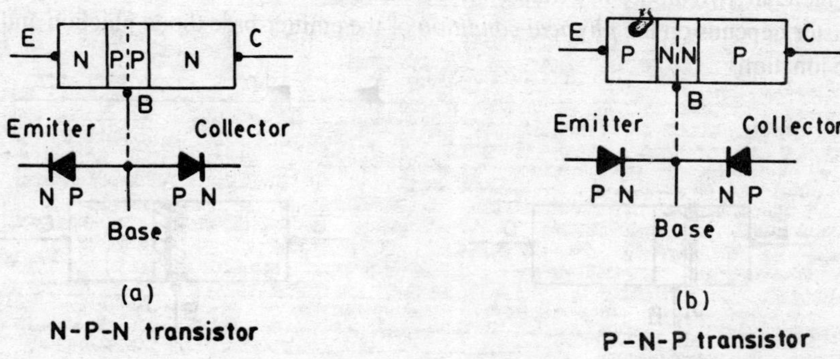


**Fig. 1.26:** Inverted connection of two p-n junction diodes.

the F-R bias is almost invariably employed. The input diode (*emitter-base diode*) of the transistor is forward biased to the extent of a few millivolts. The reverse bias given to the output diode (*base-collector diode*) is of the order of a few volts and is primarily determined by the reverse breakdown voltage of the output diode and also by the magnitude of the collector supply. F-R biasing of a transistor is shown in Fig. 1.27.

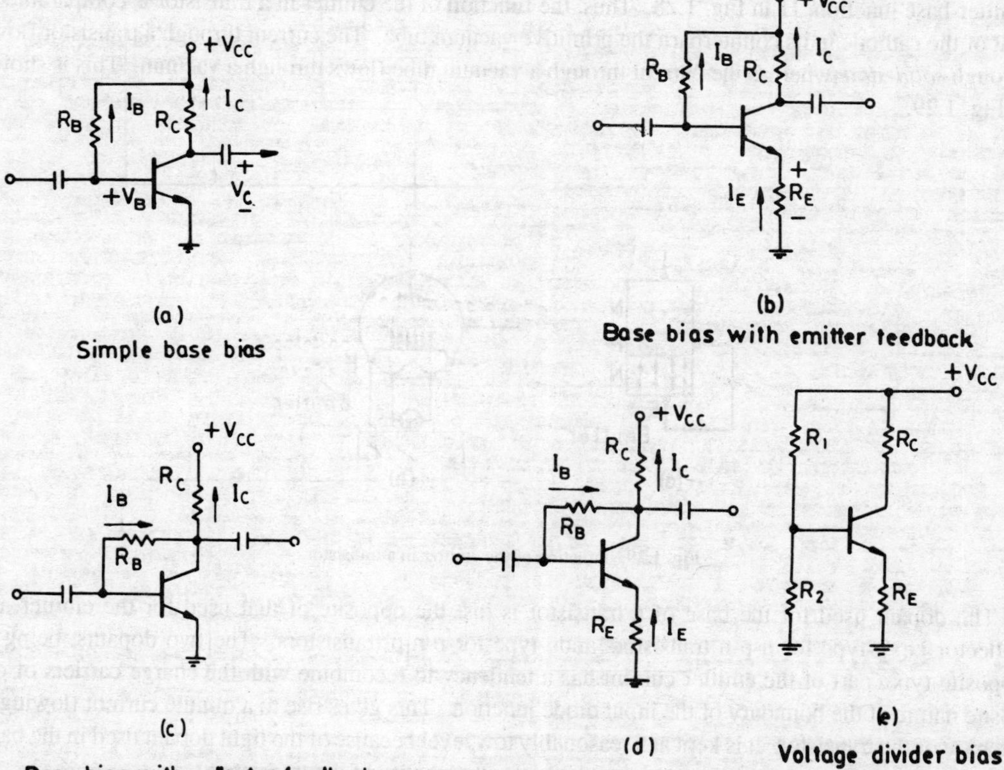


**Fig. 1.27:** Forward-reverse (F-R) biasing of a transistor for normal operation.

Because there are two diodes in a transistor, there also are *two junctions* in a transistor. The working of the transistor depends on the *physical condition* of the emitter-base diode junction and the base-collector diode junction.

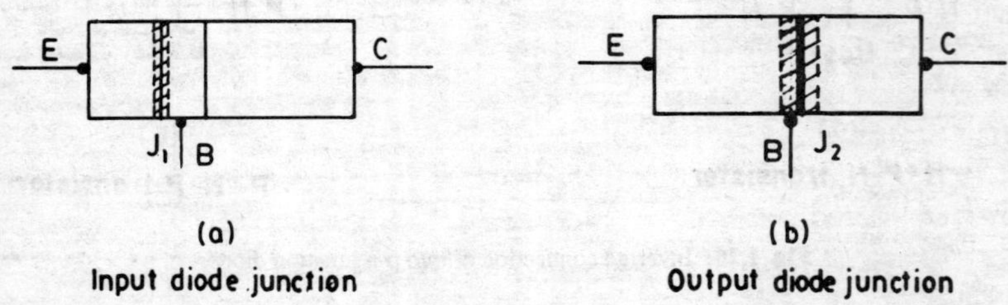


**Fig. 1.28:** Junctions in a transistor.

The emitter of a transistor is highly doped because it is the source of charge carriers, electrons in n-p-n and holes in p-n-p transistors. With normal F-R bias emitter current is set flowing towards the emitter-base junction, J1 in Fig. 1.28. Thus, the function of the emitter in a transistor is comparable to that of the cathode in its counterpart, the primitive vacuum tube. The current through a transistor flows through *solid-state* whereas the current through a vacuum tube flows through a vacuum. This is shown in Fig. 1.29.

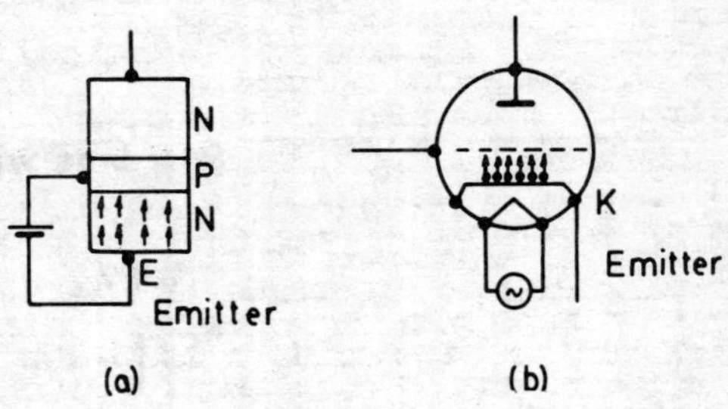


**Fig. 1.29:** Function of the emitter in a transistor.

The dopant used for the base of a transistor is just the opposite of that used for the emitter and collector i.e. p-type for n-p-n transistors andn-type for p-n-p transistors. The two dopants, being of opposite type, part of the emitter current has a tendency to recombine with the charge carriers of opposite nature at the boundary of the input diode junction. This gives rise to a minute current flowing in the base of the transistor. It is kept at a reasonably low level because of the light dopant used in the base. The lower the *base current* in a transistor, the better the quality of the transistor.

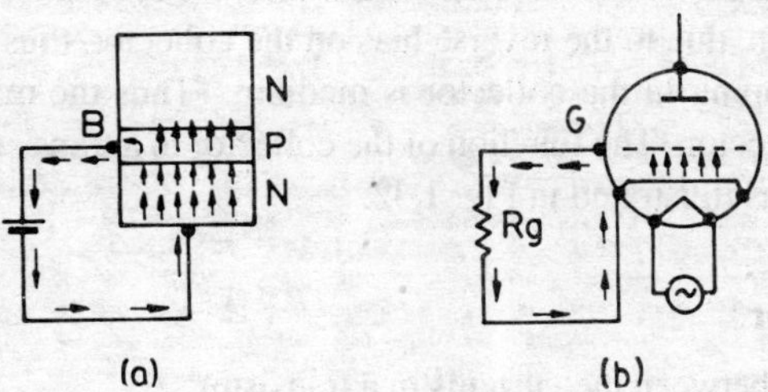


**Fig. 1.30:** Function of the base in a transistor.

The base of the transistor *controls* the collector current in the same way as the control grid of a vacuum tube controls its plate current. This is shown in Fig. 1.30. The only difference between the two being that a transistor is a *current-operated* device whereas a vacuum tube is a *voltage-operated* device.

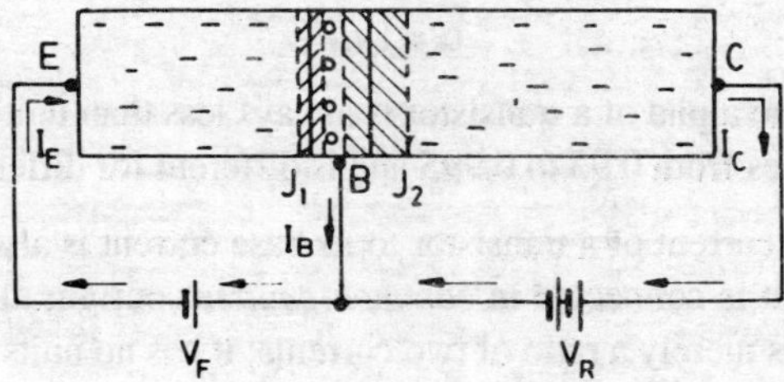


**Fig. 1.31:** The output diode junction extends deeper into the base region.

The input diode junction is extremely *narrow* because of the high level of doping in the emitter. The output diode junction is comparatively *wide,* and extends deeper into the base region to make its influence felt, as can be seen in Fig. 1.31.

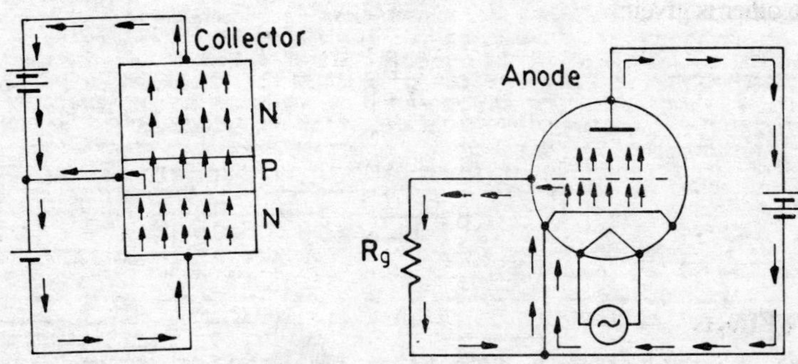


**Fig. 1.32:** Function of the collector in a transistor.

The balance of current (emitter current minus base current) flows towards the collector of the transistor. The reverse voltage on the collector is far higher than the forward voltage on the emitter. The electrostatic force of attraction, due to the reverse bias on the collector, thus *penetrates* deeper into the base region. The level of doping in the collector is medium. Thus the major portion of the emitter current flows towards the collector. The function of the collector in a transistor is comparable to that of the anode of a vacuum tube, as illustrated in Fig. 1.32.

## 1.25 Currents in a Transistor

The following relation exists between the currents in a transistor:

$$\text{Emitter current, } I_E = \text{Base current, } I_B + \text{Collector current, } I_C \tag{1.2}$$

Thus, $I_B = I_E - I_C$ $\qquad$ $I_C = I_E - I_B$

The ratio of the collector current of a transistor to its emitter current is called the *forward current gain* of the transistor and represented by the Greek letter alpha ($\alpha$). This is the current gain when a transistor is connected in *common-base* configuration. As it is merely a ratio of two currents, it has no units.

$$\alpha = I_C/I_E \tag{1.3}$$

The numerical value of the alpha of a transistor is always less than unity; it being the ratio of a part ($I_C$) to the whole ($I_E$). It varies from 0.95 to 0.995 and is different for different transistors.

The ratio of the collector current of a transistor to its base current is also called the *forward current gain* of the transistor when it is connected in *common-emitter* configuration. It is represented by the Greek letter beta ($\beta$). As it is merely a ratio of two currents; it has no units.

$$\beta = I_C/I_B \tag{1.4}$$

The numerical value of the beta of a transistor is far greater than unity; it being the ratio of a large quantity ($I_C$) to that of a much smaller quantity ($I_B$). It varies from about 19 to 199 and is different for different transistors. Transistors of the same batch number, and by the same manufacturer, may not have identical values of beta but have a range of beta values referred to as *beta spread*. This information is invariably given in the manufacturer's data sheets.

The following relation between the two current gains can be established, as it is useful in evaluating one when only the other is given:

$$\alpha = \frac{\beta}{1+\beta} \tag{1.5}$$

and

$$\beta = \frac{\alpha}{1-\alpha} \tag{1.6}$$

## 1.26 Temperature Effects

The collector current of a transistor increases with increasing temperature. The characteristics of a transistor are thus *temperature dependent*. At high temperatures (150° to 200°C) the semiconductor materials degenerate and hence their performance is unpredictable. The following characteristics of transistors are temperature dependent:

1. emitter junction resistance
2. beta
3. base-emitter voltage and
4. leakage current

***1. Emitter Junction Resistance***

The ohmic value of the emitter junction resistance varies from $25mV/1_E$ to $50\ mV/I_E$ where $I_E$ is the emitter current in milliamperes. *The emitter junction resistance increases by 1% for every 3°C rise in temperature.* If the temperature increases by 30°C, the emitter junction resistance increases by 10%. To swamp, or overcome, these changes in emitter junction resistance, and consequently to *stabilise* the gain of a transistor, a low ohmic value resistance is connected in series with the emitter.

***2. Beta***

The value of beta normally increases with increasing temperature. *The range of beta spread is about 1:4;* it can be even higher. In this range, the beta of a transistor has a minimum value and a maximum value. The *minimum value of beta* is kept in mind while designing transistor circuits.

***3. Base-Emitter Voltage***

*The base-emitter voltage of germanium transistors decreases by 1 mV per degree centigrade rise in temperature and that of silicon transistors by 2.5mV per degree centigrade rise in temperature.* Changes in base-emitter voltage ($V_{BE}$) bring about corresponding changes in the operating point of the transistor and these are swamped by increasing the collector supply voltage ($V_{CC}$).

**4. Leakage Current**

Even in the absence of emitter current, a *minute* collector current flows through the transistor. Denoted by $I_{CBO}$, this current represents the *current between collector and base with an open emitter*, see Fig. 1.33(a). *$I_{CBO}$ doubles for every 10°C rise in temperature.* At room temperature, the leakage current in silicon transistors is far less than that in germanium transistors. It is a part of the total collector current and is also undesirable. An increase in leakage current brings about a corresponding increase in collector current, resulting in a shifting of the *operating point.* Special biasing techniques keep the shift in the operating point stabilised.

Even in the absence of base current, a *minute* current flows through the collector of a transistor. Denoted by $I_{CEO}$, this current represents the *current between collector and emitter with an open base*, see Fig. 1.33(b). *$I_{CEO}$ also is temperature dependent.* Special biasing techniques are employed to swamp the changes in $I_{CEO}$.

The leakage current, an important transistor parameter, increases exponentially with temperature and can set a *limit* to the operating temperature of a transistor. Increasing leakage current results in increasing collector current, and also increasing collector junction temperature, with a still further increase in leakage current. Thus a *vicious cycle* is set up, which, if not kept under check, will result in the transistor being permanently damaged. It is referred to as *thermal runaway*.

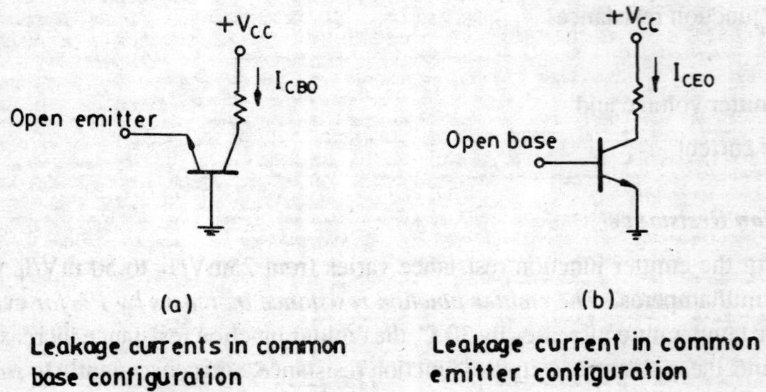


**Fig. 1.33:** Leakage currents in a transistor.

## 1.27 Collector Dissipation

The input diode junction of a transistor has to handle quite a small amount of power, which it is capable of dealing with because of the low current and voltage levels, and hence remains cool. Excess forward bias between the base and emitter of a transistor can cause permanent damage to the input diode, and the transistor reduces to a diode represented by collector-base junction with a *floating emitter*. Collector-base junctions, with floating emitter, of silicon transistors, are frequently used by TV manufacturers as *varicap diodes* in VHF tuners (BF-194B, for example). As the collector-base junction is normally operated with a reverse bias, it poses no problem.

The output diode junction of a transistor has to handle a comparatively larger amount of power which, of course, it should be capable of handling safely. To allow for cool operation of the transistor, its collector has the maximum area of cross-section. Collector dissipation, $P_C$, is given by:

$$P_C = V_R \cdot I_C \text{ watts} \tag{1.7}$$

where $V_R$ stands for the reverse voltage on the collector and $I_C$ for the collector current.

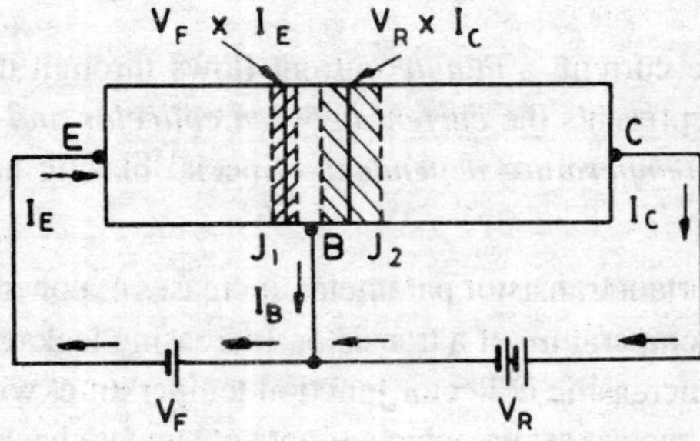


**Fig. 1.34:** Collector dissipation.

The input resistance of a transistor is *low*, whereas its output resistance is comparatively *high*. The ratio of the output resistance to the input resistance of a transistor represents a *resistance gain*. Thus a low resistance in the input is *transferred* to a high resistance in the output and hence the name *TRANSFER-RESISTOR abbreviated to transistor* (see Fig. 1.35). The same current (or almost the same current) flows through the input and output of a transistor because the collector current is only marginally less than the emitter current.

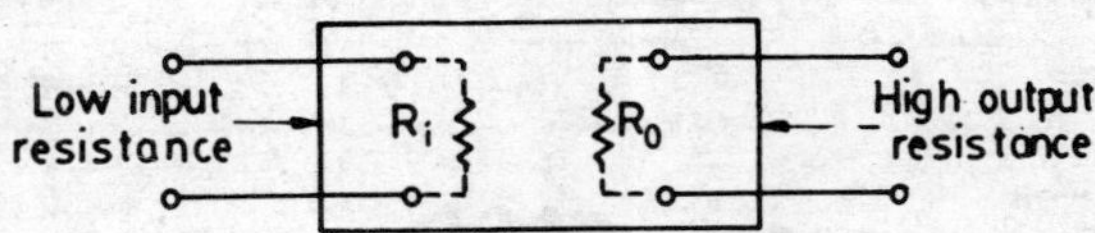


**Fig. 1.35:** A low input resistance transferred to a high output resistance.

Excess reverse voltage on the collector-base junction can also permanently destroy the output diode. The transistor now reduces to a diode represented by emitter-base junction with a *floating collector*. Because of their low turn-on voltage, germanium transistors with a floating collector are used for *signal detection*.

## 1.28 Heat Sinks

In some low power transistors and transistors of the small-signal type, sufficient increased collector heat dissipation capability may result from *mounting the transistor case directly on the metal chassis*. However, this must be done with care, since *very often the collector of the transistor is connected to the transistor case to enhance its heat dissipation capabilities.* Thus, some provision for insulating the case from the chassis (which is usually at ground potential) must be provided, unless a common collector circuit is being employed.

One method is to use a beryllium oxide washer as shown in Fig. 1.36(a). This washer has a *good thermal conductivity*, and heat transfer from the transistor case to the chassis is improved by the use of a zinc oxide filled silicon compound between the washer and the chassis. An insulated clamp over the top of the transistor may be used to increase pressure and help improve thermal dissipation.

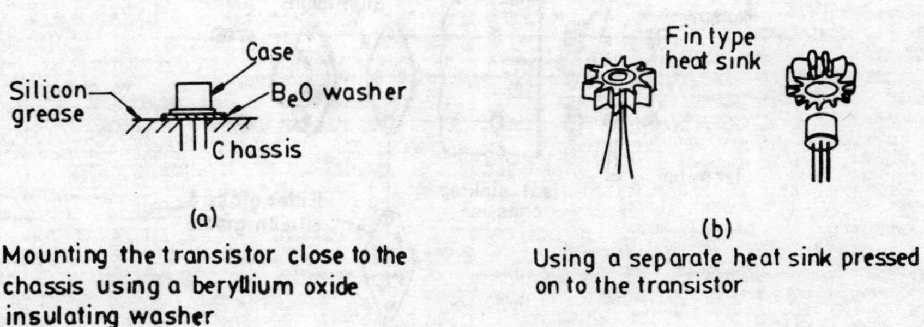


**Fig. 1.36**: Methods of mounting transistors that are encased in a metal T0-5 package.

Where transistors are mounted in teflon sockets that provide no thermal conduction from transistor case to chassis, a *press-on fin type of heat sink* may be used as shown in Fig. 1.36(b).

Popular mounting packages for power transistors are the diamond shaped T0-3 and T0-66 types, which may have dissipation in the order of 100 W. A typical heat sink that can accommodate one T0-3 power transistor and/or two stud mounted diodes is shown in Fig. 1.37. It provides cooling by conduction, convection and radiation.

**Fig. 1.37:** Typical power transistor heat sink that can accommodate a T0-3 package.

Power transistors encased in T0-3 or T0-66 type metal packages have two leads for the emitter and base, but the case, or the *mounting flange* of the case is the collector terminal so that once again insulation of the case from the heat sink is necessary. This may be accomplished by the use of an *insulating washer* as illustrated in Fig. 1.38.

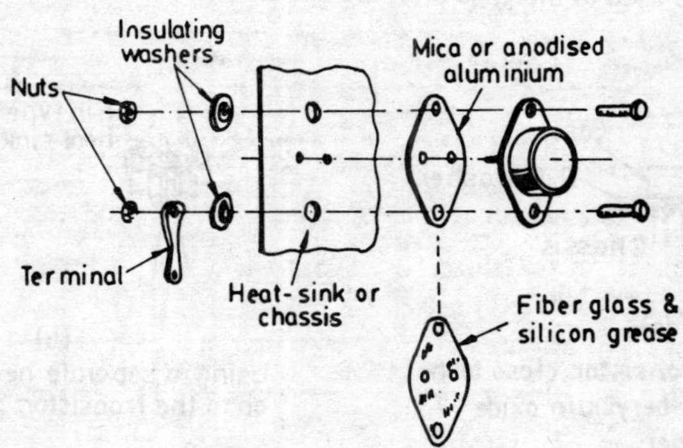


**Fig. 1.38:** Mounting of a T0-3 or T0-66 transistor package on a heat sink with an insulating washer.

## 1.29 Unijunction Transistor

The unijunction transistor is a three terminal semiconductor device with only one p-n junction. The arrangement is such that the device exhibits a *negative resistance* characteristic and so can be used for switching applications and oscillator circuits. It was originally called the double-base diode.

The physical construction of the unijunction transistor (UJT) is shown in Fig. 1.39(a). It is composed of a single bar of high resistivity silicon, contacted at the ends by base 2 and base 1 and containing a single p-n junction *located off centre* and closer to base 2.

The electrical circuit that represents this transistor is shown in Fig. 1.39(b). The junction operates by causing the conductivity of the bar to be changed. The silicon bar is doped to be n-type and a forward bias (positive applied to the emitter) applied across the emitter junction will cause positive charges to be injected into the bar of silicon. This lowers the resistance of the bar between the p-n junction and base 1.

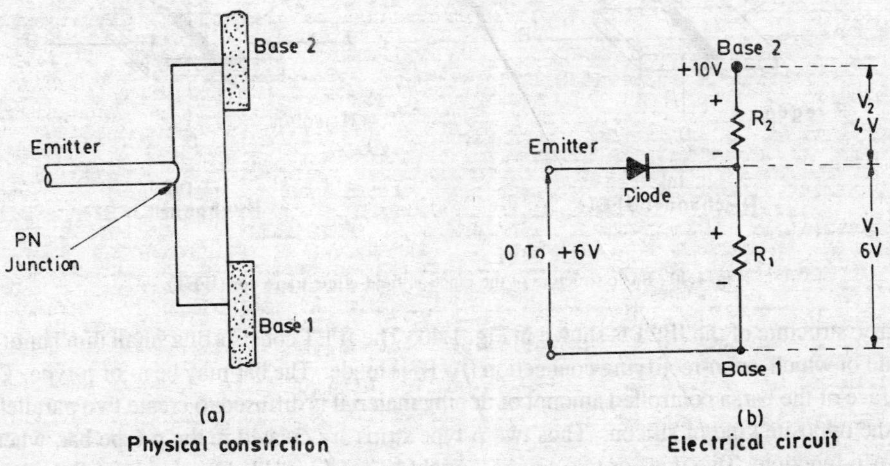


**Fig. 1.39:** Unijunction transistor.

The voltage between base 2 and base 1, in Fig. 1.39(b), is indicated as 10 V and the division of voltage across the bar is such that 6 volts exist from base 1 to the cathode of the junction diode. If the emitter is connected through an input circuit to base 1, the diode will be reverse biased. In order to inject positive charges into the silicon bar the voltage at the emitter (anode) of the diode must exceed the 6 V existing on the cathode side of the junction. When the voltage on the emitter causes the diode to conduct, the injection of positive charges into the silicon bar causes the resistance of R1 to decrease. The 6 V across R1, also decreases and this causes the number of carriers injected into the bar to increase. The voltage, V1, across R1, decreases *faster* than the emitter voltage and this gives rise to a *negative resistance* effect.

## 1.30 Junction Field-Effect Transistor

The conventional transistor employs two types of current carriers (electrons and holes) in its operation. It is a *bipolar* device. The field-effect transistor employs only one type of current carrier (either electrons or holes). It is a *unipolar* device.

The field-effect transistor has three electrodes: source, drain and gate. These correspond respectively to the emitter, collector and base of the bipolar transistor, and also to the cathode, anode and grid of the vacuum tube. One may think of the *drain* and *source* as the anode and cathode of the device, and the *gate* as the control element.

Some FETs are *single gate*, others are *dual gate*. An FET is called *symmetrical* when the source and drain terminals may be interchanged without affecting its operation. It is called *asymmetrical* when the source and drain terminals cannot be interchanged without impairing its performance.

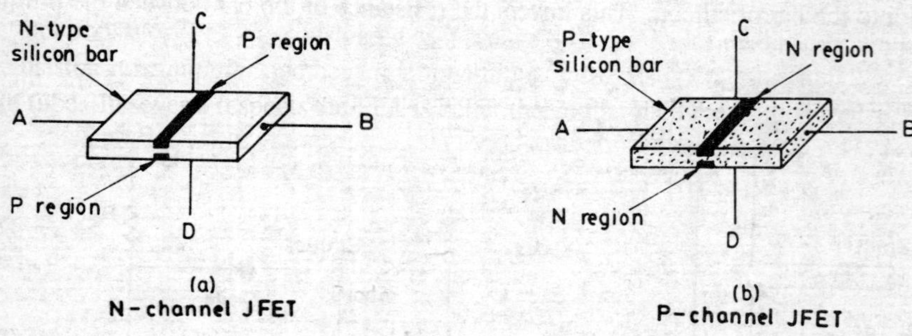


**Fig. 1.40** : Basic structure of the junction field-effect transistor (JFET).

The basic structure of the JFET is shown in Fig. 1.40. The JFET consists of a small thin bar of silicon to each end of which a non-rectifying connection (A, B) is made. The bar may be n- or p-type. On each opposite face of the bar, a controlled amount of doping material is diffused to create two parallel facing strips of the opposite kind of silicon. Thus two p-type strips are created in the p-type bar, where each forms an n-p-junction. Two dots, or two squares, might be used provided they are parallel. A connection (C, D) is made to each of the strips and this completes the basic FET. One end of the bar will be the source electrode and the other will be the drain electrode. The two strips constitute the gate electrode and are *usually tied together internally* and connected to a single terminal.

Fig. 1.41 illustrates an n-channel JFET operation. The source electrode is the common (grounded) terminal. The gate-to-source circuit forms the *input*, and the drain-to-source terminal the *output*. One dc supply voltage ($V_{DS}$ from the supply $V_{DD}$) biases the drain-source (output half) and another dc voltage ($V_{GS}$ from supply $V_{GG}$) biases the gate-source (input half).

In Fig. 1.41(a) $V_{GS}$ is zero and the drain current $I_D$ (S to D in silicon bar) is relatively high. *The current path inside the bar is called the channel.* The drain current in this instance is proportional to the resistance of the channel.

When a moderate $V_{GS}$ is applied between the gate and source, such that the gate junction is reverse biased, as in Fig. 1.41(b), a depletion layer is set up, in the normal manner for p-n junction, around each junction. The depletion layer acts like a *dielectric*. The depletion layers due to the two junctions *penetrate* into the bar towards each other and thus *reduce* the width of the channel. As fewer current

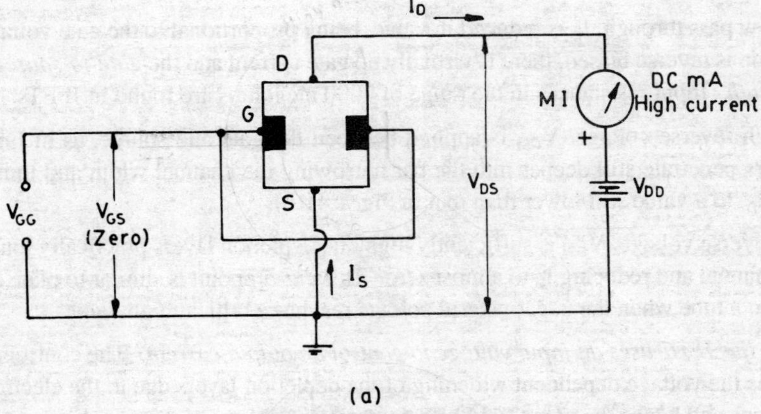


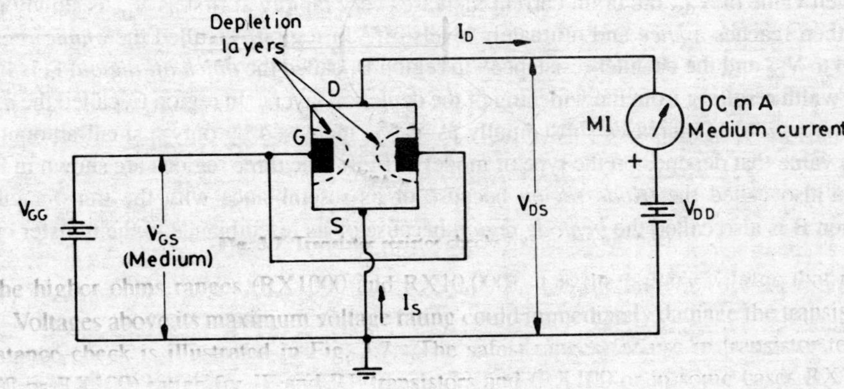


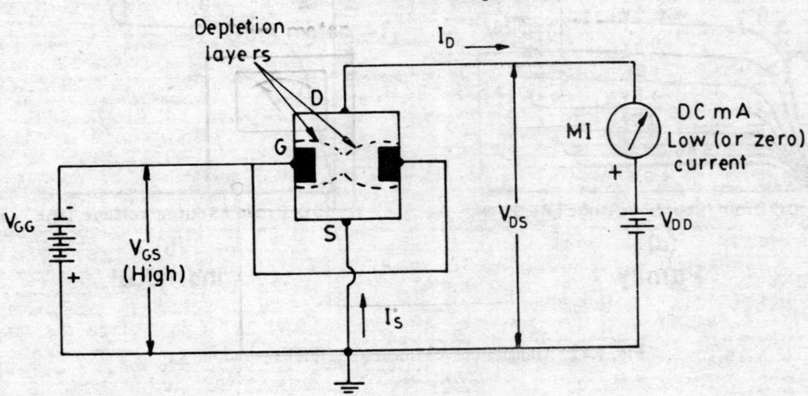


**Fig. 1.41:** JFET operation.

carriers can now pass through, $I_D$ is reduced in value, being proportional to the gate voltage $V_{GS}$. Since the gate junction is reverse biased, there is virtually no gate current and the *gate-to-source (input) resistance is very high.* Input resistances in the range of 1000 megohms are found in JFETs.

When a high reverse voltage, $V_{GS}$, is applied between the gate and source, as in Fig. 1.41(c), the depletion layers penetrate still deeper into the bar narrowing the channel width and thus reducing the drain current, $I_D$, to a value still lower than that in Fig. 1.41(b).

When the reverse voltage, $V_{GS}$, is sufficiently high, the depletion layers practically touch each other, *blocking* the channel and reducing $I_D$ to almost zero. This *cut-off* point is similar to plate current cut off which occurs in a tube when the negative grid voltage reaches a critical high value.

In this way, *the JFET uses an input voltage to control an output current.* The control mechanism in this operation is the voltage dependent widening of the depletion layers due to the electrostatic field at the gate junctions and hence the name field-effect transistor.

For a given value of $V_{GS}$, the drain current increases very rapidly at first as $V_{DS}$ is slowly increased from zero, then reaches a *knee* and ultimately levels off. In region A, called the *ohmic region,* $I_D$ is proportional to $V_{DS}$ and the channel resistance. In region B, called the *pinch off region,* $I_D$ is limited by the channel width resulting from the widening of the depletion layers. In region C, called the *avalanche or breakdown region,* $I_D$ increases substantially as $V_{DS}$ is increased by only a small amount above a critical high value that depends on the type or model of FET. The three regions are shown in Fig. 1.42. Region A is also called the *triode region* because of its resemblance with the transfer curve of a triode. Region B is also called the *pentode region* because of its resemblance to the transfer curve of a pentode.

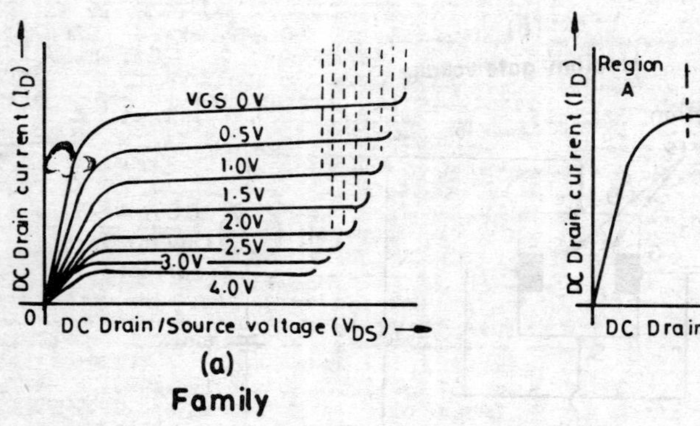


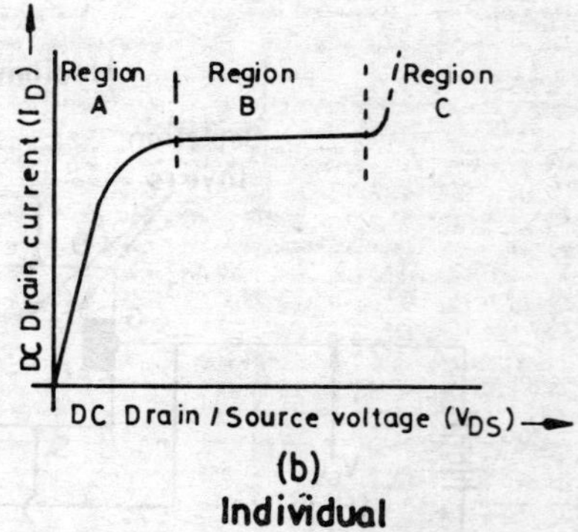


**Fig. 1.42:** Transfer curves illustrating JFET operation.

The current carriers in an FET flow through a single type of semi-conductor material and not across the junctions as in bipolar transistors. This results in the *absence of noise components caused by junction currents.*

## 1.31 Metal Oxide Semiconductor Field-Effect Transistor

Unlike the JFET, the metal oxide semiconductor field-effect transistor (MOSFET) has a metal gate electrode which is electrically insulated from the semiconductor material by a thin oxide film. *This sequence of metal oxide and semiconductor has given this type of transistor its name of MOSFET.*

Fig. 1.43 shows the cross section of a MOSFET. Two strongly doped n-type zones ($\overset{+}{n}$) are diffused into the p-type substrate for the source and drain. The metallic gate electrode forms a capacitor with the underlying substrate or channel region, the dielectric is formed by the oxide film. When the gate has a positive voltage with respect to the substrate, electrons in the substrate are attracted and the holes are repelled from the silicon dioxide interface by polarisation, thus giving an n-type inversion layer under the dielectric between the source and the drain. This *inversion layer* is called n-channel.

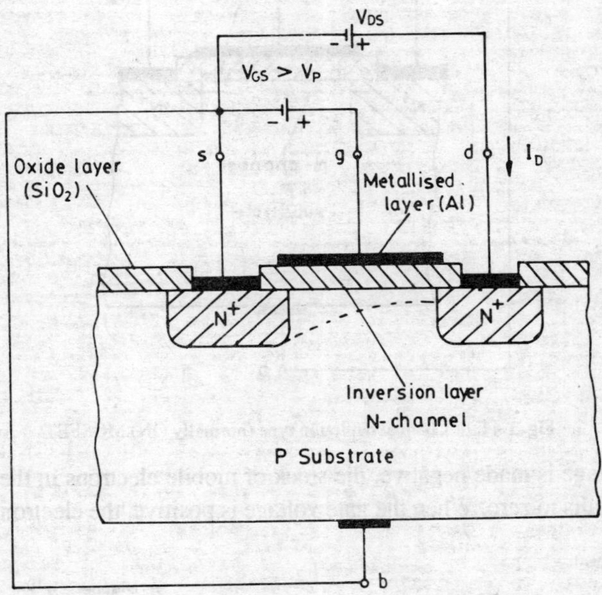


**Fig. 1.43:** n-Channel enhancement type (normally OFF) MOSFET.

When a drain source voltage is applied, a current flows through this channel. Variation of the voltage $V_{GS}$ changes the electron density, and hence the resistance of the n-channel, and therefore the current $I_D$. In contrast with the JFET, however, the gate voltage of the MOSFET can change sign without a gate current starting to flow. This type of MOSFET is called an n-channel *enhancement (normally OFF)* type because the number of mobile charge carriers (electrons) must first be increased (enhanced) by applying a positive gate voltage. There is no channel present at zero gate voltage since a positive drain voltage causes the p-n junction between the drain zone and the substrate to cut off. No current, apart from leakage current, will flow between source and drain while the gate voltage remains zero. The voltage $V_{GS}$ at which appreciable current flow starts, due to the formation of an inversion layer, is called either the *pinch-off voltage* $V_P$ (as with the JFET) or the *threshold voltage* $V_T$.

The opposite of an enhancement type is a *depletion (normally ON)* type, where a conducting channel is present even in the absence of a gate voltage. This effect is caused by immobile positive ions which are always present in the oxide layer. These positive ions give rise to an n-type inversion layer in the p-type substrate even when $V_{GS}$ is zero, so that a channel is formed. This is illustrated in Fig. 1.44.

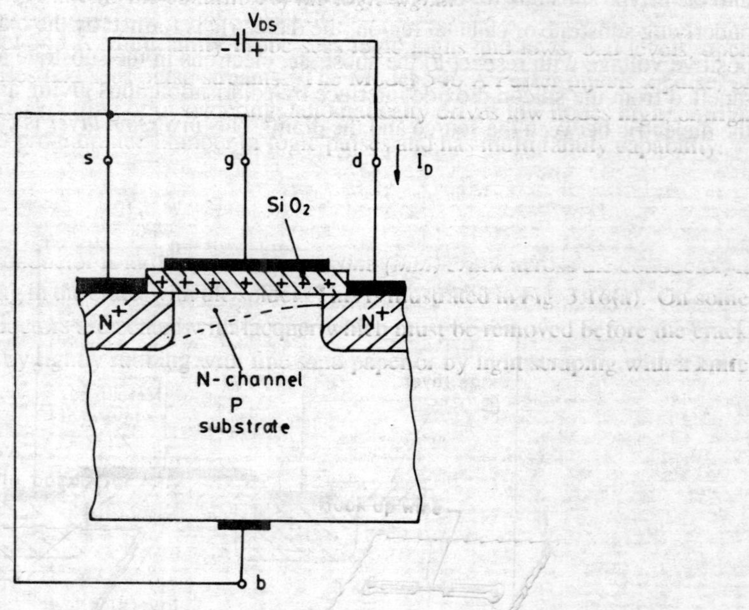


**Fig. 1.44**: n-Channel depletion type (normally ON) MOSFET.

When the gate voltage is made negative, the stock of mobile electrons in the n-channel is depleted and the drain current falls to zero. When the gate voltage is positive, the electron density and hence the

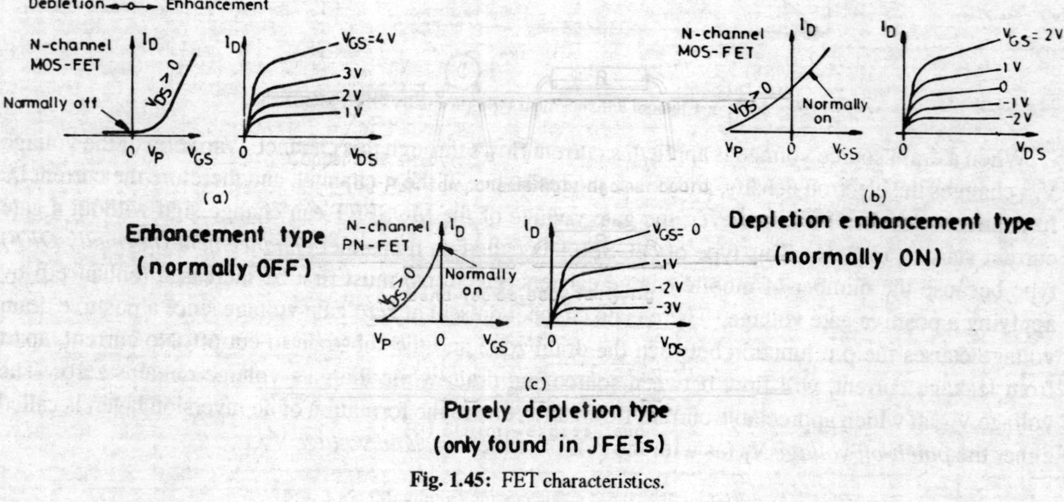


**Fig. 1.45**: FET characteristics.

drain current is increased again as in the enhancement type. This means that a *depletion type MOSFET can be used in both the depletion and enhancement mode* as shown in Fig. 1.45. The JFET is generally used in the depletion mode, because *enhancement mode in a JFET results in forward biasing of the gate-to-channel diode.*

The schematic symbols of different types of FETs are shown in Fig. 1.46.

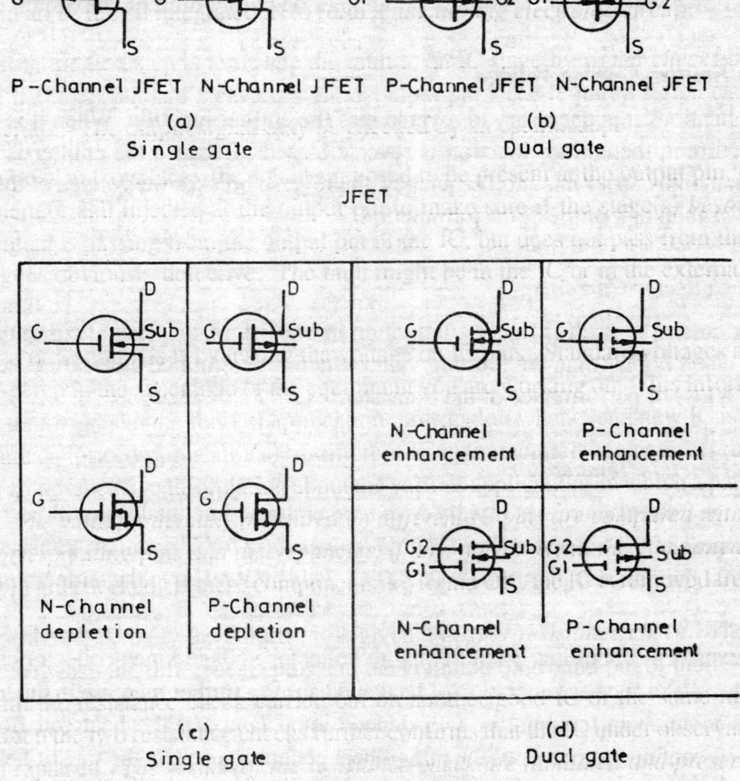


**Fig. 1.46:** Schematic symbols of field-effect transistors.

## 1.32 Operating Limits of Transistors

The range of operating conditions for safe operation of transistors is specified by the manufacturer; for example the maximum voltages, power dissipation that it can handle, maximum ambient and internal temperatures and so on. There are also some other limitations to the operation of a transistor which depend upon its design and fabrication; such as the highest frequency at which it will operate and the level of self-generated noise.

### *1. Maximum Collector Voltage*

The maximum collector voltage of a transistor depends upon the breakdown voltage between collector and base with emitter open. Denoted by $BV_{CBO}$, it is measured by applying a reverse voltage between collector and base, and increasing it until a predetermined current is reached. The manufacturer usually allows a factor of safety and quotes a value somewhat lower than the measured value.

The carrier depletion layer at the collector-base junction becomes wider as the reverse voltage is increased and it may eventually extend completely across the base region on to the emitter. Beyond this voltage, the transistor would appear as a straight resistance from emitter to collector and transistor action would be lost. This effect, known as *punch through*, can happen at a voltage lower than the collector junction breakdown voltage. This imposes yet another operating limit on maximum collector voltage.

### *2. Maximum Emitter Reverse Voltage*

In certain applications it is necessary to reverse bias the emitter junction. When it is desired to cut-off the collector current, the emitter junction is reverse biased. Because the emitter is heavily doped in order to obtain a high carrier injection efficiency, the reverse breakdown voltage of the emitter junction is usually *lower* than that of the collector junction.

### *3. Maximum Collector Current*

High collector current in itself is rarely a limitation in transistors, the normal current carrying capacity being usually *much higher* than the limiting value which is determined either by maximum power dissipation or by a loss of performance of the transistor.

### *4. Maximum Junction Temperature*

Collector leakage current increases rapidly with temperature. Eventually it will become so high as to limit the usefulness of the transistor. The *limiting temperature* for germanium transistors is of the order of 110°C and that for silicon transistors about 200°C. This limit refers to the temperature of the collector junction.

When a transistor is operating, power equal to collector voltage times collector current is also dissipated. This results in further generation of heat and causes a further increase in the temperature of the junction. The increase in temperature will depend upon *how fast* the heat can flow away and the temperature will continue to rise until the rate of heat generation equals the rate of heat flowing away.

### *5. Maximum Power Dissipation*

The property of a transistor to lose heat is expressed as its *thermal resistance*, and is quoted by the manufacturer as the junction temperature rise per unit power dissipated. From this figure the user can determine the maximum allowable power dissipation for any ambient temperature, by first determining the allowable temperature rise and dividing it by the thermal resistance. For example, a silicon transistor may have a thermal resistance from junction to ambient room temperature of 0.375 mW per degree centigrade and a maximum junction temperature of 175°C. With the ambient temperature at 20°C, the maximum allowable power dissipation is (175-20)/0.375 or 400 mW. With an ambient temperature of 100°C, the maximum power dissipation will be limited to (100-20)/0.375 or 200 mW.

The manufacturers often give a power *derating* curve as shown in Fig. 1.47.

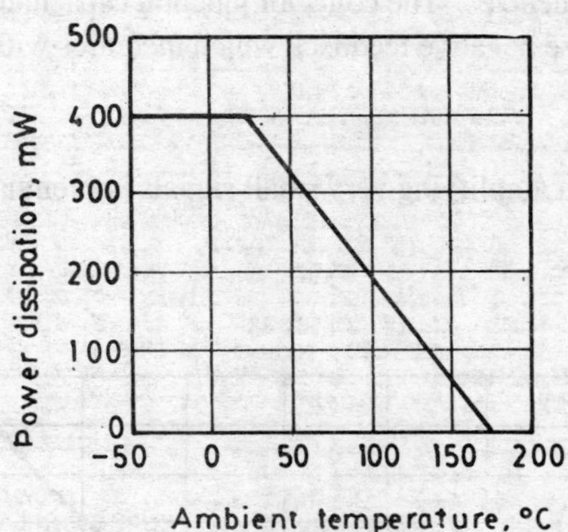


**Fig. 1.47:** Typical power derating characteristic of a silicon transistor.

### 6. Maximum Frequency

If the frequency of signal applied to a transistor is increased, the output initially remains constant, but eventually it will fall off due to *decreasing* gain of the transistor.

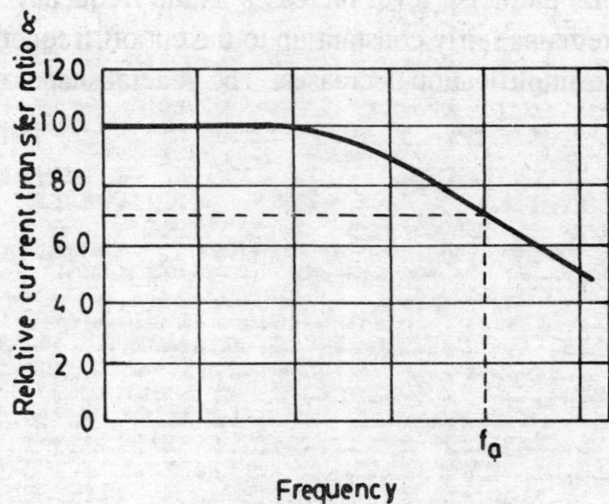


**Fig. 1.48:** Frequency/gain characteristic of a transistor.

The minority carriers take a limited time to diffuse across the base region from emitter junction to collector junction. This time is called *transit time*. It increases as the thickness of the base region increases. Thus, *for a transistor to have good high frequency performance, the base width must be very small.* The *cut off frequency* is the frequency at which the common-base current transfer ratio falls to

0.707 of its initial value and is called $f_{\alpha}$. The general form of the frequency characteristic is shown in Fig. 1.48.

Yet another limitation is imposed by the *capacitance* of the collector junction. This effectively shunts the junction at high frequencies. The collector junction capacitance in series with the resistance of the base region can also cause negative feedback which increases with frequency.

### *7. Noise*

The usefulness of a transistor in amplifying very small signals is eventually limited by noise generated *in* the transistor itself.

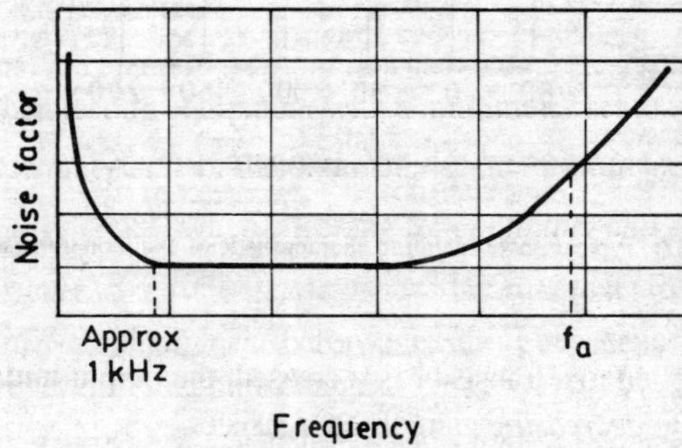


**Fig. 1.49:** Noise characteristic of a transistor.

At frequencies below 1kHz, the noise level increases as the frequency increases. At frequencies above 1kHz, the noise level remains fairly constant up to the cut off frequency. Above this frequency the noise level increases as the amplification decreases. The general shape of the noise characteristic is shown in Fig. 1.49.

## CHAPTER 2

# INTEGRATED CIRCUITS

*In conventional circuits, electrical connections are the cause of numerous failures. As electronic systems become more complex, the number of parts and connections increases, and this leads to a decrease in the potential reliability of the circuit. New devices have been discovered and old devices improved. Electronic circuits have shrunk considerably in size. Construction techniques have shifted from using discrete components to the use of plug-in integrated circuits that already contain the complete circuit. These and related topics form the subject of study in this chapter.*

### 2.1 Integrated Circuits

Since the early days of electronics, there have been continual efforts to miniaturise electronic equipment by making individual components smaller and by devising more concise assembly techniques by using printed circuits, potted modules and the like. This general trend has culminated in what is known as microelectronics and more specially from the semiconductor view point as integrated circuits. By an *integrated circuit,* we mean a circuit which has been manufactured as an *inseparable assembly of electronic parts in a single structure which cannot be divided without destroying its internal electronic function.*

There are four distinct approaches to integrated circuits. These are: (1) thin film circuits, (2) solid-state circuits, (3) multiple-chip circuits and, (4) hybrid integrated circuits.

**1. Thin Film Circuits:** Thin film microelectronic devices consist of passive elements (currently limited to resistors, capacitors and small inductors) that are bound on substrates of glass, ceramic, steatite, or other suitable insulating material. All active elements and some passive elements are added to the circuit as *discrete components.*

**2. Solid-State Circuits:** Solid-state circuits represent an *extension of silicon transistor technology.* These circuits derive their name from the fact that the circuit is formed on a single block of silicon semiconductor material. They contain both active and passive circuit elements interconnected with deposited metal patterns. The selection of circuits fabricated on the solid-state principle is at present the limiting factor in utilisation of the solid-state module. Considerable development emphasis is being given to this limiting factor. Solid-state circuit is a registered trade mark of Texas Instruments Inc.

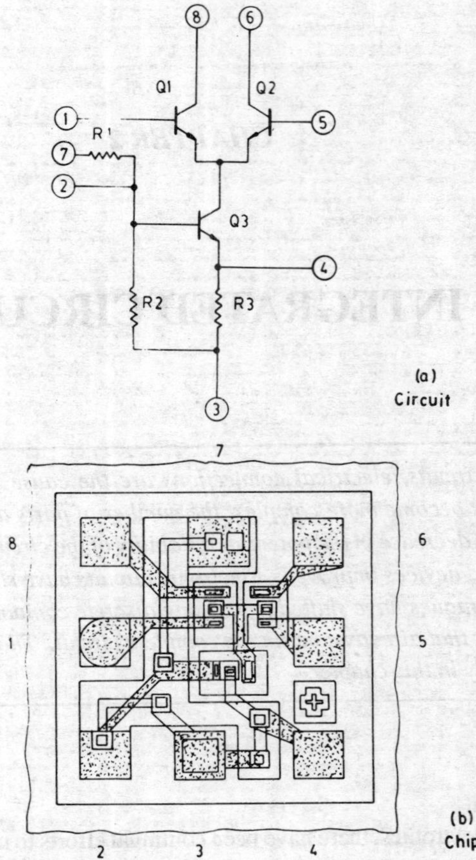


**Fig. 2.1:** Topography of an integrated circuit.

**3. Multiple-Chip Circuits:** Here the passive and active components are formed in *separate wafers*, or chips, of semiconductor material which are then mounted on to a common base and interconnected by thermo-compression bonding.

**4. Hybrid Circuits:** They utilise thin film subassemblies of *passive* components to which the *active* components are attached. Thus the number of soldering connections is reduced and the number of wafers is also decreased.

## 2.2 Solid-State Circuits

In solid-state circuits, all the components are formed on a wafer of silicon using the *diffusion planar process* which offers better control of base width and junction area than the alloy process of transistor manufacture. The various steps involved in this process are illustrated in Fig. 2.2.

To start with, a wafer of monocrystalline n-type silicon is taken. The silicon substrate, or underlayer, is oxidised (Fig. 2.2a) by heating it to a temperature of 1000°C in a flow of steam. This provides a *protective insulating layer* on the substrate. It is then covered with a photosensitive emulsion, which, when exposed to ultraviolet light, is rendered insoluble and in that form resists attack by acid (photo-resist). The emulsion is illuminated through a photographic mask so that all of the wafer except the base areas is exposed to ultraviolet light. It is then developed and the unexposed areas washed away. The silicon wafer is then washed to clear away photoresist in selected places. This completes the *photoetching process* and the slice is ready for base diffusion (Fig. 2.2b).

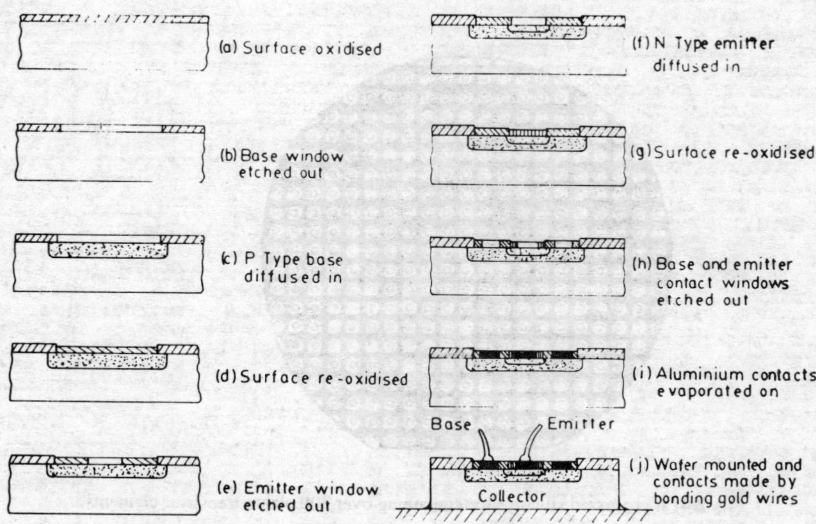


**Fig. 2.2:** Silicon diffused planar process.

The *windowed silicon wafer* is now placed in a furnace and *boron* gas admitted into the furnace. Boron is unable to penetrate the oxide layer, but readily diffuses into silicon at the windows, forming a small p-region at each window (Fig. 2.2c). The entire slice is now coated with a new layer of silicon dioxide (Fig. 2.2d). New *smaller windows* are etched out over each p-region (Fig. 2.2e). The second photoetching is similar to the first one.

The silicon slice is once again exposed to gas but this time it is *phosphorus*, an n-dopant. Thus an n-region is produced within each p-region, giving the three regions required for an n-p-n transistor. The sequence of operations is illustrated in Fig. 2.2(f) through (j).

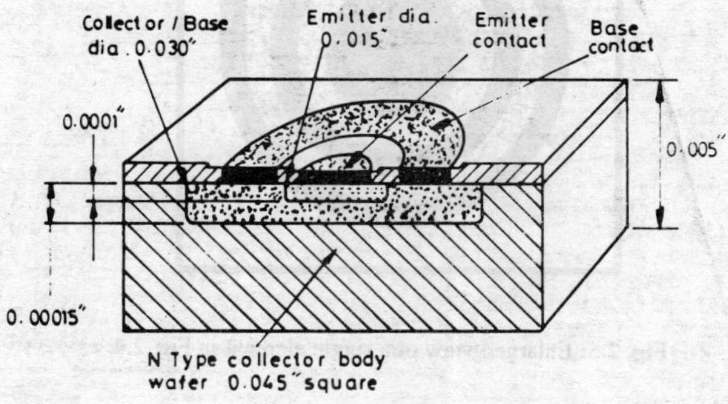


**Fig. 2.3:** Silicon planar transistor wafer.

Next, a thin layer of *aluminium* is placed over the emitter and base regions, a step called *metallisation*. This produces electrical contacts so that the different regions can be connected to the outside world. The collector metallisation is usually gold. When the silicon transistor wafer is cut into slices, several hundred transistors are formed on a single wafer.

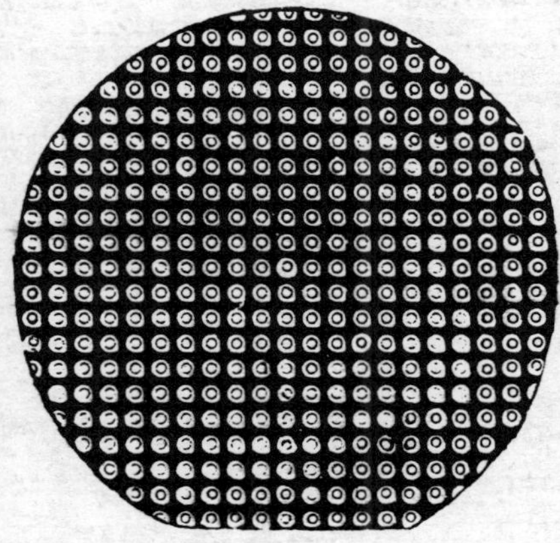

**Fig. 2.4:** A processed silicon slice containing over 300 planar transistor elements.

Gold serves as a very effective means of bonding the chips to their packages and transferring heat and collector current from the chips. Transistors produced in this way are *planar diffused* type, because, in the end result of diffusion process, the n- and p- regions all appear at the surface of the slice in one plane.

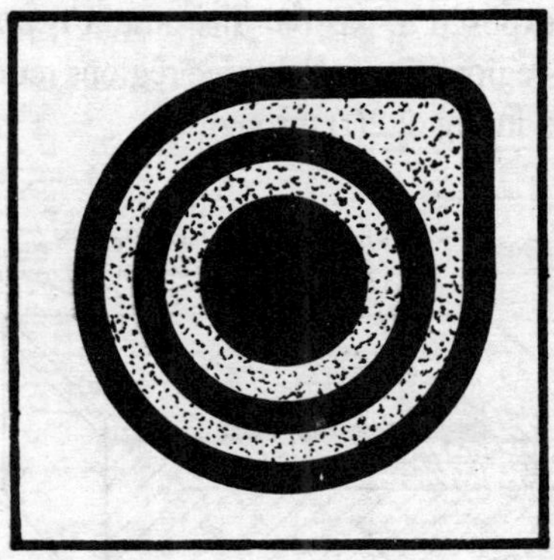

**Fig. 2.5:** Enlarged view of a single element in Fig. 2.4.

A silicon planar transistor wafer is shown in Fig. 2.3, a processed silicon slice containing over 300 planar transistor elements in Fig. 2.4, and an enlarged view of a single element in Fig. 2.4 is shown in Fig. 2.5. The individual wafers are then assembled by fusing down to a *header* which forms the collector contact and contacts are made to the base and emitter regions by compression bonding a small diameter gold wire. Cutaway view of a typical unit is shown in Fig. 2.6.

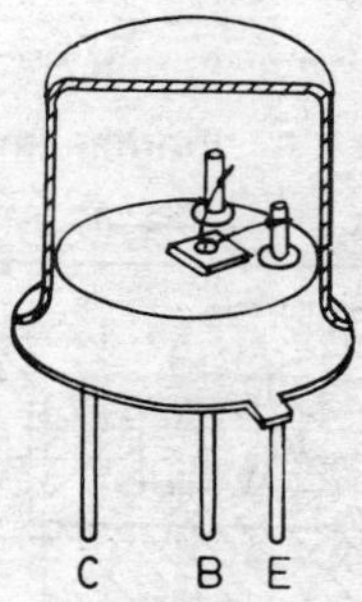


**Fig. 2.6:** Cutaway view of a silicon planar transistor.

The great advantage of the planar process lies in the protection of the collector-base junction by the oxide layer. This results in extremely low collector leakage current. It is of the order of a nanoampere or a fraction of it. Also this *low leakage current is very stable due to the insulating oxide layer over the surface.*

As is the case with all diffused transistors, very *thin base layers* are possible and this gives the possibility of *very high frequency operation.* Also, a whole range of transistors of different size can be made by this process by substituting different photographic masks in the photo-resist process.

The multiplicity of steps in the planar process are all carried out *simultaneously*, so that manufacturing cost per transistor is not high. Power transistors can be fabricated by increasing the emitter and collector areas. Base current must flow out of the base region parallel to the junctions and in so doing the base current sets up a potential drop resulting in transistor action being confined to the area adjacent to the emitter. This shortcoming imposes a limit in increasing the junction area. This is overcome by using a multiplicity of *long narrow emitter regions* to increase the effective area. A silicon diffused power transistor is shown in Fig. 2.7.

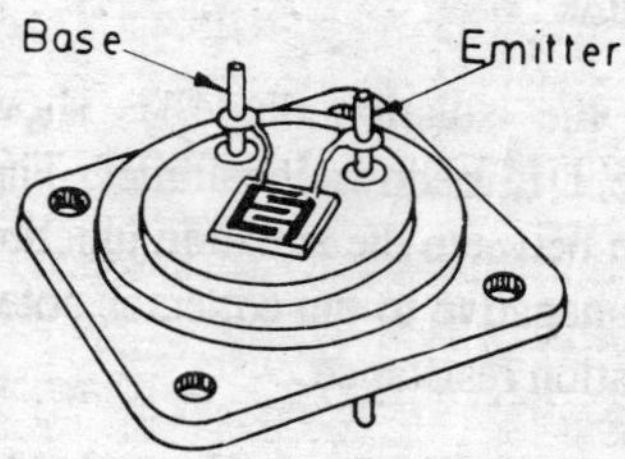


**Fig. 2.7:** Silicon diffused power transistor.

The main shortcoming of the diffused transistor is the *series resistance* between the collector junction and contact at the bottom surface of the wafer. The collector current must flow through the series resistance, thus setting up an internal voltage drop which results in a high value of saturation voltage. This is overcome in epitaxial planar transistors, Fig. 2.8, by using a composite slice consisting of very low resistance substrate on which has been deposited a thin epitaxial layer in which the transistor is formed.

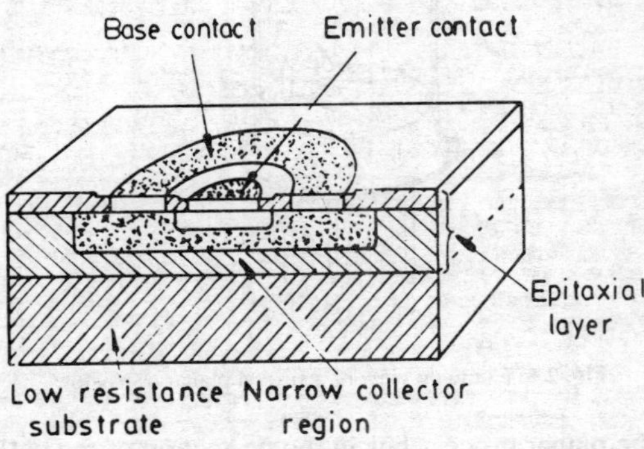


**Fig. 2.8:** Silicon epitaxial planar transistor.

## 2.3 Solid-State Component Formation

Since all of the components are formed in a single piece of conducting silicon, it is necessary to arrange them so that they are *electrically isolated.* A p-type substrate slice is used and the first diffusion of each component made n-type. The p-n junction so formed is biased in the reverse direction and so presents a high resistance. Thus *between any two components there will be two reverse-biased junctions in series, providing high resistance isolation.*

**1. Transistors:** Cross sectional view of a solid circuit transistor is shown in Fig. 2.9(a). The n-type collector region is first formed by diffusing phosphorus into p-type slice followed by diffusing boron to form the p-type base region and finally a second phosphorus diffusion to form the $\overset{+}{n}$ emitter. An $\overset{+}{n}$ region is diffused into the n-type collector region so that a low resistance collector contact can be made on the top surface to allow the interconnections to be made. This introduces an increase in the series collector resistance resulting in a consequent increase of the *collector saturation voltage*, $V_{CE(sat)}$, over that of the conventional planar transistor.

The equivalent circuit including the isolation diode, is shown in Fig. 2.9(b). The junction capacitance, C1, of the isolation diode, D1, is shown in parallel. The *series resistance*, Rs, is due to the bulk resistance of the substrate region between the isolation junction and the isolation terminal. If the latter is connected to a point slightly negative to the collector potential, the diode D1 will be reverse biased and so will present a high isolation resistance.

The equivalent circuit of two solid circuit transistors, formed side by side, is given in Fig. 2.10. The two isolation diodes appear back-to-back between the two collectors. Thus even if the isolation terminal

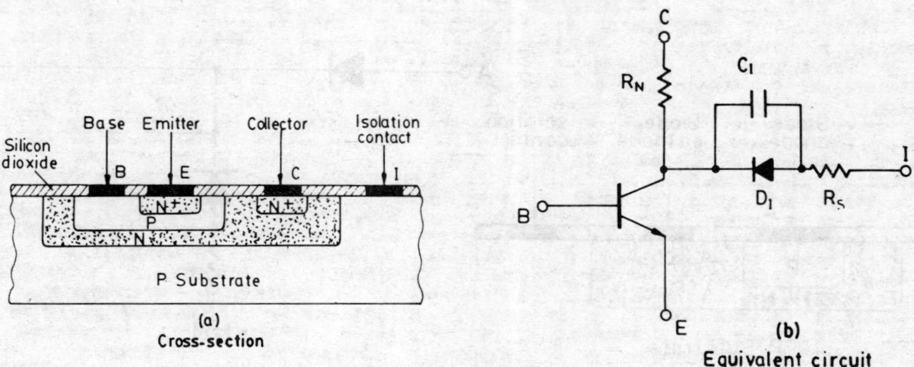


**Fig. 2.9:** Solid circuit planar transistor.

is not connected to any reference potential, the two collectors will be isolated from each other since one of the diodes will always be reverse biased. But with this arrangement, diode capacitance may result in a low isolation for ac.

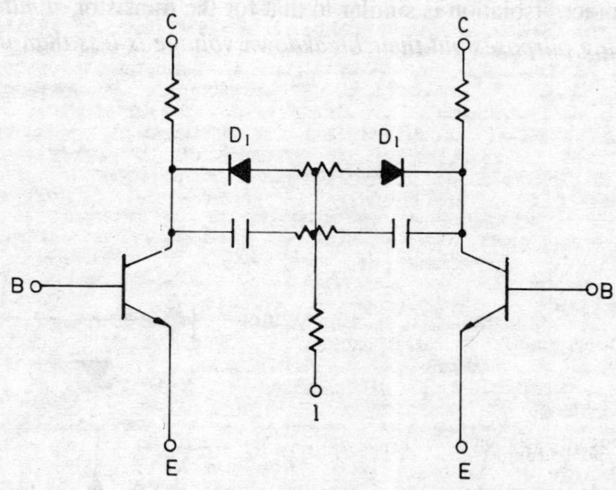


**Fig. 2.10:** Two-transistor equivalent circuit.

**2. Diodes**: Two types of p-n junctions can be formed for use as diodes; collector-base diode and emitter-base diode. Fig. 2.11 illustrates a *collector-base diode*.

The n-type cathode region is formed during the *collector phosphorus diffusion*. The p-type anode region is formed during the *boron base diffusion*. The $n^+$ cathode contact area is formed during the emitter diffusion.

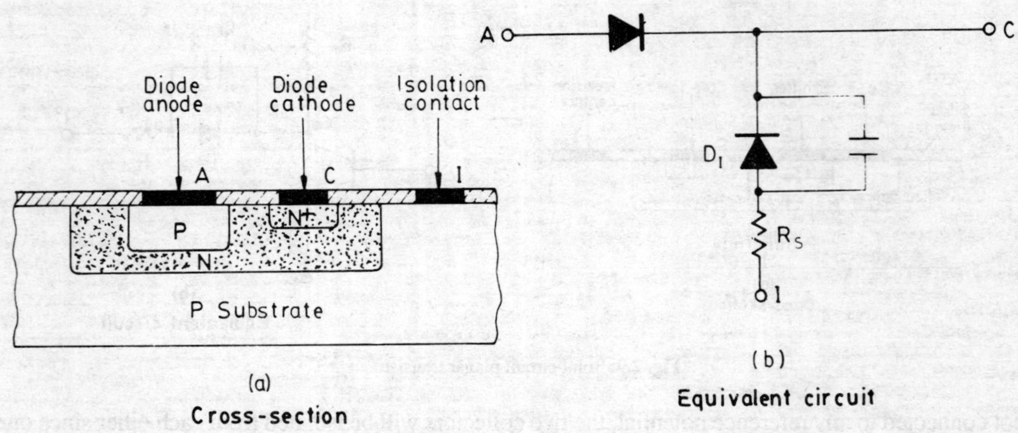


**Fig. 2.11:** Collector-base diode.

Fig. 2.12 illustrates an *emitter-base diode*. An $\overset{+}{n}$ p-n structure is formed similar to a transistor. The $\overset{+}{n}$ emitter region forms the diode cathode and the p-type base and the n-type collector regions are short-circuited by the anode contact. Isolation is similar to that for the transistor. *Emitter-base diodes offer advantage for fast switching purposes but their breakdown voltage is less than that of collector-base diodes.*

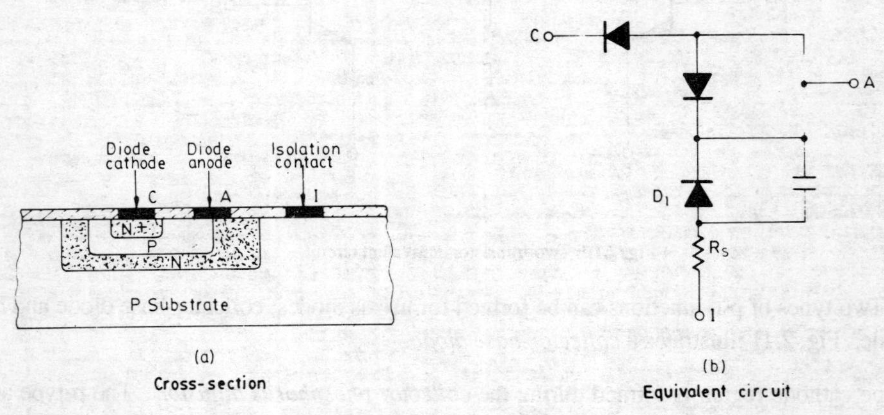


**Fig. 2.12:** Emitter-base diode.

**3. Resistors:** A resistor can be formed in a wafer of high resistivity silicon by the diffusion of an impurity into a *defined region*. The value of resistance will depend upon the dimensions of the region, the surface concentration of the diffused element, and the diffusion depth. For economic reasons, resistors are formed during one of the transistor diffusion processes. The surface concentration and diffusion depth will be fixed by the requirements of the transistor and so the dimensions of the resistor region must be designed to give the correct value of resistance. Values of resistance between 20 and 20,000 ohms can be obtained by using the p-type diffusion illustrated in Fig. 2.13.

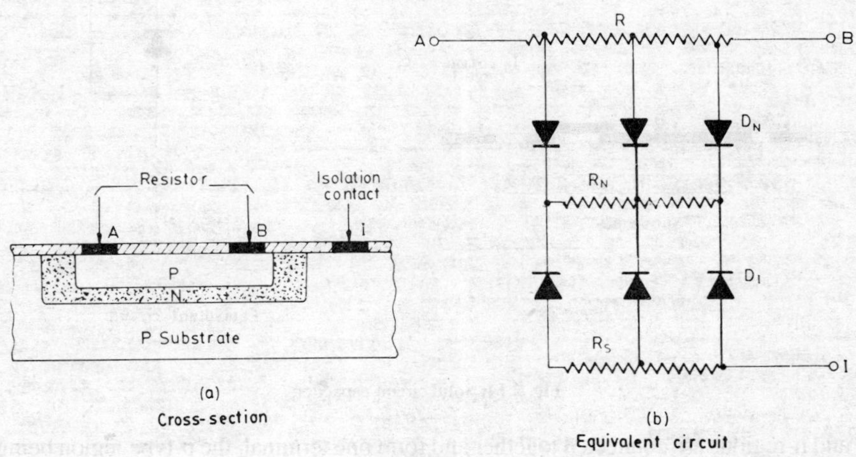


**Fig. 2.13:** Solid circuit resistor.

The main resistor R is formed in the p-type region. The resistance of the n-type region, $R_N$, does not appear in parallel with R, because as soon as there is any current through R, causing a voltage drop between the ends, one of the equivalent diodes will be reverse biased. In any case, the resistance of the n-type region is much higher than that of the p-type region, which has a high concentration of carriers near the surface of the wafer. The isolation diodes, D1, are reverse biased by connecting the substrate to the negative supply.

It is possible to form two resistors side-by-side with a very *precise ratio* of their resistance values, although it is difficult to reproduce individual resistors to closer than ± 10%. The variation of resistance with temperature is the same with each resistor. This precise ratio can be used to advantage in circuit design.

If resistance values smaller than about 20 ohms are required, the emitter diffusion can be used and the resistor formed in the $\overset{+}{n}$ region.

**4. Capacitors:** Capacitors can be formed by using the *depletion capacitance* of a p-n junction. The value of capacitance will obviously depend upon the voltage across the junction. The junction can be formed at the same time as the transistors, diodes and resistors. To realise the maximum capacitance per unit area, a structure similar to a transistor is formed, and the junctions connected in parallel as shown in Fig. 2.14.

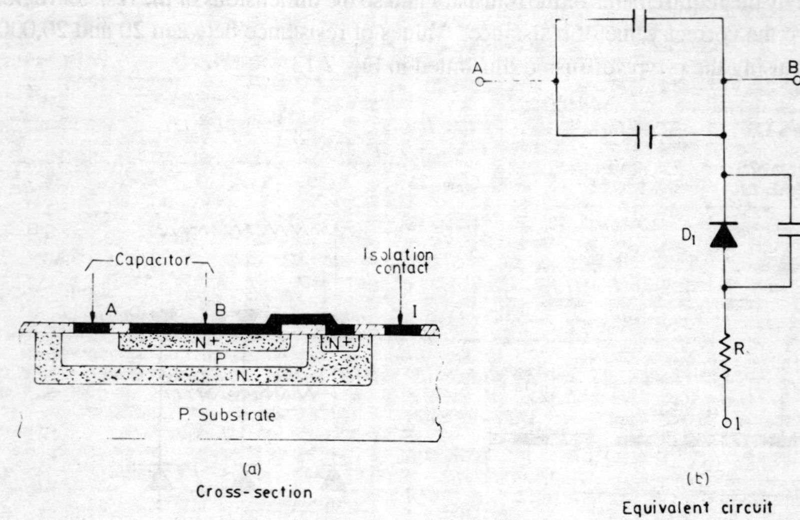


**Fig. 2.14:** Solid circuit capacitor.

The n⁺ and n-regions are connected together and form one terminal, the p-type region being the other terminal. Isolation is arranged, as for the transistor, by the junction between the p-type substrate and the n-type region.

An alternative method of forming a capacitor is to use the silicon dioxide on the surface as a dielectric by evaporating a metallised electrode on to the oxide in the required region.

**5. Circuit Formation:** The formation of various solid components into a complete circuit, or *integrating* them, is illustrated by a typical combination of components in Fig. 2.15. The illustration is, of course, only two dimensional.

The components are *interconnected* by evaporating metal connections over the silicon dioxide. Further components can be connected in a similar manner in the third dimension.

**6. Miscellaneous Features:** There are certain features of solid-state technology that tend to *dictate* the circuit arrangement. The processing cost is roughly proportional to wafer area and so for lowest cost, circuit elements of smaller area must be chosen. As resistors and capacitors made in solid circuit form take up considerably more area than transistors and diodes, so *active components become preferable to passive components.* Thus direct-coupled circuits with a minimum of resistors will offer the lowest cost circuit. By making the electrodes of two or more transistors common, considerable saving in area can be achieved since only a *single isolation region* will, in this case, be required. This is of vital importance in logic circuits where there may be multiple input or output circuits. The control of resistor values in solid circuit manufacture is not high-about ± 10%, but resistor combinations with a precise ratio are

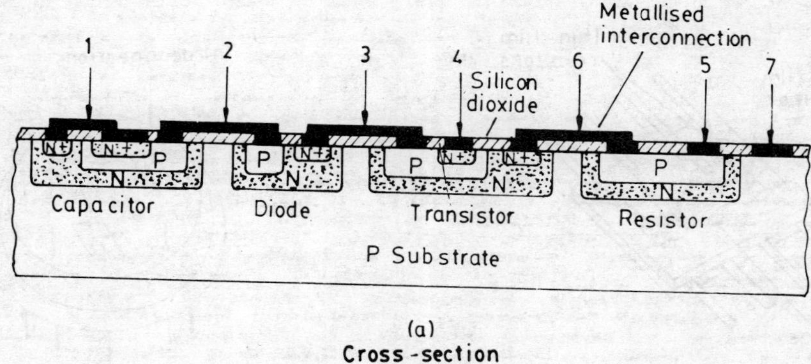


(a)
Cross-section

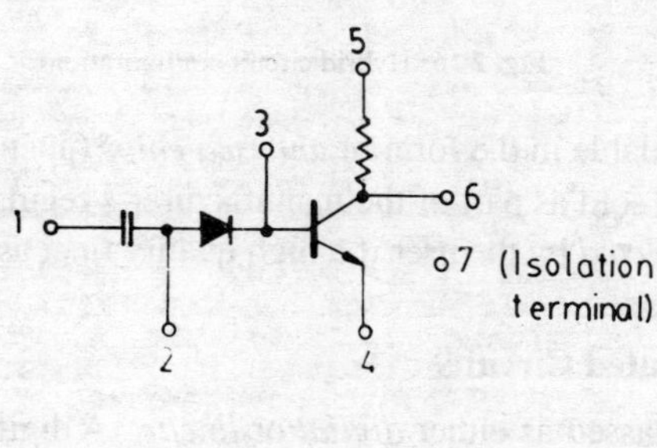


(b)
Circuit arrangement

**Fig. 2.15:** Solid circuit interconnections.

possible to ± 1%. In addition to *space saving* features, the fact that solid components receive almost identical processing enables them to be *closely matched in characteristics*. The closer such circuit elements are matched, the greater the *reliability* of the circuit.

## 2.4 Hybrid Circuits

A hybrid circuit is *one that is fabricated by combining two or more circuit types,* such as film circuits and semiconductor circuits, or even by combining one or more circuit types with discrete elements as shown in Fig. 2.16. The primary advantage of hybrid microcircuits over other microcircuits is *design flexibility*. Device connections inside a given circuit package are called *intraconnections* whereas ordinary hookups external to the package are called *interconnections*.

Several elements and circuits are available for hybrid application. These include discrete components that are electrically and mechanically *compatible* with integrated circuits. They may be used to perform functions supplementary to ICs and can be handled, tested, and assembled with essentially the same technology and tools.

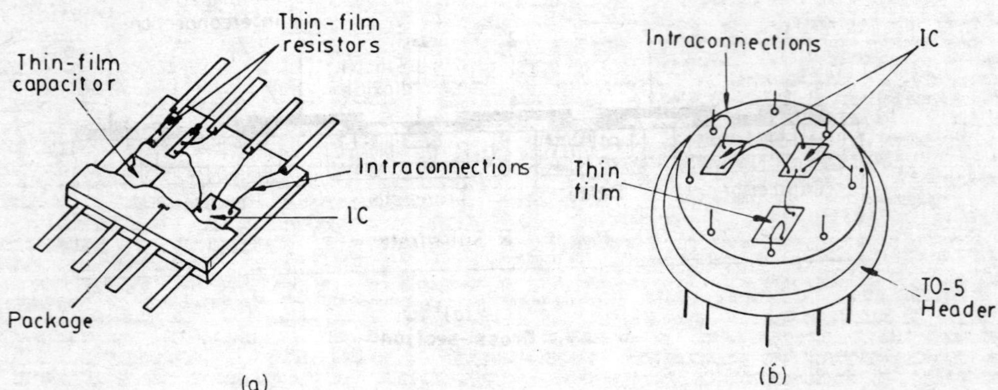


**Fig. 2.16:** Hybrid circuit configuration.

Complete circuits are available in the form of *uncased chips* (unencapsulated IC dice). These chips are usually identical to those sold as part of the manufacturer's regular production line. They must be properly *packaged* and *connected* by the user if a high quality final assembly is to be achieved.

## 2.5 Classification of Integrated Circuits

Integrated circuits may be classed as either *digital* or *linear*. A digital circuit has a discontinuous, or pulsed output. This makes it useful for applications such as computers, numerical control of machinery and data processing systems. Digital ICs may be further categorised according to how their gates are interconnected. They may thus be placed in *logic families*, each logic family having in common a certain type of interconnection and consisting of similar logic functions such as simple gates, inverters, registers, flip-flops etc.

A linear IC is an IC that is not digital. *A linear circuit element is one in which a change in the input causes a directly proportional change in the output*-double the input, double the output. Some linear ICs do not have such a response-the square law detector used in radio receivers for example. Also, such linear circuits as zener diodes are sometimes used in linear ICs. It is important not to take the term too literally, and to understand its limitations. The basis of most linear circuits is the *linear transistor amplifier*.

## 2.6 Large Scale Integration

Much of the reliability credited to ICs arises from the fact that *interconnections are made within the chip* and are, hence, not subject to the problems arising from discrete connections such as vibration, fatigue, cold solder joints, etc. Thus, it was only natural to build on the same chip many interconnected circuits to provide some overall system or *subsystem*. This has been especially effective in the digital areas and has been enhanced by the use of MOS devices instead of bipolar transistors.

*Large scale integration*, (*LSI*) refers to applications where more than 100 gates are fabricated on the same chip. If the number of gates is more than 12 but less than 100, this is referred to as *medium scale integration* (*MSI*). The number of components may be much higher than the number of gates. For example, 100 inverters consist of 200 components. Or 100 NOR gates, each of which may be obtained by interconnecting four COS/MOS inverters, would consist of 400 components.

## 2.7 IC Packaging

Integrated circuit packages have evolved from the still used transistor type (modified to include more leads) to the widely used *flat pack*, and to the *dual-in-line package* (DIP). Generally these packages are limited to 14 or 16 leads, but some configurations are available with many more. Packages with up to 160 leads are available with many large-scale integrated devices.

**1. Modified T0-5 Package:** Modified T0-5 is a term used to describe a package having the same general appearance as a standard T0-5 transistor package but which has been modified by increasing the number of external leads and the dimensions of the package.

Since the modified T0-5 package typically has 8,10 or 12 leads compared with the 3 or 4 leads of the standard transistor package, *its pin-circle diameter is slightly larger.*

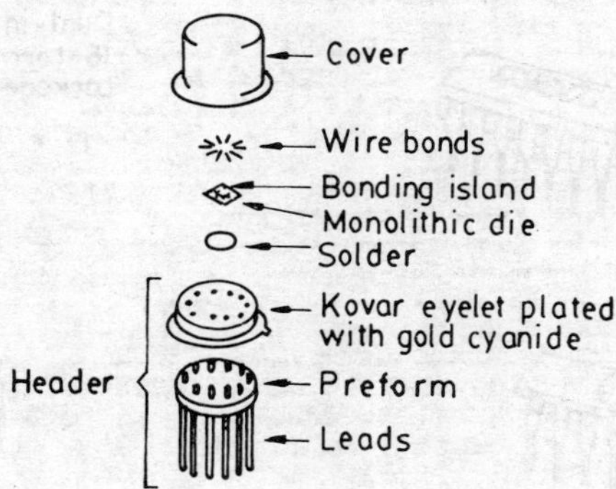


**Fig. 2.17:** Exploded view of modified T0-5 package.

The modified T0-5 package can be either *pin-plugged* or *embedded* into the printed circuit board. It can also be plugged into a socket which has been soldered into the printed circuit board.

**2. Flat Pack:** Many types of integrated circuit flat packages are being produced in various sizes and materials. These packages are available as square, rectangular, and circular configurations with 10 to 60 external leads, and may be fabricated of metal, ceramic, epoxy, glass or combinations thereof. The *ceramic flat pack*, shown in Fig. 2.18, is representative of all flat packs with respect to general packaging requirement.

**3. Dual-In-Line Package:** The dual-in-line package (DIP) was designed primarily to overcome the difficulties associated with handling and inserting packages into mounting boards. Dual-in-line packages are easily inserted by hand or machine and require no spreaders, spacers, insulators or lead forming. It is also possible to field service the devices with standard hand tools and soldering irons.

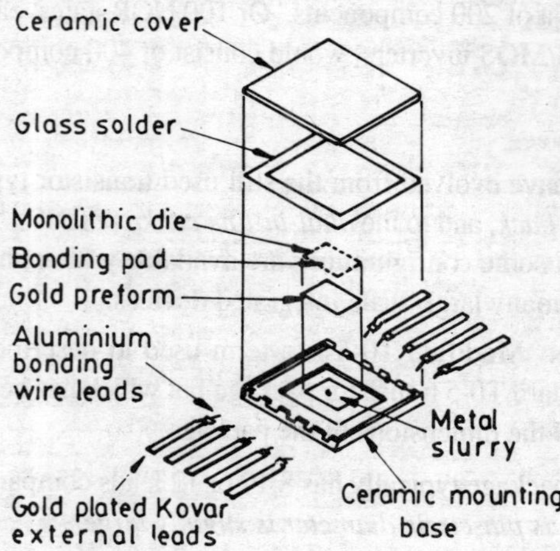


**Fig. 2.18:** Exploded view of ceramic flat pack.

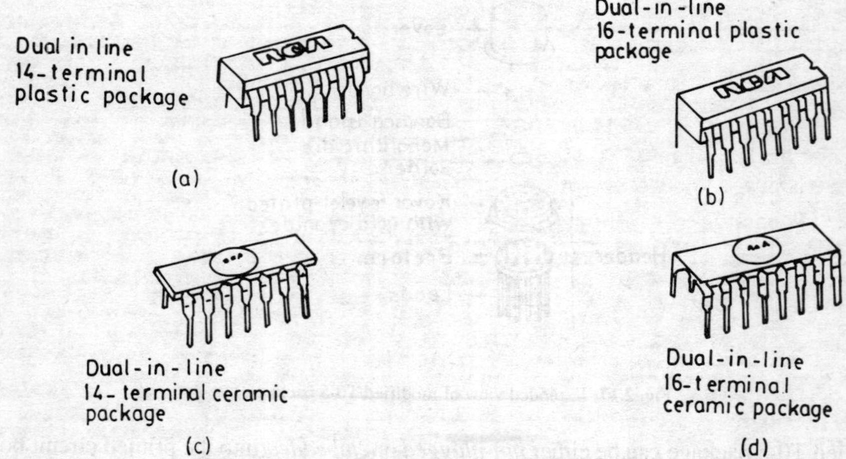


**Fig. 2.19:** Dual-in-line package (DIP) assembly.

Plastic dual-in-line packages are finding wide use in commercial applications and a number of military systems incorporate ceramic in-line packages. At present *ceramic dual-in-line packages* are the most common of the two package types to be found in microelectronic systems.

## 2.8 Printed Circuits

Printed circuits are appearing more and more in electronic equipment, and have proven to be equal in or perhaps better than, conventional type construction. A few simple machine operations automatically produce circuits that formerly required a production line of workers performing the same jobs by hand.

One method of manufacturing a printed circuit is the *photoetching process*. A plastic or phenolic sheet with a thin layer of copper coating (copper clad) may be used. The copper coating is covered with

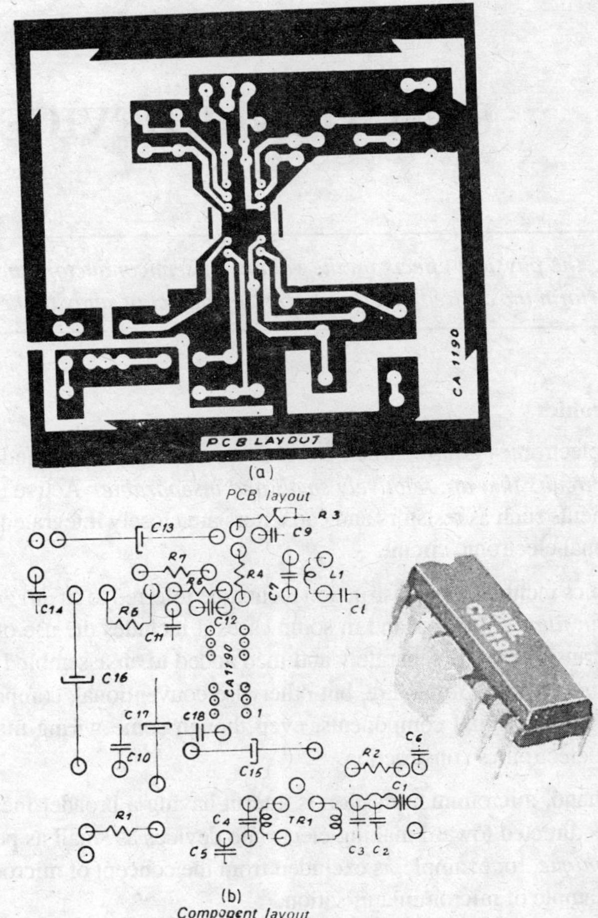


**Fig. 2.20:** Printed circuit board.

light-sensitive enamel. A *template* of the circuit that will ultimately appear on the plastic sheet is placed over the sensitized surface. The entire sheet is exposed to light. The area of the copper that is exposed, reacts to light and can then be removed by an etching process.

The exposure of the printed circuit is similar to a photographic exposure. The enamel on the unexposed circuit protects the unexposed copper. After the etching bath, the enamel is removed from the printed circuit by a solvent solution. This leaves the metallic surfaces in a condition that is suitable for soldering parts and connections. The printed circuit board layout and component layout of a typical integrated circuit are shown in Fig. 2.20.

*CHAPTER 3*

# FUNCTIONAL DEVICES

---

*Employing the physical effects on the volume of semiconductors and harnessing these effects to perform the desired functions form the subject of study in this chapter.*

---

### 3.1 Microelectronics

The term microelectronics is used when we refer to those processes and techniques involved in the construction of *circuits that are relatively small and inseparable*. Active elements such as transistors, and passive elements such as resistors and capacitors are closely integrated on a single wafer, or block, to form a functional electronic circuit.

Microelectronics includes the construction technique referred to as *integrated circuitry*; it also involves the use of *thin film techniques* and in some cases it includes the use of *discrete microcomponents* (for example, a transistor made separately and then added to an assembled unit). The term is not used to include construction using miniature, but otherwise conventional, components with interconnecting wiring between the individual components, even though some wiring may be required between assemblies in microelectronics construction.

On the other hand, microminiaturisation is a term having a broader meaning. The term is used to refer to all efforts directed toward making electronic devices as small as possible. The *micromodular construction technique*, for example, is excluded from the concept of microelectronics, although it represents a good example of microminiaturisation.

### 3.2 Functional Microelectronics

Two basic trends are predominant today in the field of microelectronics; one towards *further miniaturisation* using improved techniques for the fabrication of each element of the integrated circuit on the chip as a discrete element; the other is towards *employing the physical effects* in the volume of semiconductors and harnessing these effects to perform the desired functions. The latter trend received the name *functional microelectronics*.

*It is the physical effects that impose a limit on the scale of integration.* Of these, the factor responsible for complication of *heat removal* is the most serious one. As the size of microcircuits decreases and the scale of integration grows, the problem of heat removal becomes difficult. In an effort to remove the heat dissipation in the chip, the design engineer can decrease the power consumption, but only to a certain limit at which the circuit must operate properly. The result is a lower packing density. What complicates the problem of heat removal is the fact that with a decrease in the size of microcircuit the resistance of conductor grows, so the circuit dissipates more power.

The phenomenon of *ion migration* in film conductors at high current densities also places a limitation on the packing density and reduces the reliability of circuits.

Integrated electronics today, in essence, faces the same problems as those which spurred its development at the beginning; the problems of dispensing with the dominance of components, simplifying interconnection problems, and improving the reliability.

The only way to solve the problems is to abandon the classical concept of component based circuitry design and to evolve the devices that would use the properties of solids to perform the functions of, say, a generator, amplifier, etc., in the system. Some devices of this type, called *functional units*, are already known in practice. A distinguishing feature of these devices is that *they have no discernible circuit elements equivalent to the elements in a conventional circuit*. There is reason to believe that further progress in microelectronics can be made precisely through the use of functional units.

Examples of the simplest functional units can be quartz crystal vibrators and negative resistance devices. Other examples include Gunn diodes and charge-coupled devices featuring considerable functional capabilities. These devices rely on the *volume effects* in solids and dissipate energy almost in the entire *bulk* of the crystal, so that heat removal problems are less severe than they are in classical devices with p-n junctions.

A search for the ways of evolving functional systems based on the direct utilisation of physical phenomenon has led to the appearance of new branches in radio electronics. These are cryoelectronics, magnetoelectronics, thermoelectronics, molectronics, acoustoelectronics and optoelectronics.

## 3.3 Four-Layer Functional Devices

A specific example of semiconductor functional devices is a four-layer pnpn structure which displays the regions of low, high, and negative resistances in its current-voltage characteristic. By properly selecting the *desired operating conditions and loads*, it is possible to construct various devices for many applications using a considerably less number of circuit elements.

Fig. 3.1 illustrates a four-state current switch composed of two thyristors in which two layers, n and p, are common. The common control layer has a metal ring brought to ground potential. The common

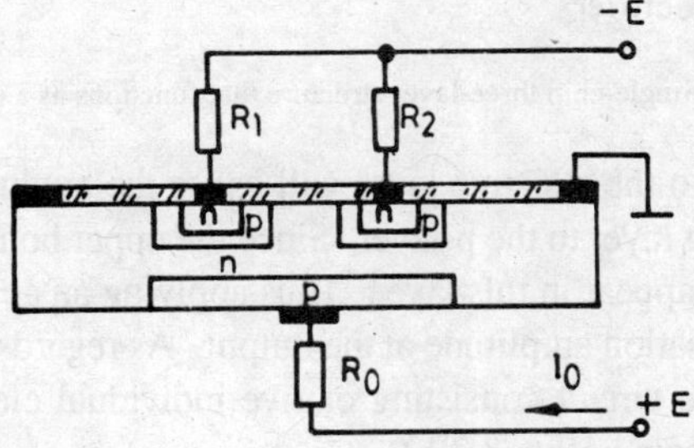


**Fig. 3.1:** Four-layer current switch.

p-layer receives a positive voltage via a resistor of high value to keep control current $I_0$ constant. *The value of the current determines the state of the switch.* Separate p and n layers are taken to negative potential via resistors of different values to bias the upper p-n junctions in the forward direction.

The switch can be in one of the four stable states depending on the value of $I_0$. When the two thyristors go off and the current is small, the switch shifts to the *first state*. When the left thyristor of high gain goes on, the switch swings to the *second state*. For its biasing to conduct, the left thyristor requires a lower current than the right one. When the right turns on and the left off, the switch moves to the *third state*. This happens when the current becomes sufficient to bias the right thyristor into conduction. The process is as follows:

Since R2 < R1 and the current through the right thyristor is much higher than it is through the left, while the current $I_0$ remains constant, the holes injected into the n-region from the lower p-region travel to the right thyristor. As a result, the entire current flows through the right thyristor, and the left one stays off. In the *fourth state*, both thyristors are made conductive by applying a sufficient current to the output of the switch. This switch is suitable for use in various digital devices.

## 3.4 Rectifiers

Fig. 3.2(a) illustrates one more functional microcircuit. This is a single-chip three-layer structure that functions as a rectifier. The upper layer is a resistor separated by a thin dielectric layer from a semiconducting layer performing the function of a *thermo-EMF generator.* The insulating layer must be a good dielectric with an enhanced thermal conductivity. Since the thermal constant for the structure is much greater than the period of alternating current, the temperature during the period will not change in essence and thus the output voltage will not change too.

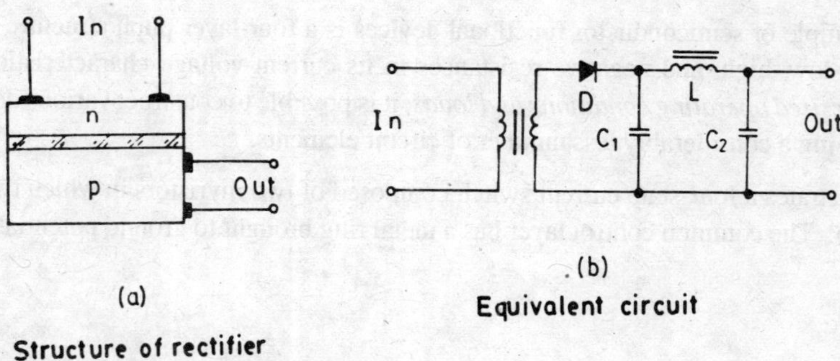


**Fig. 3.2:** Single-chip three-layer structure that functions as a rectifier.

Alternating current applied to the resistive layer will cause the evolution of thermal energy which flows through the thin insulating layer to the p-layer. Since the upper boundary of the p-layer gradually heats up the *Seebeck EMF* will appear in this layer. Thus applying an ac voltage to the input results in a dc voltage of a very small pulsation amplitude at the output. As regards the functions it performs, the microcircuit is equivalent to the circuit consisting of five individual elements: a transformer, choke, diode and two capacitors, as shown in Fig. 3.2(b).

### 3.5 Piezoelectricity

The application of an electric field to a dielectric specimen produces two results; (i) *it induces dipole moments* in atoms or ions and (ii) *it displaces ions* relative to each other. Hence, the dimensions of the specimen change very slightly. On the other hand, mechanical stresses applied to a specimen do change the dimensions of the specimen but, in general, do not produce dipole moments. Thus we find that in most of the materials, *dielectric polarisation causes mechanical distortion*; but the reverse is not true-mechanical distortion does not cause polarisation. Thus electromechanical conversion is restricted to one direction only, namely, from electrical to mechanical. This electromechanical effect is called *electrostriction* and is present in all materials. Further, in a purely electrostrictive material, the mechanical distortion caused by any polarisation in a given direction is the same as that caused by polarisation of equal magnitude but opposite direction.

In yet another category of solid dielectric materials, *the mechanical deformation produced by polarisation in a given direction reverses when the direction of applied polarisation is reversed.* Such materials get polarised on application of a mechanical stress and thus permit conversion of mechanical energy into electrical energy and vice-versa. These materials are called *piezoelectric materials.*

Substances, such as quartz, exhibit what is known as the *piezoelectric effect*, that is to say, a mechanical strain applied to a suitable piece cut from the substance causes an EMF to be developed between two surfaces of that piece; and conversely, an EMF applied between two opposite faces of the piece causes a mechanical deformation.

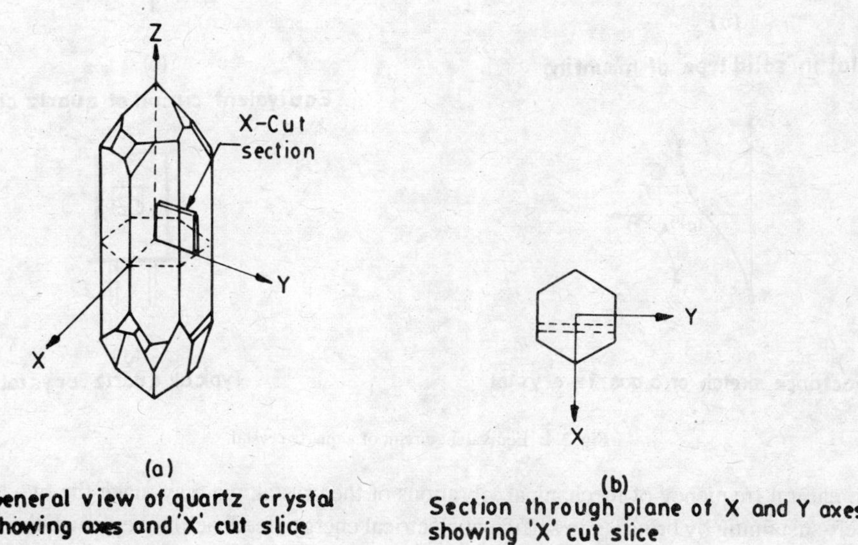


**Fig. 3.3:** Quartz crystal showing x-cut slice.

Slices cut from a quartz crystal also have another important property, they behave electrically as a *resonant circuit of extremely high quality factor*, resonant at the natural frequency of mechanical vibration of the slice.

The complete quartz crystal in its natural state is of the general shape shown in Fig. 3.3(a), and for convenience of reference, three axes through the crystal are defined as follows: the vertical axis passing through both points of the crystal is called the *optical or z-axis*; the line at right angles to the z-axis parallel to any major face of the crystal is called the *electrical or x-axis*; and the line at right angles to both these axes, and perpendicular to any face of the crystal is called the *mechanical or y-axis*.

Slices cut from the crystal for various purposes may then be described in terms of the angles between them and the three axes of a crystal; thus a slice cut with its faces perpendicular to the x-axis is called the *x-cut crystal* as shown in Fig. 3.3, and this cut is frequently used in filters.

If such a slice be mounted between two flat metal plates to which electrical connections can be made, it is found to behave like the equivalent circuit given in Fig. 3.4. C, L and R are merely the *electrical equivalent of the corresponding mechanical properties of the crystal*, and their values are determined by the dimensions and physical constants of the crystal material. C′ represents the electrical capacity between the faces of the crystal, and in practice its value may be increased by the capacity of the wiring. The series circuit composed of C, L and R resonates at the natural (mechanical) resonant frequency of the crystal, $f_R$. In addition to this series resonant frequency, the circuit of Fig. 3.4(b) also has an antiresonant frequency, $f_A$. The reactance sketch of the crystal is therefore of the form shown in Fig. 3.4(c).

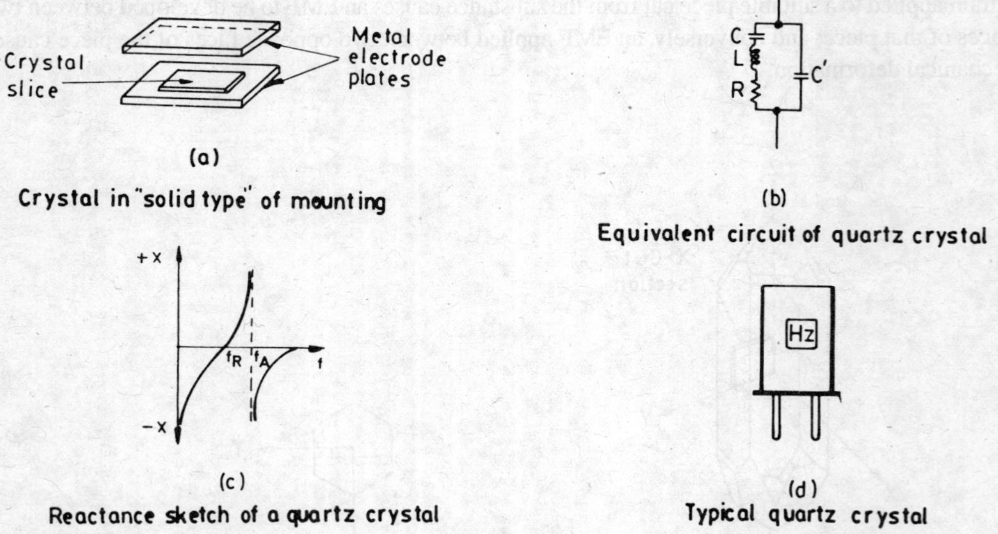


**Fig. 3.4:** Equivalent circuit of a quartz crystal.

At the natural frequency of mechanical vibrations of the crystal, we may make the two actions mutually self-sustaining by bringing in sufficient electrical energy (feedback) to replenish the energy that is lost as heat during each cycle. The effect of the crystal, thus, is to produce an oscillating voltage whose frequency is determined by the *natural frequency* of the crystal.

## 3.6 Acoustoelectronics

This branch of science and engineering deals with the *excitation and propagation of ultrasonic waves in solids*, the frequency range of which extends from tens to hundreds of megahertz. The discovery of acoustoelectronic effect has made it possible to harness the range of unltrasonic frequencies for practical applications and to evolve unique types of solid-state devices.

*These devices work on the principle of conversion of electric energy to elastic waves which travel in solids at a comparatively low speed, which is five orders of magnitude lower than the speed of electromagnetic waves.* This feature opens up new possibilities for further microminiaturisation of radio and television equipment.

*Surface acoustic wave* (SAW) devices can practically handle all the tasks of data processing (delaying, filtering, amplifying etc.). Moreover, the technology for the fabrication of acoustic systems is *compatible* with IC technology. The possibility thus exists for producing hybrid form acoustic and electronic microcircuits on the market. These are delay lines, amplifiers, memories etc.

### 3.7 Delay Lines

The velocity of sound in any medium is very low compared to the velocity of propagation of electromagnetic radiation, so that the utilisation of the *transmission delay* of sound waves is attractive for storage systems.

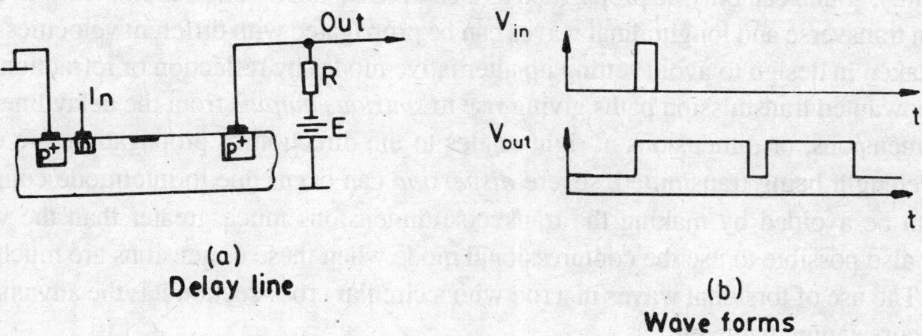


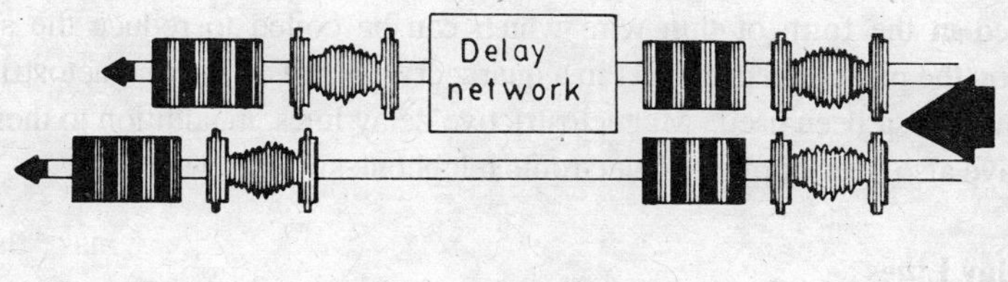


**Fig. 3.5:** Delay line waveforms.

An acoustic delay line consists of the essential parts shown in Fig. 3.6. The electrical input energy is transformed by the input transducer to acoustic energy, which propagates through the delay medium with the velocity of sound to the output transducer, where the acoustic energy is converted back to the electrical output signal.

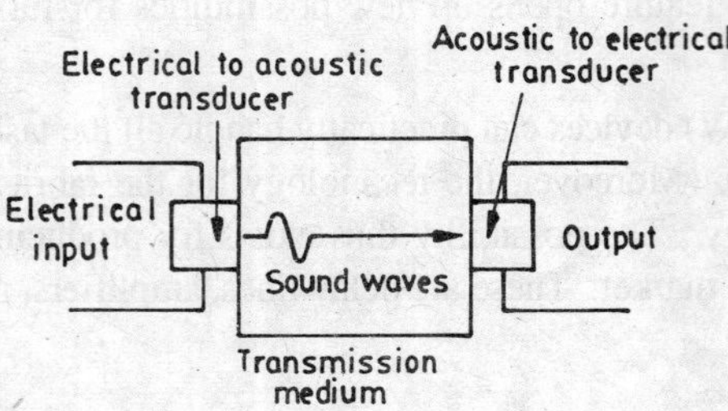


**Fig. 3.6:** Essential parts of an acoustic delay line.

Liquid media have the following advantages over solid media. Firstly, the velocity of sound in a solid is several times its velocity in a liquid, so that the *path length* for a given delay line is less in a liquid. Secondly, sound can only be propagated as a compressional or *longitudinal wave* in a liquid, but in a solid both transverse and longitudinal waves can be propagated with different velocities. Care must therefore be taken in design to avoid setting up alternative modes by reflection or refraction, since they may lead to unwanted transmission paths giving rise to *spurious outputs* from the delay line. When the transverse dimensions, or dimensions at right angles to the direction of propagation, are comparable with the wavelength being transmitted, severe *dispersion* can occur due to intermode coupling. This dispersion can be avoided by making the transverse dimensions much greater than the wavelength, although it is also possible to use the compressional mode when these dimensions are much less than a wavelength. The use of torsional waves in a rod with a circular cross-section has the advantage that the dispersion occurs at any diameter.

Water had been used as the transmission medium but the most successful units with a liquid medium have employed mercury. Fused quartz solid delay units have also been built, in which *multiple internal reflections* were used to reduce the volume required to contain a given transmission delay. Solid media have also been used in the form of thin wire which can be coiled to reduce the space occupied. Transducers based on the piezo-electric effect in a quartz crystal and on the magnetostrictive effect in a magnetic material have also been used. Magnetostrictive delay lines, in addition to their use in several early computers, have also been applied in electronic telephone switching systems.

### 3.8 Ultrasonic Delay Lines

The connection of an ultrasonic delay line is shown in Fig. 3.7. The transducers at the input and output ends are connected by an acoustical waveguide which may be fabricated from steel, quartz, glass, water or mercury. When the signal is to be delayed by 64 μs, it is preferable to use ultrasonic delay lines with a solid waveguide.

The input signal is applied to an amplifier 1. From the amplifier, the signal is then routed via a matching transformer to excite a piezo-electric transducer, *a*. For its operation, this transducer depends

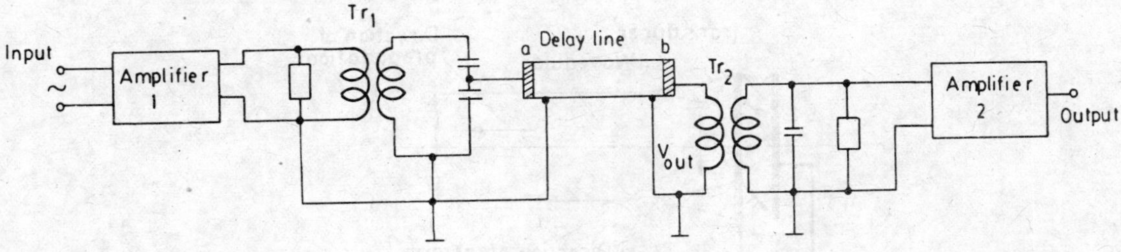


**Fig. 3.7:** Connection of an ultrasonic delay line.

on the well known piezo-electric effect. The transducer converts electric signals to ultrasound which is then propagated in a sound conducting medium. Another transducer, *b*, at the receiving end converts the ultrasonic waves back into electric signals. The delayed signal is then fed via a matching circuit to an amplifier 2.

**Fig. 3.8:** Delay line.

If the delayed signal is to be free from distortion, it is important that the piezoelectric transducers have a flat frequency response, that is, produce about the same signal at any frequency of the spectrum. Unfortunately, piezoelectric transducers show good performance at the high frequency end and a poor performance at the low frequency end of the range. Because of this, the signal spectrum is usually *modulated* into the high frequency end of the range.

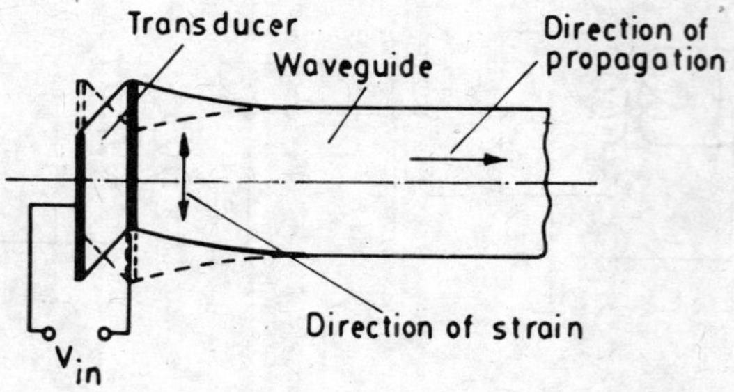


**Fig. 3.9:** Excitation of transverse mode in a delay line.

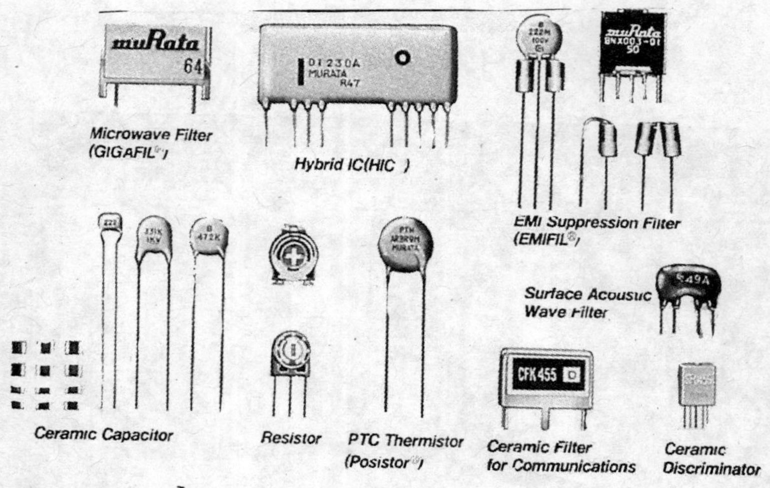


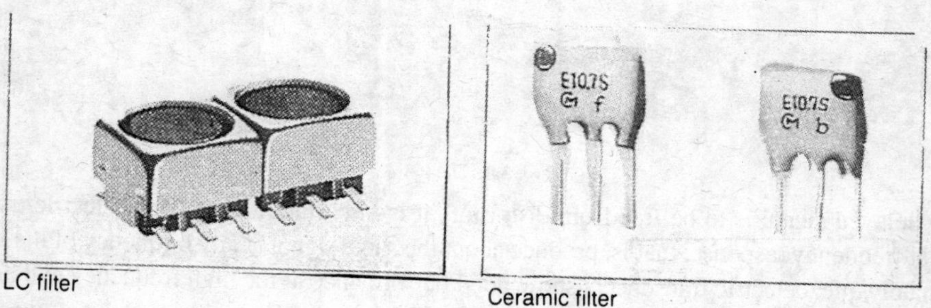


**Fig. 3.10:** Filters.

The delay line of a colour tv receiver is usually excited in a *transverse acoustical mode*, see Fig. 3.9. The applied electric field causes the transducer to deform in the transverse direction, at right angles to the direction of propagation of waves.

### 3.9 L C Filters

Filters are electronic components that *pass or cut certain frequency bands*. Analog filters practically used since old time include a variety of kinds which differ in structure and the principle of operation — like the LC filters, RC active filter, vibrator filter, crystal filter, mechanical, ceramic and other filters shown in Fig. 3.10.

The LC filter has a simple construction that consists of capacitors and coils, but has wide applications in both low and high frequency ranges. This is due to the *large degree of freedom* in the centre frequen-

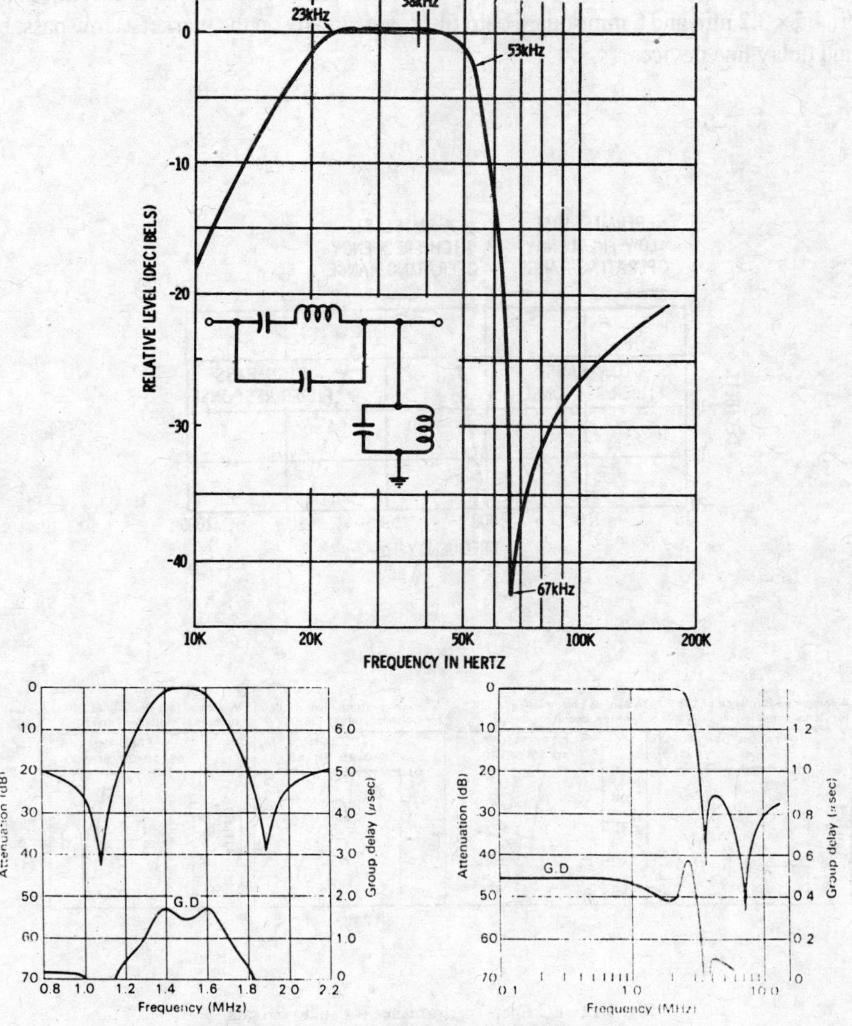


Fig. 3.11: LC filters' performance for video circuits.

cy and band width., *small size* resulting from the performance of core material of coils and *low cost*. Many LC filters are used in AV equipment such as video tape recorder (vtr), television (tv), radio and other audio equipment.

LC filters used in video signal are getting smaller as vtr cameras also get smaller. Starting at 7 mm square, moving to 5 mm, chips now are 4 mm square. The assortment of 4 mm chips is expected to grow as SMT (Surface Mounting Technology) becomes more popular.

Audio filters also are following the same footsteps as their video counterparts. As audio products use more digital technology, so will filters for digital circuits have a wider variety.

LC filters themselves are evolved from a traditional construction of wound wire and capacitors to a new concept. This can be seen with the *multilayer chip LC filter* with a monolithic construction, a type of multilayer inductor. An inductor is built with laminated ferrite and conductive material. Such inductors are further laminated with conductive material to form capacitors, and then the chip is baked into an LC filter; 4.5 × 3.2 mm and 5 mm square chip filters are already on the market as low pass, band pass, equaliser and delay line devices.

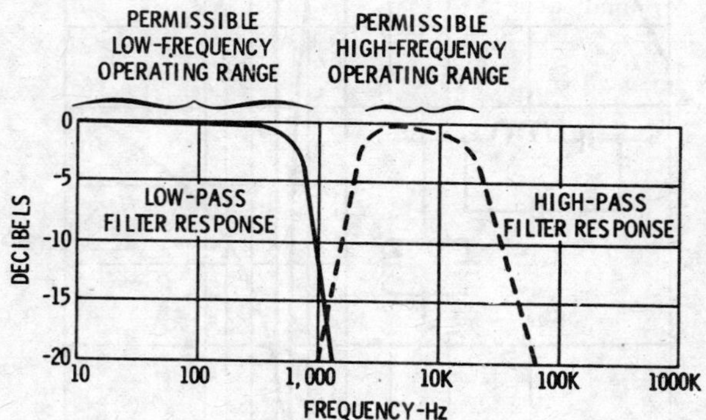


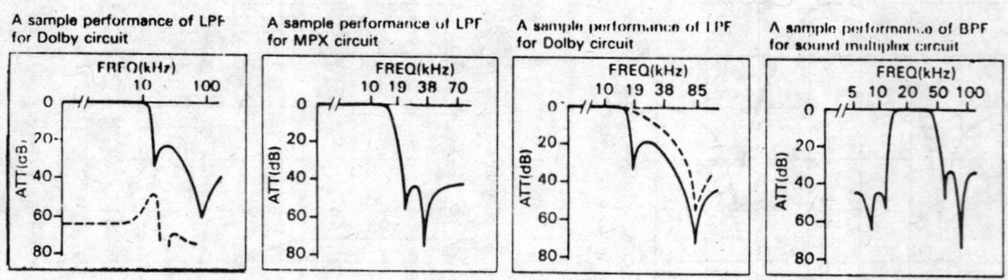


**Fig. 3.12:** LC filters' performance for audio circuits.

### 3.10 Surface Acoustic Wave (SAW) Filters

Surface acoustic wave filters are compact, highly reliable solid elements not only used as if filters in tv sets, but are beginning to be used widely in new *media devices* such as optical communication, catv, mobile radio, space communication and are becoming the *key component* that determines the performance of these devices.

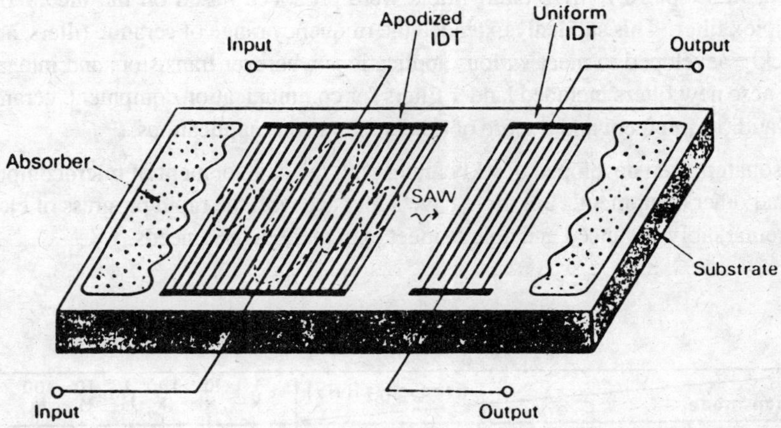


**Fig. 3.13:** Fundamental configuration of SAW filter.

Generally, the SAW filter is formed by input and output *comb-shaped transducers* (IDTs - inter-digited) on a piezo-electric substrate, as shown in Fig. 3.13. When an electric signal is applied to the input side electrode, a SAW is generated. This SAW is received by the output electrode and transferred to an electric signal. Thus, the filter works as a transverse filter. Generally, in the SAW filter, the *overlapped length* of electrode fingers in one IDT is varied in order to obtain the required filter characteristics.

**Fig. 3.14:** SAW Filter.

The SAW filter production process is fundamentally the same as the LSI manufacturing process but the absolute accuracy of electrode dimensions is more important in the former than in the latter. Since one of the objectives of the adoption of SAW filters is *to make circuit adjustment unnecessary*, the problem of reproducibility is extremely important.

### 3.11 Ceramic Filters and Resonators

The ceramic filter began its history with the appearance of lead zirconate titanium in 1955. Since then many developments have occurred beginning with the first production of ceramic filters for AM radios in 1962. Then, in 1966, 10.7 MHz radio filters were produced based on the theory of the energy enclosed multiplex filter. This suddenly extended the frequency range of ceramic filters, and new types were then quickly developed to meet various applications wherever transistors and integrated circuits were in use. These new filters included ladder filters for communication equipment, ceramic filters for television and audio equipment and a score of others for various applications.

Ceramic resonators then developed quickly along with the development of microcomputers, remote control units and other equipment. But these have kept pace with the rapid progress of electronics and have become remarkably advanced and able to meet extremely diverse needs.

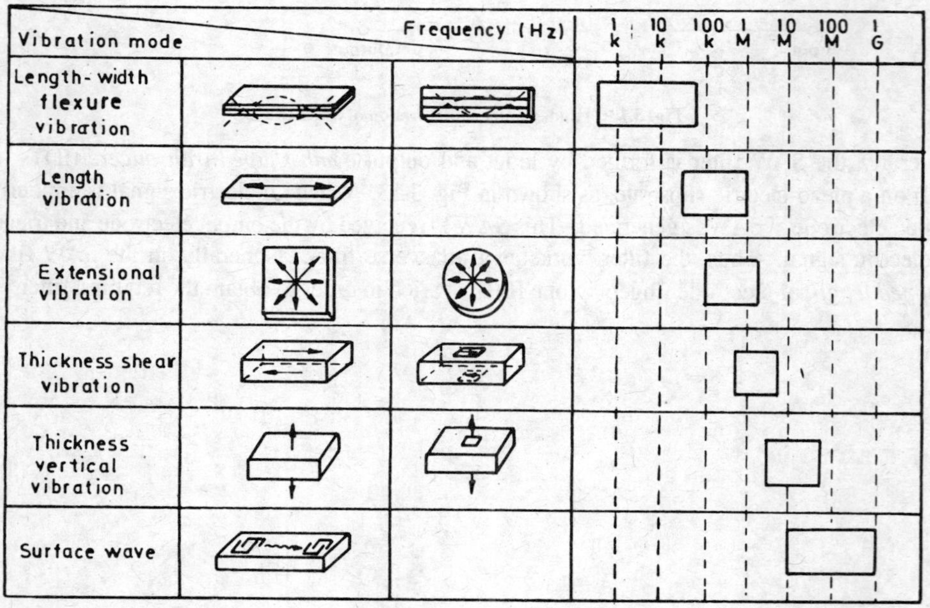

| Vibration mode | | Frequency (Hz) | 1 k 10 k 100 k 1 M 10 M 100 M 1 G |
|---|---|---|---|
| Length-width flexure vibration | | | |
| Length vibration | | | |
| Extensional vibration | | | |
| Thickness shear vibration | | | |
| Thickness vertical vibration | | | |
| Surface wave | | | |

**Fig. 3.15:** Vibration modes and their frequency ranges.

Electronic equipment is becoming *smaller and lighter* day by day, and ceramic filters and resonators have also followed this trend. For these components, size and vibration modes are closely related. Fig. 3.15 shows vibration modes and frequency ranges.

AM and FM filters, and small ladder filters are available for many types of communication equipment. All of these contribute greatly to *miniaturisation*.

Chips are the smallest, low profile components available and innovative techniques are now making automatic mounting of these components increasingly easy and reliable.

### 3.12 Wave Absorbers

Electric wave absorbers *attenuate* electromagnetic waves by using energy loss to prevent reflection and external leakage. They are used in *electromagnetic shields* and for preventing tv ghosts.

*Ohmic, dielectric and magnetic loss are used to attenuate and absorb electric waves.* Therefore, electric wave absorbers are required to have a large electromagnetic loss to effectively attenuate electromagnetic waves. In the past, carbon materials were commonly used for ohmic losses; such materials were contained in foam urethane or mixed with rubber materials.

Ferrite electric wave absorbers take advantage of the large loss ferrite has in the high frequency range and a phenomenon where magnetic characteristics change in inverse proportion to wavelength. They are used as electric wave absorbing panels for *tv ghost prevention.*

## *CHAPTER 4*

# CHARGE-COUPLED DEVICES

*Invented in 1969, charge-coupled devices have offered a base on which to build solid-state photoelectric imaging devices with a number of scanning elements answering the requirements of broadcast TV standards. These are discussed in this chapter.*

### 4.1 Introduction

The charge-coupled device (CCD) is not a MOS process, rather it uses MOS processing to make potentially very fast (say upto 100 MHz) low power *dynamic shift registers*. Some special features of this device are:

(1) there are no *transistors* on the silicon substrate

(2) *no windows* are required to be cut for diffusing dopant

(3) there are no *p-n junctions* in the structure.

Charge-coupled devices operate on the principle of generation and storage of minority carriers (a charge packet) in *potential wells,* transfer of charge packet together with potential wells along the dielectric (oxide)-semiconductor interface and its subsequent detection.

### 4.2 Charge-Coupled Devices

Charge-coupled devices have a three layer structure produced by the common MOS technique. They are extremely simple; (see Fig. 4.1). After oxidation of the surface of a semiconductor wafer to form a

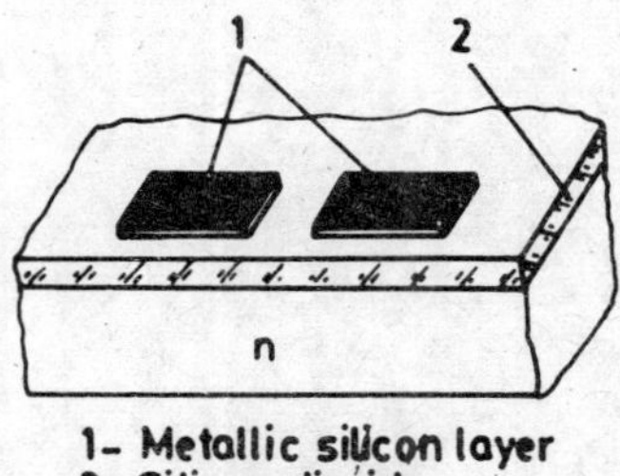


**Fig. 4.1:** Charge-coupled device.

thin insulation layer, metal conductors and electrodes are suitably formed on this layer. In comparison with the process required for the manufacture of bipolar and common MOS transistor ICs, the process for the production of a CCD involves one-fourth, and half as many operations respectively. *CCDs can form the basis for memory, delay, logic and image transfer devices.*

Consider the operation of a device with two electrodes on the oxide layer, Fig. 4.2. A negative voltage applied to the n-type substrate must be higher than the threshold voltage required to form a homogeneous depletion layer at the substrate-insulator interface. Increasing the negative voltage (storage voltage) on an electrode 1 (gate) can produce directly under this gate a deeper layer called a potential well (see Fig. 4.2a). In the *storage mode*, the device can receive and store charge (minority carriers) generated in the semiconductor by an appropriate source. Because the minority carriers in the n-type silicon are holes, and the gate is negative with respect to the substrate, the gate attracts the holes and retains them in the potential well.

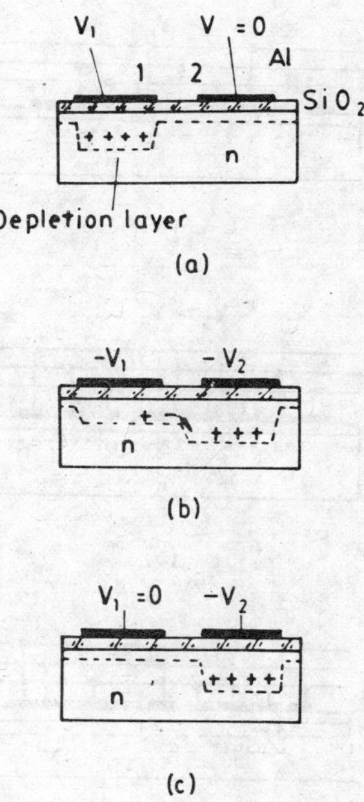


**Fig. 4.2:** CCD structure.

In the *transfer mode*, the charge packet moves to a gate 2 held at a more negative voltage (transfer voltage) to form a yet deeper potential well under the gate. Now the holes stored under the gate drop into the deeper well and flow to the gate 2. The initial conditions of storage can be restored after removing the storage voltage from the gate and decreasing the transfer voltage on gate 2 to the storage voltage.

As soon as the storage mode is set up under the gate 2, the process of charge transfer can be extended to move the charge packet to the next gate and hence along the substrate. Thus, using only two levels of voltage permits shifting the charges from under one gate to the next through the entire device. *Control over the transfer of charges is made by clock pulses applied to CCD gates.*

## 4.3 Shift Registers

CCDs are most commonly used to perform the function of shift registers. The control gates of a register form an electrode system with one to four phases. The principle of a *three phase CCD shift register* is illustrated in Fig. 4.3. Each third gate is connected to one and the same phase line. First, the storage voltage $-V_2$ is applied to one group of electrodes, say 1, 4, 7 etc.; with the other electrodes kept at a lower bias voltage. The bias $-V_2$ produces potential wells under gates 1, 4, 7 etc.; with the result that the charges injected into the semiconductor flow to these wells. Assume that the charges resulting from the injection are present under the gates 1 and 7 and are absent under the gate 4.

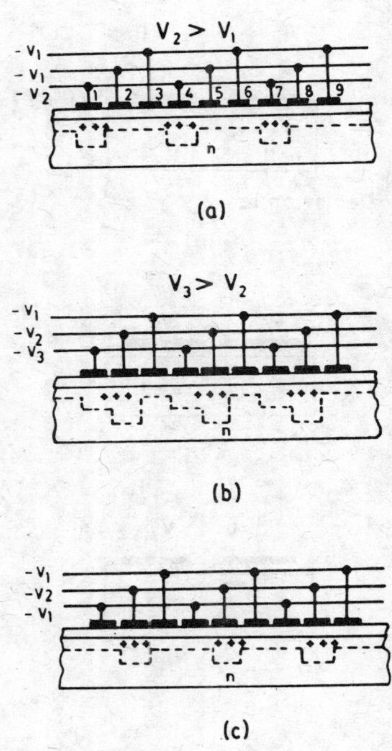


**Fig. 4.3:** CCD as shift register.

At the next stage, the transfer voltage $|V_3| > |V_2|$ is set on the other group of electrodes, such as 2, 5, and 8, Fig. 4.3(b). Since the potential wells under these gates become *deeper*, the charges under the gates 1 and 7 along with the vacant state under the gate 4 will *shift* one step to the right. Further, after restoring the initial conditions of storage, Fig. 4.3(c), the potential on electrodes 2, 5, and 8 decreases to the storage voltage $-V_2$, and on the others to $-V_1$. Thus *charges (or a no charge state) can shift along through the device.* Charge carriers move from one potential well into the next by virtue of three mechanisms: self-induced drift, diffusion, and edge effect.

The *self-induced drift* results from the repulsion of charge and enables their fast transfer, but this drift occurs only at high charge densities. *Diffusion* leads to an exponential decay of charge under the transfer gate. *The edge effect* acts in the direction of charge transfer and can appreciably speed up the process. This field also causes an exponetial decay of charges.

## 4.4 Transfer Effciency

The process of charge transfer in a CCD can be defined by the transfer efficiency, η, which specifies a relative portion of charge being shifted in one *transfer cycle*.

$$\eta = 1 - q_l/q_r + q_l = 1 - \varepsilon \qquad 4.1$$

Here $\varepsilon = 1 - \eta$ is the *loss factor* (transfer inefficiency); $q_l$ is the charge lost during transfer, and $q_r$ is the residual charge transferred. The *transfer efficiency* largely depends on the rate of motion of charge carriers from one potential hill to the next; therefore, *electrons are more preferable charge carriers than holes*. A substantial decrease in the transfer efficiency results from the trapping of charge carriers in surface states.

## 4.5 Transfer Time and Storage Time

Another important parameter is the transfer time required to shift the charge between the neighboring gates. This time is far less than 1 μs.

The *storage time* can, in fact, be in the order of a few seconds; while the clock rates of transfer voltages can typically be in the range of a few megahertz. As the storage time elapses, the potential wells get filled with minority carriers appearing through thermal generation. To increase the time of data storage, use should be made of wide-gap semiconductors.

## 4.6 Structure of a CCD as Shift Register

The transfer of charges from gate to gate is the primary aim one strives to achieve in all applications. But in many cases there is a need to produce a *generator of charges at the input and a detector at the output*.

It is possible to *generate* charges by several methods; for example, by forward biasing the diffused p-n junction in n-type silicon under the action of a surface avalanche breakdown in the MOS structure.

Detectors can be built around a reverse biased p-n junction or Schottky diode. The charge can be *detected* by changing the capacitance according to the charge stored or by changing the potential on the surface electrode with the charge change.

A shift register with 24 gates is available which can act both as a circulating storage (*memory*) and as a *controlled delay line*. This is shown in Fig. 4.4. The width of a CCD gate is not more than 10 μm, the gate capacitance is 0.4pF, and the channel length (the centre-to-centre distance between two neighboring gates) is equal to 10μm. The charge transfer time is below 1 μs; the register operates in the range of a few megahertz. The CCD can also store information over a period of several seconds. The transfer voltages (voltages on the gates) range to about 10V, which is comparatively easy to secure in silicon devices.

The *response* of a MOS IC can be raised two to ten fold using circuit engineering methods in combination with the current manufacturing methods. The former methods can increase the response by

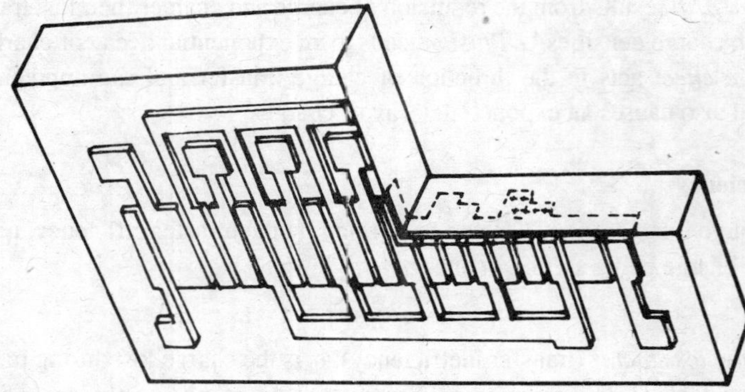

**Fig. 4.4:** Structure of a CCD as shift register.

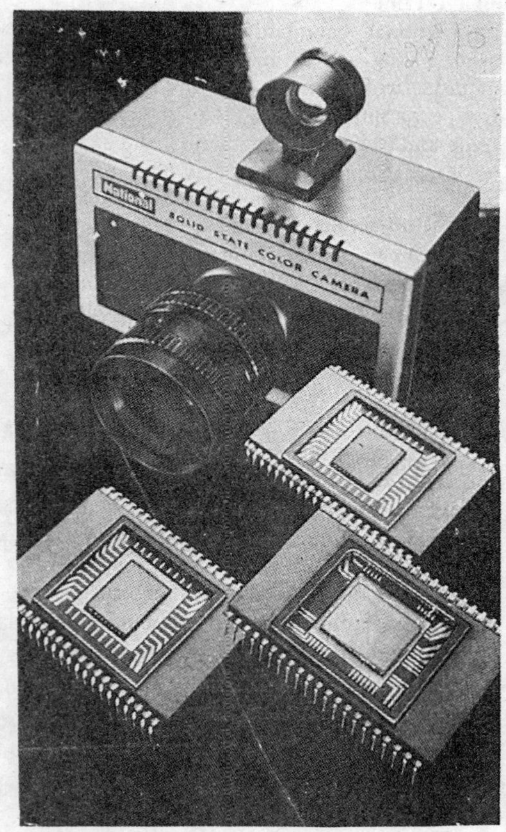


**Fig. 4.5:** CCD Imager.

introducing additional control signals. Some circuits employ the principle of four-phase, or *multiphase locking*, to speed up the response without the power consumption or the dimensions of transistors. For example, using the design of a four-phase control circuit of somewhat increased complexity, it becomes possible to fabricate by common diffusion techniques, the shift registers with a clock rate of 10 MHz and RAMs (random access memories) with an access time of less than 10 ns.

## 4.7 CCD Imager

Reticon's D-series photodiode arrays are charge-coupled photodiode devices (CCPD) with video rates typically up to 20 MHz. The typical dynamic range is 64 dB and the picture elements are 13 × 13 μm. Clock rates for the diode arrays can be up to 10 MHz and a diode reset is provided. The D-series are available in four configurations: 256, 512, 1024 and 2048 elements. They are packaged in a 22 pin ceramic dual-in-line package and pin-compatible across the series.

*CHAPTER 5*

# OPTOELECTRONIC DEVICES

*The relation between light and electricity was first observed by Hertz in 1887 and was termed the photoelectric effect. The devices that resulted from photoelectric emission were vacuum and gas photo tubes and were given the name photoelectric devices. Modern solid-state devices, which include emitters, sensors and couplers are called optoelectronic devices. We study these devices in this chapter.*

## 5.1 PHOTOELECTRIC EFFECT

In 1887, Hertz observed that a spark passes between two metal electrodes at a lower voltage when ultraviolet light is shining on the electrodes than when they are not illuminated. It was then discovered by J J. Thomson in 1893 that negative charges are emitted by a metal surface on which ultraviolet light impinges.

J.J. Thomson was able to show that the negative electric charge that leaves the zinc plate, Fig. 5.1, under the influence of ultraviolet light, or X-rays, consists of electrons. The emission of electrons by the action of ultraviolet light, is called the *photoelectric effect.* The electrons that are given off by the metal plate are called *photo-electrons*, they are not different in character from other electrons.

## 5.2 OPTOELECTRONICS

Vacuum and gas photo tubes, that resulted from the discovery of photoelectric effect, were commonly called photoelectric devices. Modern solid-state devices, which include emitters, sensors and couplers, are called optoelectronic devices and have a much wider scope

*Optoelectronics may be defined as the technology that makes use of the interactions between photons (small packets of light energy) and electrons.* Optoelectronic products may be categorised into *sensors* (responders to light), *emitters* (of light) and *users* (of light) which are often a combination of the first two.

There are two other categories in the area of optoelectronic sensors besides the *photoemissive* type. The *photovoltaic* devices generate a voltage, while the *photoconductive* devices, which require an external source, may be further subdivided into two branches. One is the bulk type often referred to as *photoresistors*; the other uses reverse biased depletion junctions and includes *photodiodes and phototransistors.*

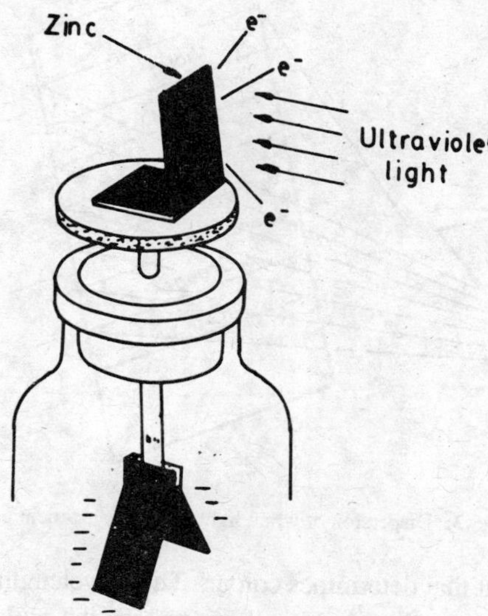


**Fig. 5.1:** A simple experiment showing the photoelectric effect. A negative charge is emitted by a zinc plate upon which ultraviolet light impinges.

Optoelectronic solid-state emitters are of three main types. They are *light emitting diodes* (LEDs) that emit visible radiation for display purposes, *infrared* (IR-type) *emitters* and *laser diodes.*

Finally, emitters and sensors are often combined in one package to provide a *coupling function*, such as providing a high degree of *isolation* between low powered integrated circuits and ac power line voltages, etc.

## 5.3 THE VISIBLE LIGHT SPECTRUM

Light occupies only a small portion of the electromagnetic spectrum. Light wave frequencies are extremely high and since frequency and wavelength are inversely related — the shorter the wavelength the higher the frequency — we can expect the wavelength of light to be small. The spectrum of visible light is shown in Fig. 5.2.

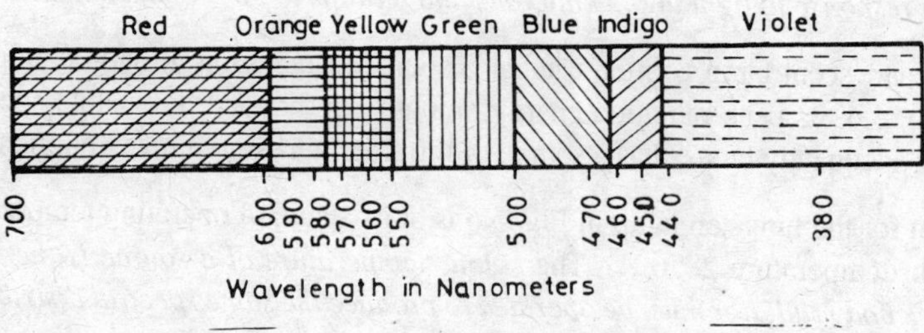


**Fig. 5.2:** The spectrum of visible light.

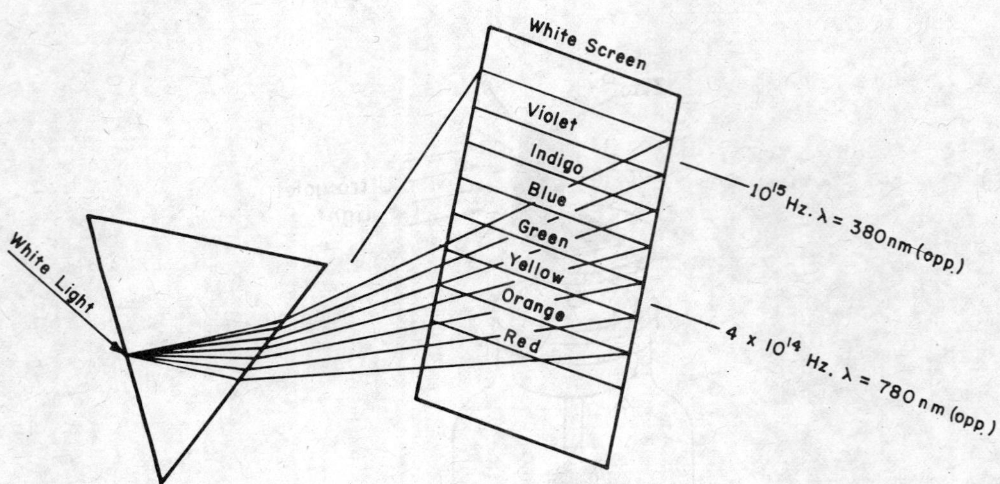


**Fig. 5.3:** Dispersion of white light into its constituent colours.

It is the wavelength of light that determines colour. The wavelength decreases as we go from left to right, that is, as we proceed through orange, yellow towards the violet end of the spectrum. Thus a colour such as red has a longer wavelength than blue. Red, for example, has a wavelength of 700 nanometers ($10^{-9}$ m), while violet has a wavelength of 400 nanometers.

To produce the visible light spectrum we take white light (sunlight) and pass it through a prism, Fig. 5.3. The prism will then separate the white light into its constituent colours, a process known as *dispersion*. We can reconstitute white light by sending coloured light in proper proportion through a prism, a process known as *recombination*.

The colours of light can also be separated by *filters*, as shown in Fig. 5.4. Every colour has a particular wavelength. The colour may also be a mixture of different wavelengths. One of these may be made stronger than the others and it is this one, the *dominant wave-length* of light, that represents the colour of an object.

## 5.4 SPECTRAL RESPONSE

Optoelectronic devices utilise energy in the visible and infrared regions. Fig. 5.5 shows the response of average human eye to the visible radiation that extends from 0.4 to 0.76 microns in wavelength. *The eye has a maximum response to light at 0.55 microns and it drops to 50% of maximum at 0.61 microns.*

The tungsten incandescent lamp is often used as a source of radiation in photoconductive devices. As illustrated in Fig. 5.6, its peak radiation energy output is far beyond the visible spectrum into the infrared, and it may adequately be used as a source of infrared radiation in many applications.

The curve drawn for the tungsten lamp in Fig. 5.6 is at its standard operating temperature, and corresponds to a colour temperature 2870°K. The *colour temperature of a source is the temperature at which an ideal black body radiator must be operated to produce the same spectral distribution or visual effect as the source in question.*

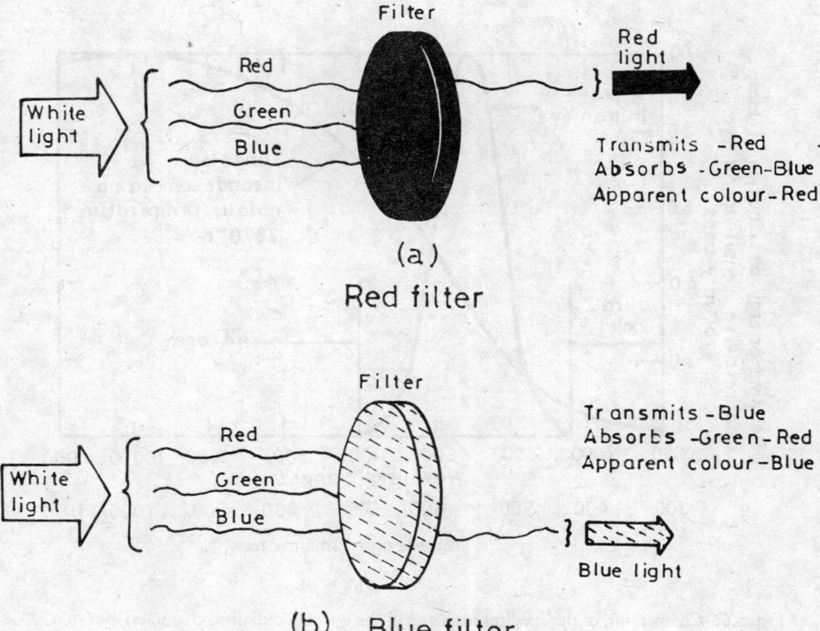


**Fig. 5.4:** Selective colour filtering.

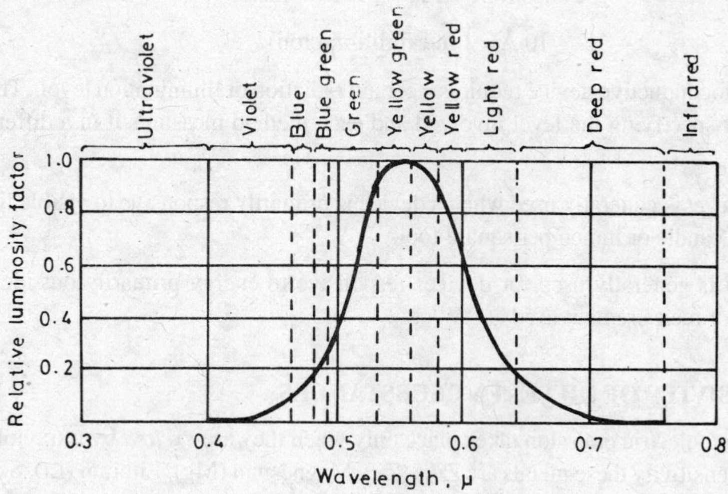


**Fig. 5.5:** Spectral response of the eye to visible radiation.

For lower colour temperatures, the peak moves more to the right with very little visible radiation at 1600°K The sun's radiation (colour temperature 6000°K) has its peak in the middle of the visible

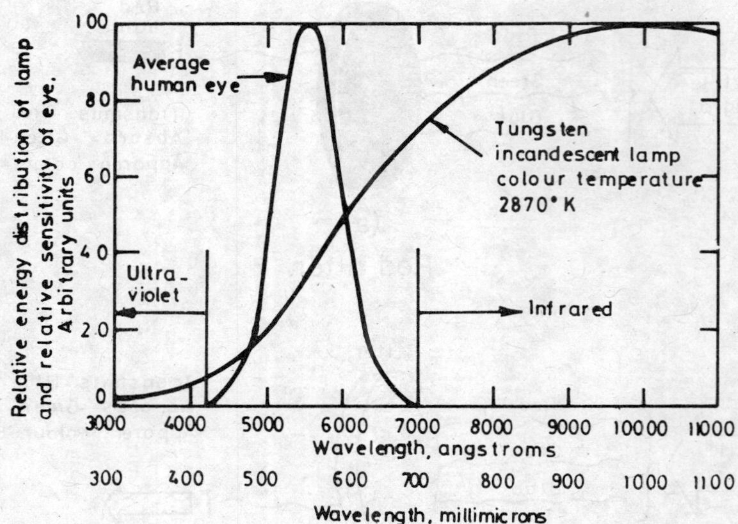


**Fig. 5.6:** Comparison of the spectral response of the eye and radiation characteristics of a tungsten lamp working at a standard colour temperature.

spectrum, but its energy extends deep into the infrared. The wavelength in Fig. 5.6 is also given in *angstroms*, Å, a frequently used unit in optics.

$$1 \text{ angstrom} = 1 \text{ Å} = 1 \times 10^{-10} \text{ meter} \tag{5.1}$$

$$10 \text{ Å} = 1 \text{ m}\mu \text{ (milli-micron)}$$

To activate a photoconductive device requires a certain radiation or illumination level. There are two different methods of specifying the level involved and each method measures it in a different type of unit.

An *illumination level* is generally used when a device is primarily responsive to visible light. In this case, the unit is foot candle or lumen per square foot.

A *radiation level* is generally used for devices responsive to energy primarily outside the visible spectrum. The unit of measurement used is mW/cm$^2$.

## 5.5 PHOTOEMISSIVITY OF DIFFERENT SUBSTANCES

For most metals, photoelectric emission takes place only when they are *excited* by ultraviolet light. In order of increasing sensitivity these metals are Zinc (Zn), Magnesium (Mg), Lithium (Li), Sodium (Na), Rubidium (Rb), Cesium (Cs) ... etc. In this respect *alkali metals* are extremely sensitive and in some cases these metals respond to light rays from visible part of the spectrum. *The frequency at which electrons are just liberated from the metal is called the threshold frequency of the metal.* When the frequency is increased beyond the threshold frequency, more and more energetic electrons contribute to emission. Thus the emission current increases with an increase in frequency till a state of *saturation* is

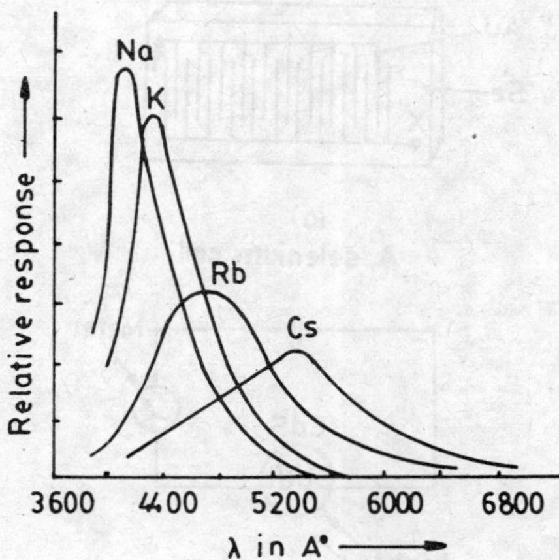


**Fig. 5.7:** Photoemissivity of different substances.

reached. The curves in Fig. 5.7 form the basis for selection of materials used in the construction of instruments based on photoelectric effect. In the case of metals, the curves are helpful for measuring or computing the number of photo electrons emitted per incident quantum as a function of the light employed and also the energy distribution of the emitted electrons.

Photoelectric emission is exhibited by non-metals, liquids and gases as well. The characteristics like *fluorescence* and *phosphorescence* of liquids and gases are due to photoelectric mechanism.

## 5.6 PHOTOCONDUCTIVE CELLS

Photoconductive cells are made of a thin layer of semiconductor material such as cadmium sulfide, cadmium selenide, or lead sulfide. The semiconductor layer is enclosed in a sealed housing. A glass window in the housing permits light to fall on the active material of the cell.

The resistance of a semiconductor material changes when radiant light energy is incident on its surface. Such materials have a high *dark resistance* (when not illuminated) and low *irradiated resistance* (when illuminated). *A photoconductive cell simply acts as a conductor whose resistance changes when illuminated.*

A photoconductive cell using selenium as the semiconductor material is shown in Fig. 5.8(a). To get a large dark-to-light resistance ratio, the semiconductor material is made in a particular shape, *interdigital pattern.* Fig. 4.8(b) illustrates a practical circuit using cadmium sulfide as the semiconductor material. CdS cells provide a very high dark to light resistance ratio and give maximum response to radiations of 5000Å.

## 5.7 PHOTOVOLTAIC CELLS

Photovoltaic cells are also light sensitive semiconductor devices but *they produce a voltage, when illuminated,* which increases as the intensity of light falling on the semiconductor element of this two element cell increases. The basic material most commonly used for these cells is *selenium.*

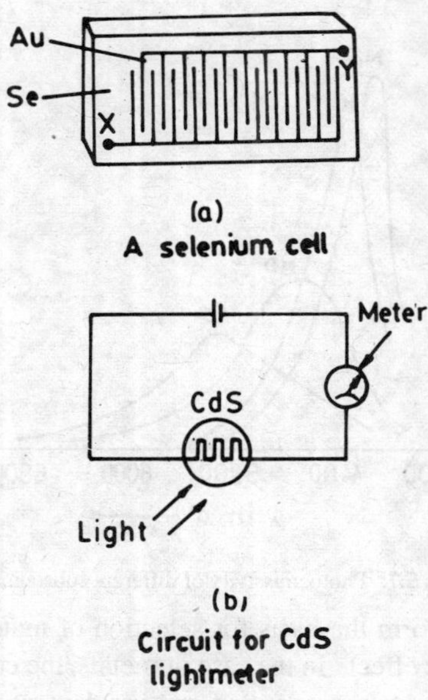


**Fig. 5.8:** Photoconductive cell.

Photovoltaic cells convert light energy into electric energy, which may be used directly to supply small amounts of electric power for electrically powered devices. Because of the *low levels of power* which photovoltaic cells generate, they have been used in the past in low power devices such as light meters and photographic exposure meters.

Photovoltaic cells consist of a single semiconductor crystal which has been doped with n- and p-type materials. When light falls on the p-n junction, which is the boundary of these dopants, a voltage appears across the junction. About 0.6 V is developed by the photovoltaic cell in bright sunlight. The amount of power the cell can deliver depends on the extent of its *active surface*. An average cell will produce about 30 mW/cm$^2$ if operating into a load of 4 ohms. To increase the power output, large *banks*

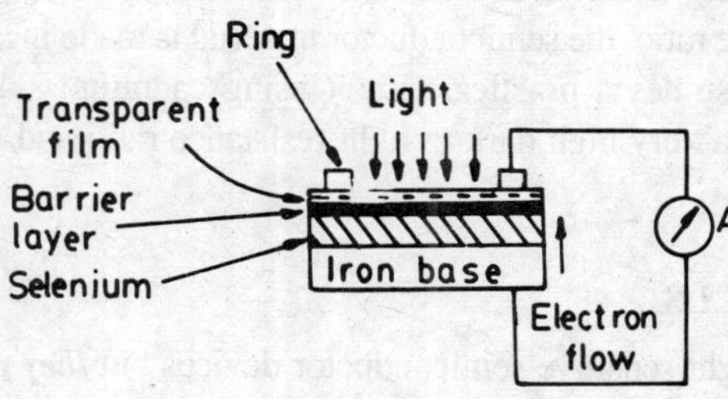


**Fig. 5.9:** Iron-Selenium photovoltaic cell.

*of cells* are used in series and parallel combinations. An example is the use of solar cells to power the experimental circuits in lunar and space modules.

Photovoltaic cells may be used directly to trigger sensitive relays or where required their output may be amplified electronically.

## 5.8 PHOTOEMISSIVE CELLS

Photoemissive cells are more appropriately known as *photoelectric cells* (PECs) and are used in talking motion pictures, television, photoelectric relays, photomultiplier tubes ,..., and so on.

The cell may be made by depositing a thin layer of an alkali metal on the inner surface of a small vacuum tube, as shown in Fig. 5.10. The collecting electrode is positively charged, so that photoelectrons are attracted to it. Illumination of the metal surface by any radiation with wavelength shorter than the *photoelectric threshold* causes emission of photoelectrons and a consequent flow of electric current through the circuit. The current maybe registered on an ammeter. It is found that *the magnitude of the current is proportional to the intensity of light*.

Photoelectric cells are of two types, *vacuum type* photo cell and *gas filled* photo cell. Vacuum type photoelectric cell consists of a sealed glass or quartz bulb which is completely evacuated. Two electrodes are provided in the bulb, one electrode forming the cathode and the second electrode forming the anode or collector. These cells find wide application in television and photometry. *The emission current of a vacuum type photoelectric cell is very small. To increase the current, the tube is filled with an inert gas like neon or argon at low pressure*. The photoelectrons *collide* with the molecules of the inert gas and produce *ionisation* causing the current to increase. When emission takes place by collision, there is an *abrupt increase* in current. The gas filled photo-tubes are used in sound reproduction in films, burglar alarms, counting machines, street lighting, ..., etc.

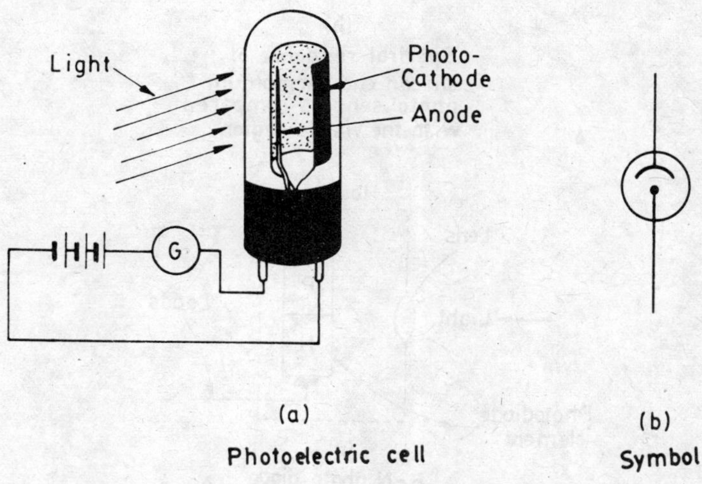


**Fig. 5.10:** Photoelectric cell.

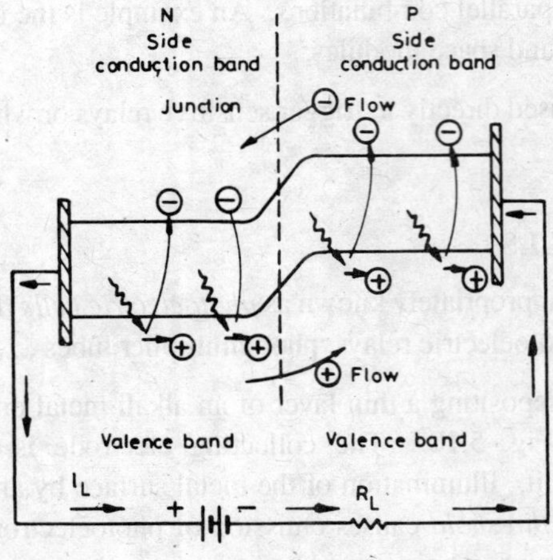


(a)

**Photoeffect in a reverse biased P-N junction**

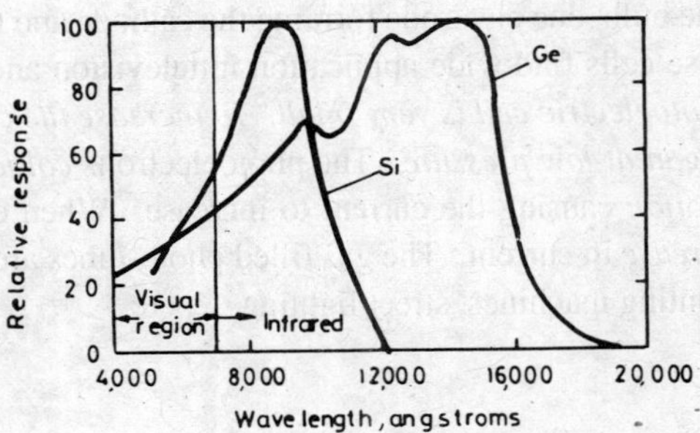


(b)

**Spectral response of silicon and germanium photo sensors compared with the visible region**

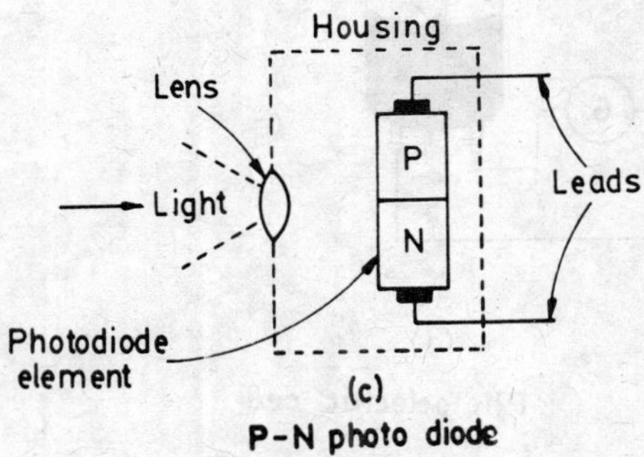


(c)

**P-N photo diode**

**Fig. 5.11:** p-n Photodiode.

### 5.9 p-n Photodiodes

Bulk type photoconductive cells could also be fabricated from silicon or germanium, but the *current sensitivity to radiation can be made much larger by the use of a reverse biased p-n junction.* The mechanism of current control through radiation is similar to that explained for the photoconductive cell, whereby photons create electron-hole pairs on both sides of the junction. With no light applied, the reverse current is simply the reverse saturation current due to the minority carriers, holes in the n-type and electrons in the p-type. When light is applied, the *photoinduced electrons* in the conduction band of the p-type will move across the junction down the potential hill with the thermally generated minority carriers. Similarly, holes produced in the valence band of the n-type (by electrons being photoexcited into the conduction band) are available to add to the current flow by moving across the junction to the p-side, see Fig. 5.11(a). The spectral response of the diode to wavelengths is shown in Fig. 5.11(b). *The photoinduced current is a function of irradiance.*

The active diameter of these devices is only 0.1 inch, but they are mounted in standard TO-5 packages with a window as shown in Fig. 5.11(c). They can operate at frequencies of the order of 1 MHz and are used in character recognition, optical communication equipment and encoders.

### 5.10 p-i-n-Photodiode

Although the p i n photodiode provides a much higher frequency response than the CdS bulk type photoconductors, an even faster response may be achieved in the p-i-n photodiode. This consists of adding a layer of *intrinsic* silicon between the heavily doped p- and n-type silicon materials. This has the effect of reducing the transit or *diffusion time* of photo induced electron-hole pairs. That is, carriers created by photons incident on the middle of depletion region have less distance to travel than if generated at one side or the other of the depletion layer. The relatively thick layer (2.5 $\mu$) of high resistivity intrinsic silicon ensures the absorption of most of the incident photons.

Irradiated voltage-current characteristics exhibit a linear relation. The device is essentially a *constant current generator* at a given light level.

### 5.11 Avalanche Photodiodes

Current sensitivity may be increased from 30 to 100 times by operating a photodiode in its reverse breakdown or avalanche breakdown region. In the avalanche photodiode the construction is such as to provide a very *uniform junction* that exhibits the avalanche effect at voltages between 30 and 200 V for the TIXL 55-59 series. The *dark current* is typically 10 $\mu$ A. The electron-hole pairs that are generated by incident photons are accelerated by the electric field to *kick* new electrons from the valence to the conduction band. In this way, a typical *photomultiplication* of 50 is obtained, so that light current may be 0.5 to 1 mA.

This mechanism does not reduce the fast response found in the p-i-n photodiode. The avalanche photodiode may be operated at a modulation frequency of 2 GHz with an excellent signal-to-noise ratio that rivals the photomultiplier tube. Their present disadvantage is their expense.

### 5.12 n-p-n Photo Duodiode

The current flow in the photodiode is relatively small. In an attempt to obtain an *inexpensive device with higher current capabilities*, another junction is added to form a duodiode. Resembling a transistor, but

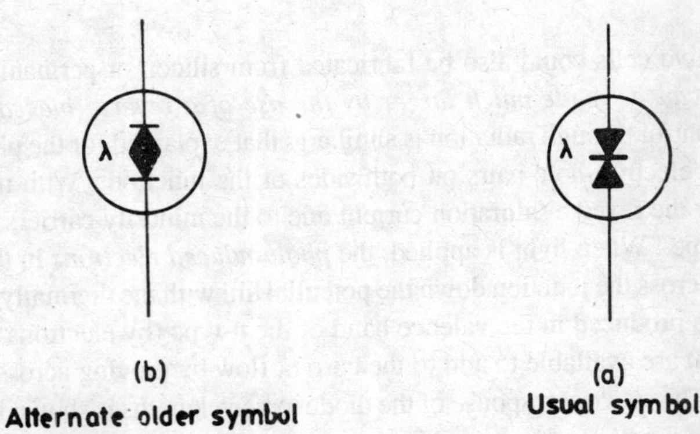


**Fig. 5.12:** Schematic symbol of n-p-n photo duodiode.

22 μA/mW/cm$^2$, roughly ten times that of the previous devices. It has a typical dark current of 0.01 μA, a rise time of 2μs and fall time of 45μs with a power dissipation of 250 mW. Because of its two symmetrical junctions it can be used with either a.c. or d.c. bias voltages.

## 5.13 Photo transistor

The phototransistor is a junction transistor constructed so that light can fall on the structure near the collector junction. The action of the light increases the current through the reverse biased collector junction, and, if the base terminal of the transistor is left open circuited, the increase of current is *amplified* by the transistor to result in a much higher collector current, in much the same way as the collector/base leakage current is amplified in the common-emitter connection. After amplification, the photocurrent may be high enough to be used in a control system, for example, to set a relay. *The phototransistor is a light detector which combines a photodiode and a transistor amplifier.*

One of the important features of the photoconductive diodes and phototransistors is the very *small active area of the devices.* This enables them to be used very successfully for position indicators. Also, if incident light is focused from a large area on to the active region, very high sensitivities can be obtained.

A *lens* concentrates the light on a very thin p-type wafer, sandwiched between an n-type collector and emitter, as shown in Fig. 5.13(a). Although the phototransistor has three sections, only two leads may issue from the housing. In such a phototransistor base bias is provided from an external circuit, on which the photodiode current is superimposed.

Current in a phototransistor is dependent mainly on the intensity of light entering the transistor window and is little affected by the voltage applied to the external circuit. Fig. 5.13(b) illustrates the graph of collector current, $I_C$, as a function of collector-emitter voltage, $V_{CE}$, and function of illumination H. It is evident that the phototransistor acts as a *constant current source*, and that current depends almost entirely on the illumination level and very little on the voltage level.

Fig. 5.13(c) depicts the frequency response curve of a phototransistor for comparison with the response of the human eye. *Peak sensitivity of the phototransistor is at about 900 nm.*

The *angular adjustment* of a phototransistor and its source of illumination are important considerations. The reason is that the illumination of the photo junction is proportional to the cosinc of the angle

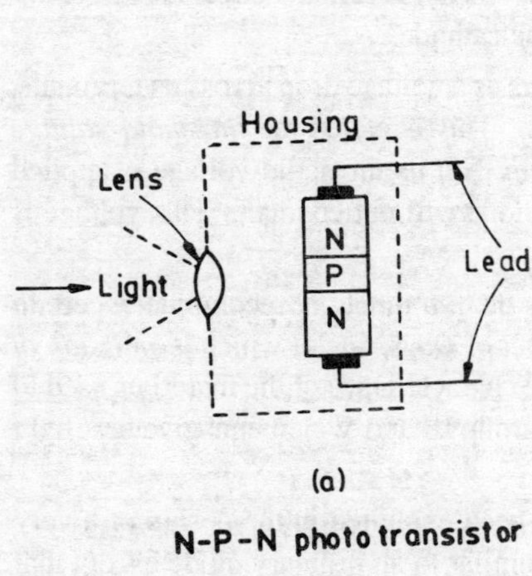


(a)

**N-P-N photo transistor**

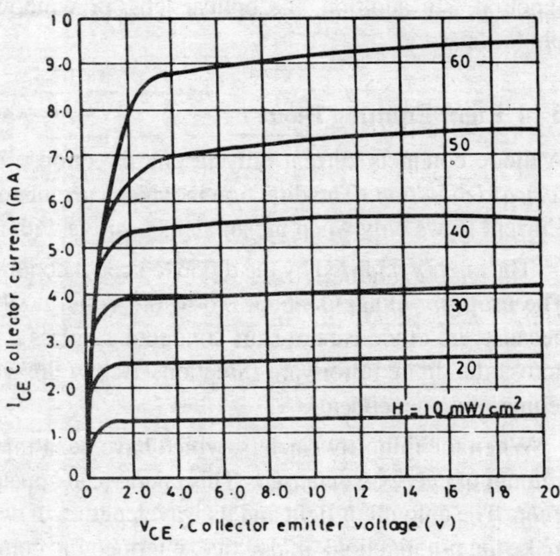


(b)

**Collector characteristic for photo transistor**

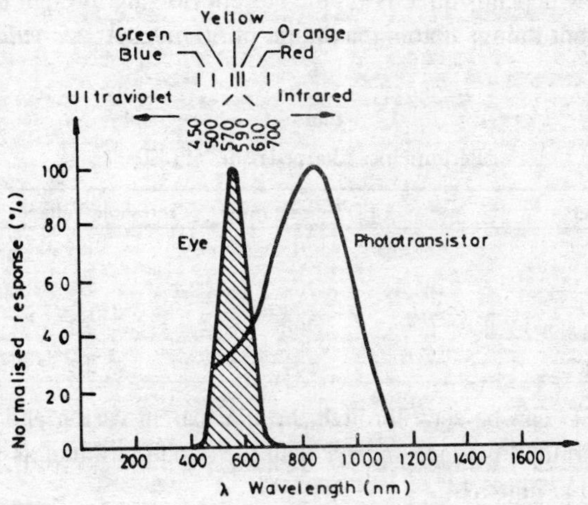


(c)

**Spectral response of a photo transistor and the human eye**

**Fig. 5.13: Photo-transister** Because of its two symmetrical p-n junctions it can be used with either ac or dc bias voltages.

between the direction of radiation of the light source and the perpendicular to the surface of the photo junction. In addition, the optical lens, or window, and its size further affect the response of the phototransistor.

## 5.14 Light Emitting Diode

A diode conducts current only in one direction. The direction in which current flows is called the *forward direction*. The direction in which current does not flow (blocks) is called the *reverse direction*. Current flows only when the anode is positive with respect to the cathode.

The energy required by the diode to keep it conducting appears as a voltage drop across its terminals. The drop in voltage is about 0.6 to 0.7 volts in silicon diodes. This is called the *threshold voltage* because the diode cannot start conducting unless a voltage more than its threshold voltage is applied across the diode in forward direction. Below this voltage the diode will not conduct. This voltage is temperature dependent.

When the minority carriers, which have been injected across the p-n junction, recombine, a certain amount of *energy is released*. This energy can appear in a number of ways, one of which is *emission of light*. The amount of light and its wavelength will depend on the characteristics of the materials used to make the p-n junction. In practice, intermetallic compounds of group III and V elements give best light emission.

The energy that is dissipated as heat in ordinary diodes has been exploited in the design of a very popular and useful device called the *light emitting diode*. It is similar to an ordinary diode except that *the energy required by the diode to keep it in conduction is given out in the form of light rather than heat*.

LEDs are made of materials like gallium arsenide, gallium phosphide, and can have different colours depending on the type of material. *Threshold voltage for an LED varies from 1.6 to 2.2 V* (see Table 5.1). The intensity of glow depends directly on the current flowing through the LED. Commercially available LEDs have current ratings upto a maximum of 50 mA. *A safe value to operate an LED is between 15 to 20 mA.*

**Table 5.1**

**Performance Characteristics of LEDs**

| Colour | LED Current (mA) | Threshold Voltage (V) |
|---|---|---|
| Red | 20 | 1.6 |
| Red | 20 | 2.2 |
| (High glow) Yellow | 20 | 2.2 |
| Green | 20 | 2.2 |

The cathode of the LED can be seen through the transparent casing and is a broad disc shaped electrode. The cathode terminal is made *shorter* than the anode terminal as a physical indication of polarity. This is illustrated in Fig. 5.14.

By doping the compound with other elements, the wavelength of light emission can be changed. For example, gallium phosphide doped with silicon and zinc gives a green light, and doped with oxygen and zinc gives a red light.

Although most LEDs have a peak spectral response at (red) 6600 Å and (yellow) 6100 Å, (green) 5600 Å devices also are available. High output types used for solid-state lamp indicators (such as

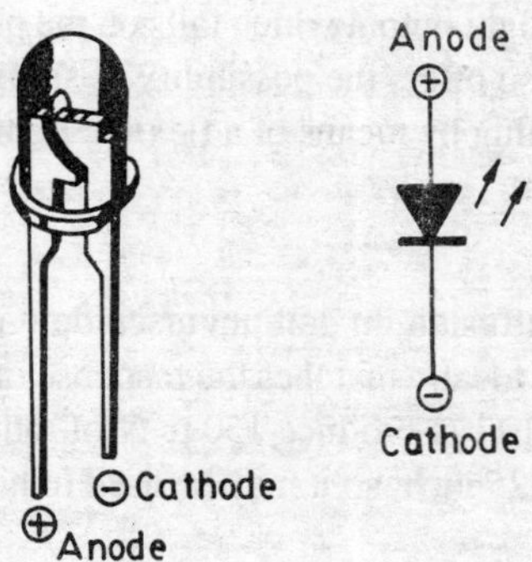


**Fig. 5.14**: LED leads are always connected in the forward direction. The cathode can be recognised by three features: (1) the shorter terminal; (2) flattened body; and (3) wider of the two inside electrodes.

MLED 600) may have a brightness, $B_L$, of 200 foot lamberts (a measure of source intensity) at only 10 mA and 1.6 V. This represents only 10 mW of input power. The brightness may be increased to 1100 foot lamberts at 50 mA.

LEDs used as indicators have *solid-state reliability*, typically $10^6$ hours of life time. They may also be used in optical switching circuits, since, unlike an incandescent lamp, they can be switched 'on' and 'off' in 1 ns.

### 5.15 Semiconductor Relay

An interesting application of the light emitting diode is its use in conjunction with a silicon *photodetector* to form a four terminal device with the output completely *isolated* from the input. The system is *similar to a relay* in that an alternating signal applied to the input will change the impedance between the output terminals from high to low. The arrangement is illustrated in Fig. 5.15.

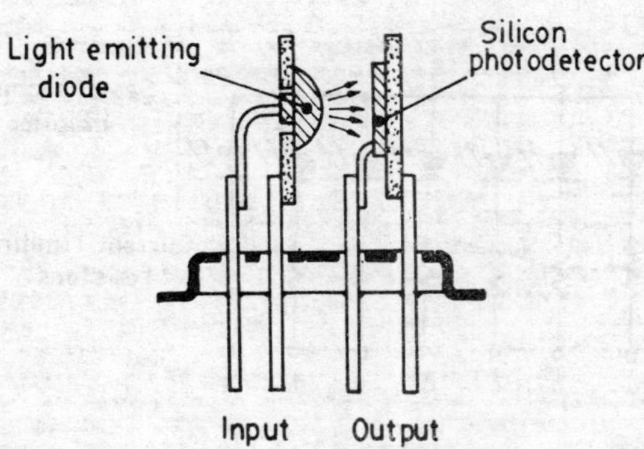


**Fig. 5.15**: Semiconductor relay.

A silicon photodetector — either a photodiode or a phototransistor — is mounted in close proximity to the light emitting diode and the assembly encapsulated in a transistor case. An input signal applied to the light emitting diode causes a light output which falls on the photodetector and results in an output signal. The use of *light coupling* also offers the possibility of siting the two elements apart in different sections of an equipment, and coupling by means of a flexible light pipe.

## 5.16 Infra Red (IR) Emitters

If the p-n junction results from a diffusion on gallium arsenide (Ga As) the radiant energy is infrared with a typical peak at 9000 Å. This ideally matches the response of silicon photodiodes and phototransistors. For example, the MLED 910 can produce 150 μ W of radiant energy at 50 mA and 1.2 V. Its diameter is 0.06 inch and length 0.125 inch, so it may be used in high density applications.

## 5.17 Alpha Numeric Display

Fig. 5.16(a) depicts a seven-segment display. It contains the seven rectangular LEDs, 'a through g'. *Each LED is called a segment, because it forms part of the character being displayed.* A typical character display is shown in Fig. 5.16(b).

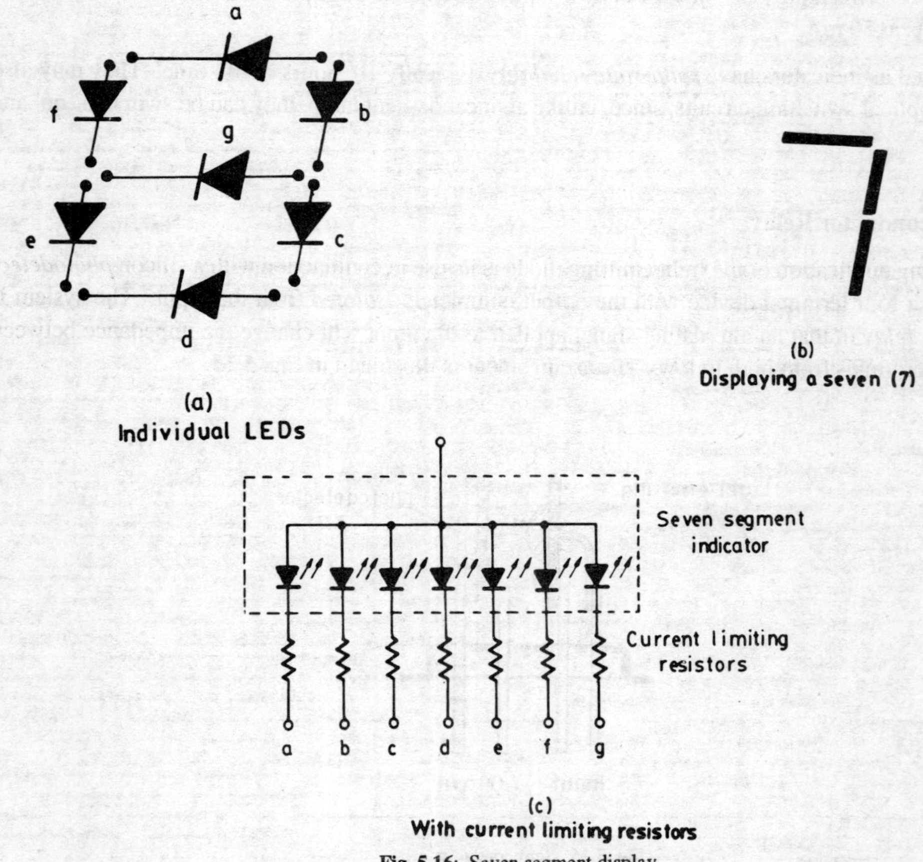


Fig. 5.16: Seven segment display.

External current limiting resistors are included to keep the LED currents at safe levels. By grounding one or more resistors, we can form any digit from 0 through 9. For instance, by grounding 'a', 'b' and 'c', we can get a 7 as shown in Fig. 5.16(b). Grounding 'a', 'b', 'c', 'd' and 'g' produces a 3. The schematic diagram of a *seven-segment display* is given in Fig. 5.16(c).

One way of producing an alpha numeric display is to make a $5 \times 7$ array of LEDs (plus one for the decimal point) in a monolithic structure. By driving the appropriate horizontal and vertical inputs, 64 different characters can be displayed, including the full alphabet, all numerals, and various signs.

Another method that can display all numerals and nine letters is the *seven-segment monolithic device* shown in Fig. 5.17. Mounted in a standard DIP package, each of the seven-segments consists of five LEDs, interconnected by evaporated aluminium on the surface of the chip. Each segment provides a brightness of 200 foot lamberts at 5 mA and 1.7 V. The compactness of the display allows numerous packages to be mounted side by side to give any degree of readout. The seven-segment displays are driven from a decoder/driver that accepts binary coded decimal (BCD) input.

Yet another type of display that uses a seven-segment system is the electroluminescent or *liquid crystal display* (LCD). It consists of normally transparent liquid crystal material between two electrodes. When an electric field is applied (15 to 60 V), the material becomes turbulent, reflecting and scattering ambient light. It provides excellent brightness under high ambient light conditions and requires only 50 μW of power per segment. This power is much lower than that for the LED, but the life expectancy is not as high as that of the LED (minimum 10,000 hours).

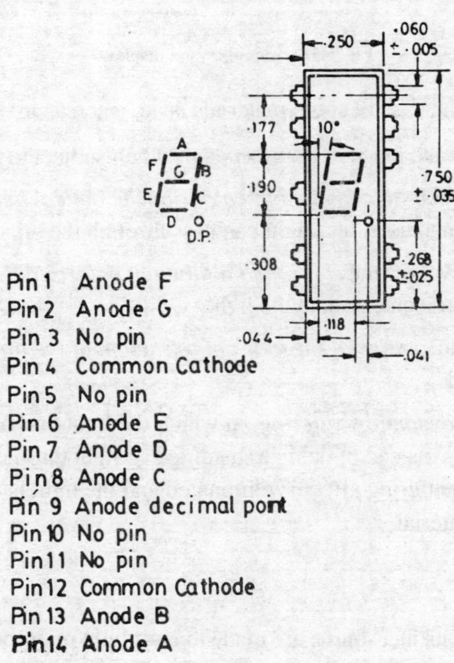


**Fig. 5.17:** Showing a 0.19 inch high LED, seven-segment display mounted in a standard DIP package.

## 5.18 Liquid Crystal Display

Liquid crystal materials find wide application in *digital display consumer products*, such as electronic watches and calculators. Fig. 5.18 depicts a typical liquid crystal display. It consists of a two piece transparent case with the liquid crystal material filling the space in between. Very thin electrodes are deposited on a transparent material (such as tin oxide) on the inner surface of both the base and the cover. The electrodes are shaped to give the desired shape for display. The exterior of either the cover, or the base, may be coated with a mirror like material.

Liquid crystals are liquids in which the inter molecular bonds have definite relationships, like in crystals. *When a liquid crystal material is subjected to an electric field, its optical characteristics change.* The electric field in the case of some materials is ac, whereas in other materials it is dc.

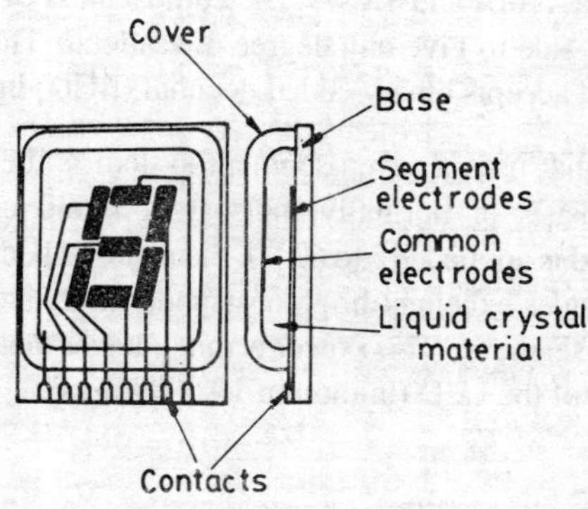


**Fig. 5.18:** Liquid crystal display.

Four main phenomena can take place in liquid crystals when subject to an electric field.

(1) The first mode is based on *a change in colour or the optical density of the material.* When the optical density increases, less light can pass through the crystal material.

(2) The second mode involves the *controlled rotation of polarised light.* As bias is applied to the liquid crystal, the molecules rotate the light.

(3) The third mode is the *change of the index of refraction of the liquid crystal material* with an applied electric field.

(4) The fourth mode is *dynamic scattering*, in which the light entering the liquid crystal is scattered in all directions instead of being transmitted straight through. When a bias is applied to the electrodes, the scattering effect is eliminated and the light is transmitted directly through the liquid crystal material.

## 5.19 Laser

In lasers billions of excited atoms in a source are made to emit light in proper agreement with the phase and direction. This agreement is called *coherence* (see Fig. 5.19) and it includes *space coherence* (i.e. direction) and *time coherence* (i.e. phase).

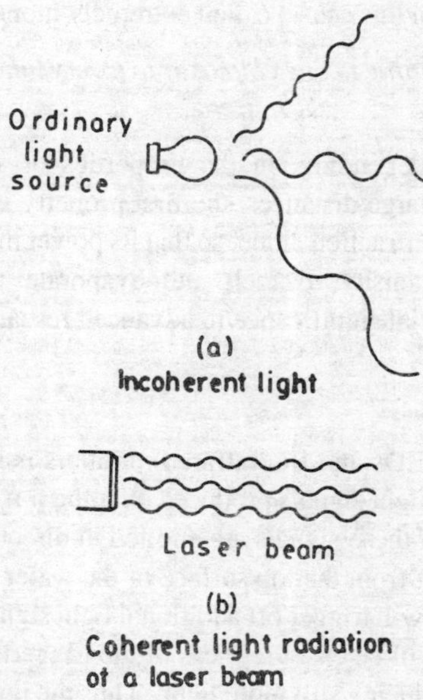


**Fig. 5.19:** Patterns of light radiation.

Excited atoms may give spontaneous emission or induced emission of light. In the *spontaneous emission*, the emissions of different atoms are random in direction and in phase. If the light, of just the wavelength which the atoms can emit, is passed through the assembly, the excited atoms may give *induced emission.* This light is the signal and the induced emissions from each atom agree with the signal in direction, polarisation and phase. The result is that the induced wave becomes stronger and stronger as the signal passes over successive excited atoms.

Actually the beam has to be repeatedly reflected back and forth between two highly polished mirrors so that the coherent radiation (i.e. the induced one) may become quite powerful in comparison with the incoherent (i.e. the spontaneous) one. An important aspect is that the same light may also be absorbed by the unexcited atoms to raise them to the excited state. In any normal system, the unexcited atoms are much larger in number than the excited ones. Hence, the induced emission would not buildup. If we desire the induced emission to be dominant, an essential condition to be achieved is that the excited atoms should be in excess of the unexcited ones. This is called *population inversion.*

Initially some spontaneously emitted light acts as signal to produce induced emission. During its successive passages, the light is both absorbed and emitted. But because of population inversion, the induced emission goes on dominating over the spontaneous one. This is called buildup of *laser action.* In effect, the total result is that the beam becomes:

(1) extremely directed one (2) the angular spread is limited only by the diffraction of the aperture, and (3) is extremely coherent one-the wavelength spread becomes extremely narrow.

A *monochromatic* source of wavelength λ really has some wavelength spread. It is about $10^{-2}$ Å for the best sources. Angstrom unit, Å, is equal to $10^{-10}$ m. In a laser the wavelength spread becomes as small as $10^{-10}$ Å. A laser beam is, in this sense, called extremely monochromatic.

*The optics of producing or pertaining to one colour, or to a very limited range of wavelength is called monochromatic.*

Lasers are put to many uses depending on the properties of small angular spread and small wavelength spread. In measuring large distances, the first property is used. If a lens is used to focus laser light, the image is limited by diffraction alone, so that its power may be focused on an area as small as $10^{-6}$ $cm^2$ giving sufficient intensity to melt and evaporate the material there. Its highly monochromatic character is of special significance in advanced research.

### 5.20 Invisible Light Laser Diode:

The energy emitted from visible LEDs and IR (infrared) emitters is spontaneous or *incoherent* i.e. as holes and electrons recombine, the light signals produced over the p-n junction area are random in time, direction and polarisation. Most of these signals are emitted in directions which take them away from the junction to form the light output from the top surface of the wafer. However, some light will move along the plane of the junction, and will trigger off additional light signals by a process called *stimulated emission.* The stimulated signals will be exact replicas of the triggering signals and so they also move along the plane of the junction, producing still more light. Thus the initial light signal moving along the junction can be *multiplied many fold* before it reaches the surface. If now some of the light is reflected at the surface, travels back to the other surface and is partially reflected there, so that it 'bounces' to and fro, *it will be amplified further with each bounce, and coherent light output will be produced*, with the radiation concentrated from the junction region. This is *laser action* and the diode is then called a semiconductor laser or a junction laser.

The basic form of semiconductor laser is shown in Fig. 5.20. The p-n junction is formed by diffusing zinc into n-type gallium arsenide. A wafer is cut from the slice and the front and back faces precisely ground and polished to be parallel to each other and perpendicular to the junction. The reflectivity of the gallium arsenide to air interface is sufficient to give the required reflection, and so no further surface treatment is needed. The other two faces of the wafer are intentionally made irregular to give in-

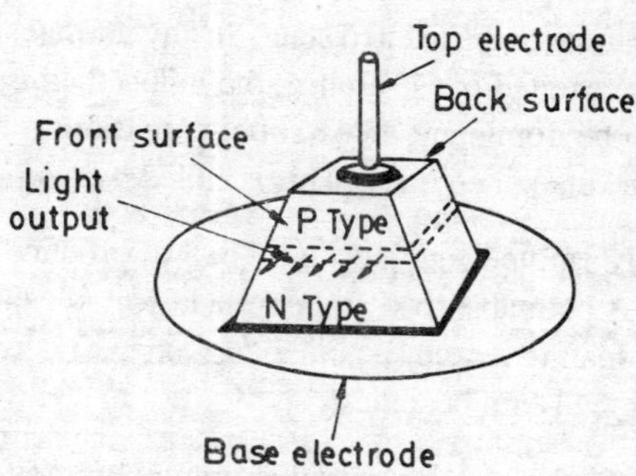


**Fig. 5.20:** Semiconductor laser. Note that the front and back surfaces must be parallel.

coherent reflection. The *distance* between the two reflecting faces will determine the frequency at which the laser will operate-the radiations must have a wavelength such that the distance between the two faces is equal to an integral number of half-wavelengths. Typical laser wafers are between 0.2 and 0.4 cm. square at the junction.

With such a semiconductor laser, a current density of several thousand amperes per square is neccessary to initiate laser action at liquid air temperature, and the light output is concentrated over a very small area in the region of the junction-about 0.005 sq mm-with a peak intensity of the order of $10^6$ watts per sq. cm.

LASER is an abbreviated form of "light amplification by stimulated emission of radiation."

## 5.21 Visible Light Laser Diode:

Light sensitivity of the human eye is illustrated in Fig. 5.21. The graph is called the *photopic luminosity curve.* It shows that visible light ends and near infra red begins at or near a wavelength of about 720 nm. However, the curve is asymptotic at both extremes since the actual point at which the eye no longer responds is determined by the intensity of radiation.

The new aluminium-gallium-arsenide (Al-G-As) light emitting diodes emit a spectrum of wave lengths *centered* at 880 nm. In other words, some of the radiation is closer to the visible wavelength than the 880 nm figure suggests. Les's look at the evolution of visible light laser diodes.

The first laser diode was made in 1962 from gallium-arsenide (Ga-As) semiconductor alloy still used to make many such lasers. Fig. 5.22 compares the simplest possible light emitting diode chip with an early Ga-As laser diode. The laser is a p-n diode alike in nearly all respects to the LED. Unlike ordinary LED, however, the laser version has a very smooth, flat junction, and two opposing ends of the chip have been cleared to form perfectly flat, parallel reflective facets that function as end mirrors. *The mirrors provide the optical feedback necessary to establish and sustain laser operation.* Radiation is emitted from the chip at the junction region of both facets.

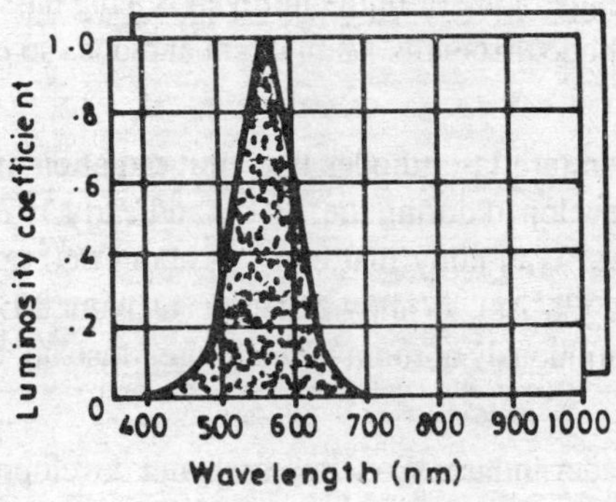


**Fig. 5.21:** The photopic luminosity curve of the typical human eye.

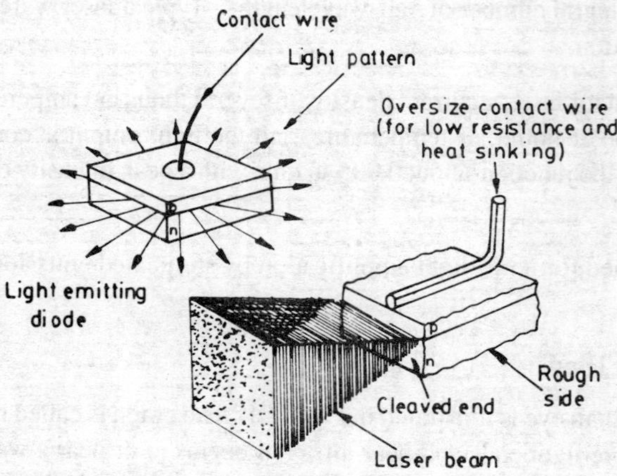


**Fig. 5.22:** Physical structure of a simple LED and junction laser.

Todays' sophisticated laser diodes have structures much more complex than the simple p-n junction version shown in Fig. 5.22. Nearly all are made by forming multiple layers of semiconductor alloys in a sandwich like configuration. This confines to the junction region radiation that would otherwise be in the crystal or escape out of the sides of the chip.

At room temperature, Ga-As lasers emit invisible near infrared radiation at a wavelength centered at about 904 nm. The wavelength range of these lasers is much narrower than similar diodes operated as non-laser infrared emitting diodes.

However, even the very first laser diodes emitted red light because they could only be operated when cooled to the temperature of liquid nitrogen (–196˚C). Light is shifted downward in wavelength from near infrared to visible red because *the emission wavelength of any light emitting diode laser, or otherwise, decreases with the temperature of the diode junction.* A liquid nitrogen cooled Ga-As laser, for example, emits a wavelength of about 845 nm. At this temperature, the laser operates continuously. Operating a laser diode at the temperature of liquid nitrogen is a big hassle. That's why scientists have long sought to develop a laser diode that emits visible light and does so continuously at room temperature.

Several kinds of room temperature laser diodes that emit radiation at the far end of the visible red portion of the spectrum were developed during the 1960's and early 1970's. One was made from gallium-arsenide-phosphide (Ga-AsP), an alloy that emits at about 860 nm. A second was made from aluminium-gallium-arsenide (Al-Ga-As), an alloy that emits at about 805 nm. These early visible light lasers could not be operated continuously at room temperature. Instead, they were driven by brief (200 ns) current pulses.

As early as 1970, RCA, a leader in laser diode research and development, had produced lasers that emitted at 690 nm in a pulse mode, but they were *unreliable* due to high current density required to attain laser operation. During the late 1970's, however, RCA scientists produced laser diodes of Al-Ga-As that operated continuously at room temperature and emitted wavelengths as low as 740 nm. The race to develop practical visible-light laser diodes has intensified. The main incentives are laser-semiconductor discs and video discs and laser printers. In both cases laser diodes could be better utilised as compact low-power consuming radiation sources *if only their emission wavelengths could be reduced.* This would permit the laser beam *to be focused to a smaller spot* than that available from a conventional near infrared emitting laser diode.

The small spot size of focused light from visible laser diodes would mean that audio discs and video discs could carry as much information as competing discs designed for use with 632.8 nm radiation emitted by helium neon (He-Ne) laser. And the laser non-impact printer could achieve better resolution. Furthermore, the laser printer would require less power since the same energy could now be focused into a small space.

Until a few years ago, the very best visible laser diodes emitted in the visible red above about 700 nm. Scientists at Japan's Sharp Corp announced the first laser diode to emit continuously at wavelengths below 700 nm. Various versions of their new Al-Ga-As laser deliver from 5 to 10 mW (about the same power as a small penlight) at 683 nm.

Recently, the development of indium-gallium-arsenide-phosphide (In-Ga-As-P) laser diodes, that emit at 621 nm at room temperature, has been announced. This wavelength is lower than that emitted by the He Ne laser.

Will laser diodes that emit visible hight replace He-Ne laser? In some applications they might. While laser diodes produce a much more broadly diverging beam than helium-neon laser, a simple convex lens can focus the light from a laser diode into a beam as narrow as that from a He Ne laser. Of even more significance, a laser diode can be operated from a compact battery-powered supply and its light output can be pulse or intensity modulated.

*The laser diode's biggest drawback is its temperature sensitivity.* The current supplied to the laser must exceed a certain threshold point before the laser will begin to function as a laser and not merely as an LED. The threshold point, however, decreases with temperature. To avoid destroying the diode by pumping too much current through it, the laser chip must be cooled to a constant temperature (perhaps by a thermo-electric module). Or it must be powered by a temperature regulated power supply. Alternatively, the supply can be regulated by a photo diode circuit that *monitors* the radiation from one end of the laser. The set up automatically compensates for temperature changes since the radiation from a laser diode biased by a constant current, increases in power as temperature decreases. This guarantees that the laser will not be operated above the limit where excessive optical radiation can cause damage to the chip facets.

Its temperature sensitivity not withstanding, the era of the *visible light emitting laser diode* has finally arrived. Hopefully, the demand created by the video disc and laser non-impact printer industries will warrant mass production of these important new lasers. When that occurs, prices will fall. You can then expect to see various kinds of penlight size laser diode assemblies, complete with battery powered, regulated supplies, and perhaps, self contained modulator circuits.

# CHAPTER 6

# SENSORS

*In general, electrical quantities are easy to measure and it is not necessary to convert voltages, currents, or frequencies in order to measure them. Such a conversion when it is done, is carried out by a device called transducer. Sensors are information measurement elements that have functions with information conversion as their object. Sensors can be thought of as acting in place of the five human senses of sight, hearing, smell, taste and touch. Sensors are also able to detect phenomena such as ultrasonic sounds and other forms of energy that are not detectable by the human senses. Transducers differ from sensors in that the signals for the input energy are converted into electrical signals. In the modern age, the terminology for measurement has shifted from the transducer to the sensor and the act of measuring something has come to mean "the ability to feel". These operations and the results of these operations form the subject of study in this chapter.*

## 6.1 Thermistors

Thermistors can be classified as *NTC* (negative temperature coefficient) and *PTC* (positive temperature coefficient) thermistors. The PTC thermistor is an oxide semiconductor of n-type titanic barium family. Conductivity is produced by adding small amounts of rare earth elements to titanic barium, the *main component* of a PTC thermistor. The conductivity of titanic barium (BaTiO) changes notably from the Curie Point by phase transition. In other words, when the element reaches a certain temperature, its resistance increases abruptly. This is shown in Fig. 6.1(a). It has a temperature coefficient opposite to the negative one of the NTC thermistor. PTC thermistors, Fig. 6.1(b), featuring a simple structure with electrodes on both sides of a disc, are used as temperature sensors, current limiters, fuse elements and constant heating elements. They are also receiving attention in the communication field, which is demanding increasingly higher reliability.

### *Characteristics of PTC Thermistors*

*1. Resistance-Temperature (R-T) Characteristics*

If we measure the resistance of a PTC thermistor with a small voltage in a range that does not induce Joule heat by gradually raising the surrounding temperature, resistance-temperature characteristics as shown in Fig. 6.1(c) are obtained. The specific temperature where resistance increases abruptly is called Curie temperature (Tc). It is defined as the *temperature corresponding to twice the minimum resistance*

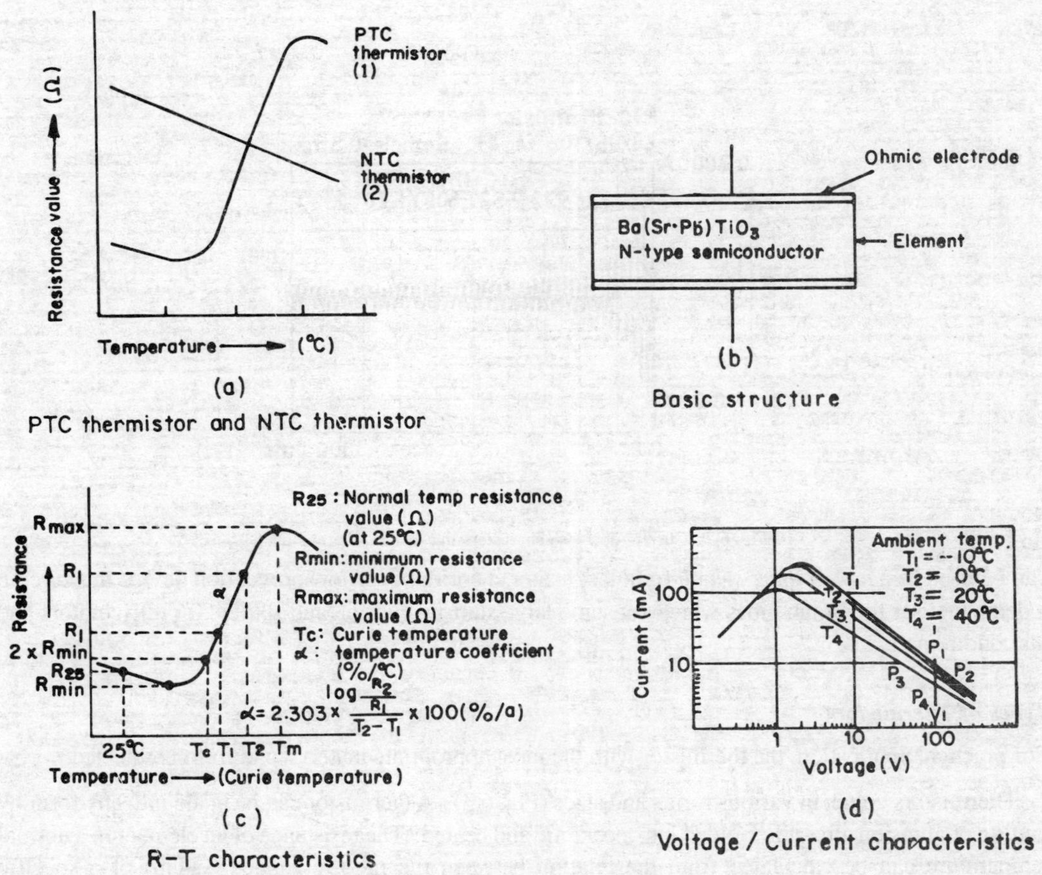


$$\alpha = 2.303 \times \frac{\log \frac{R_2}{R_1}}{T_2 - T_1} \times 100 (\%/a)$$

**Fig. 6.1** Characteristics of PTC thermistors

*value* ($R_{min}$) and it is an important parameter of material characteristics. If we raise the temperature even higher, the resistance reaches the maximum value, and then drops.

### 2. *Current Voltage (iv) Characteristics*

When we raise the voltage across the terminals of a PTC thermistor, it shows an iv characteristic as shown in Fig. 6.1(d). After the PTC element nears the Curie temperature through self heating by Joule heat, it shows negative current characteristics—the current decreases while voltage rises. When we plot current and voltage on a double logarithmic scale, we can see that this *negative current area* remains constant. If a voltage applied to a PTC element exceeds a certain value, the current increases with the voltage until it breaks down. This is called the *breakdown voltage*, and is used to show a *safety level* against a rated voltage. It also shows a characteristic change against change of surrounding temperature as shown in Fig. 6.1(d).

### 3. *Current Decrease Characteristics*

The current of a PTC thermistor decreases when a voltage of 10—100 V is applied, as shown in Fig. 6.2. This characteristic, in which a large current is transferred into a small current at a certain temperature,

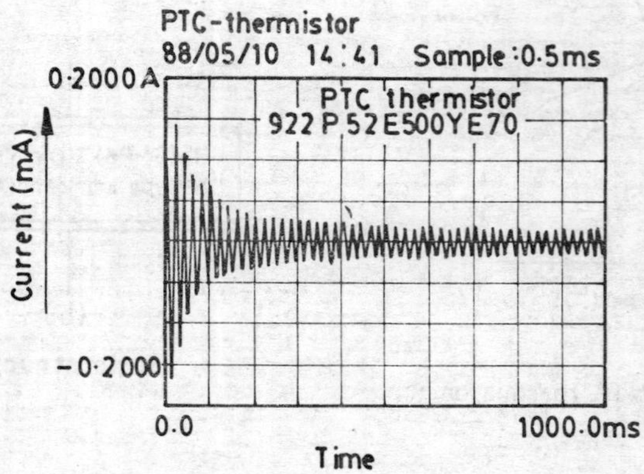


**Fig. 6.2:** Current decrease characteristics of PTC thermistor

can be used as a *non-contact switch* for a fast heater, a current overflow prevention device in a circuit, a demagnetiser for Braun tubes which consume large start-up current, and starters for refrigerators and air conditioners.

### *Types of Thermistors*

For practical application, the thermistor with the most appropriate material and form is selected.

Thermistors come in various forms and sizes (Fig. 6.3). A thermistor can be made into any form by cutting or abrasion after it is molded into a ceramic and heated. The resistance of an element in a normal temperature can be calculated from the relation between the ratio-resistance (Ω-cm) of a specific material and its size.

Heat resistant paint is usually applied to an element with lead wires soldered directly on it for mechanical protection, insulation, humidity resistance and weather durability.

Besides the basic forms, any form can be produced to meet a specific demand. For example a large heat value of several hundred watts can be obtained by shaping a thermistor into a *beehive* by humid extrusion molding, then sending air through the holes.

### *Applications*

When a PTC thermistor is exposed to the air for a long period, its resistance changes by about 5-10 per cent for 10,000 hours with a small variation according to the material. For this reason *PTC thermistors are considered inappropriate for preventing small resistance changes that correspond to the specific temperature for which the NTC thermistor is the most appropriate.* However, when a PTC thermistor is used as a *heat sensor* it can work around the Curie temperature where the resistance of a PTC thermistor changes considerably. The Curie temperature itself does not change over a long period since it is specific to the material used.

PTC thermistors in series are connected to a power transformer and are used as a ballast for electric discharge lamps or the casing of power transistors that can be heated when something goes wrong.

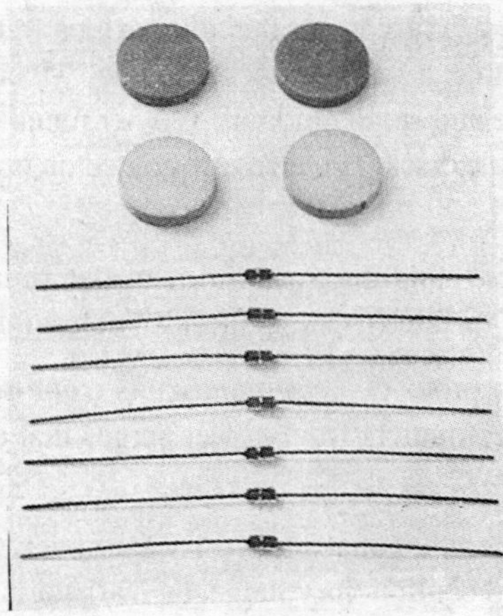

**Fig. 6.3:** Thermistors

PTC thermistor may deteriorate at low temperatures if it is used in a reducing gas or with an oily material. Since the terminals of thermistors are plated, they can be *corroded* by high humidity, chlorine gas or other gas containing sulfur.

## 6.2 Varistors

The varistor, a *nonlinear resistor* that does not obey ohms law is one of those wonderfully simple components whose operating principles are not yet fully understood. What is understood is that *the varistor is well suited for protecting electronic circuits from transient voltage surges.*

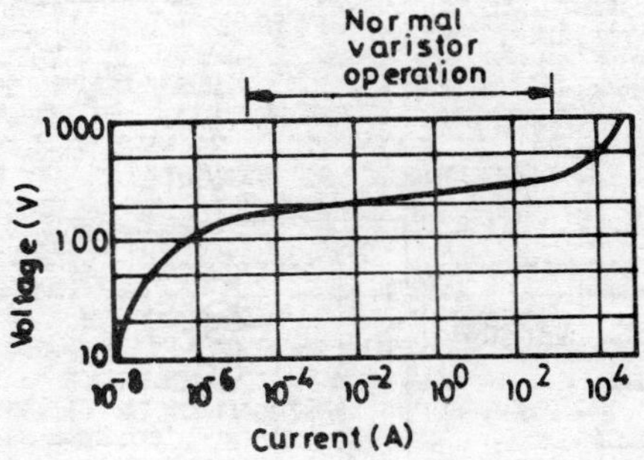


**Fig. 6.4** Varistor v-i characteristic

The current through a standard resistor is directly proportional to the voltage across the resistor. The current through a varistor, however, varies according to the *nth power of the voltage*, where n varies from 2 to 25. Fig. 6.4 shows the voltage versus current characteristics of a typical varistor.

Varistors are made from silicon carbide or metallic oxides yet their construction very closely resembles that of a resistor rather than a semiconductor device. Metal varistors, for example, are made from powdery grains of zinc oxide and small amounts of other metal oxides. The zinc oxide powder is pressed and sintered to form ceramic discs. The discs are coated on opposite sides with silver to provide solderable contact regions.

The varistor's voltage and current handling capabilities are determined, respectively, by the thickness and width of the disc. The size of the zinc oxide grains also determines the varistor's *voltage rating.*

Varistors are primarily used to protect electronic circuits from transient voltage surges. They are particularly useful for protecting computers from power surges that cause memory errors or even loss of memory contents.

The operation of a varistor resembles that of two back-to-back zener diodes. When the voltage across the varistor is below its rated continuous voltage, the varistor is in a *high impedance state.* When the voltage exceeds the rated level, the varistor suddenly begins to conduct and its formerly very *high impedance falls* to a few ohms.

Voltage transients can be caused by power line surges, lightning, switch arcing, component failures, and the sudden removal of a voltage across a transformer inductor, or relay coil.

Fig. 6.5, for example, shows a relay coil connected through a switch to a power supply. When the switch is opened, the field in the coil collapses and induces a series of high voltage spikes across the coil terminals. The varistor serves to short to ground the spikes having an amplitude above the varistor's voltage rating.

Because of its nonlinear v-i characteristic, a semiconductor diode will also function as a varistor. The lower the voltage drop across the diode, the higher the resistance. The widest *resistance variation* is obtained when the diode is forward biased by the applied voltage. An example is the 1N 34 (see Fig. 6.6a). At +0.05 V its resistance is 50,000 ohms and at +2.5 V its resistance is 70 ohms. Thus a resistance change of 714:1 corresponds to a voltage change of 50:1.

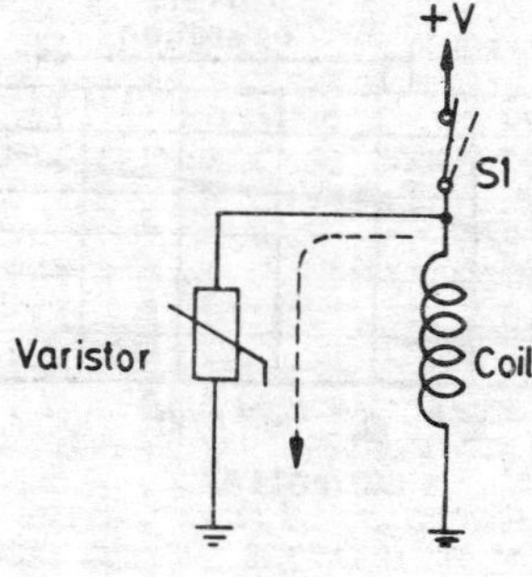


**Fig. 6.5** Varistor voltage clamp

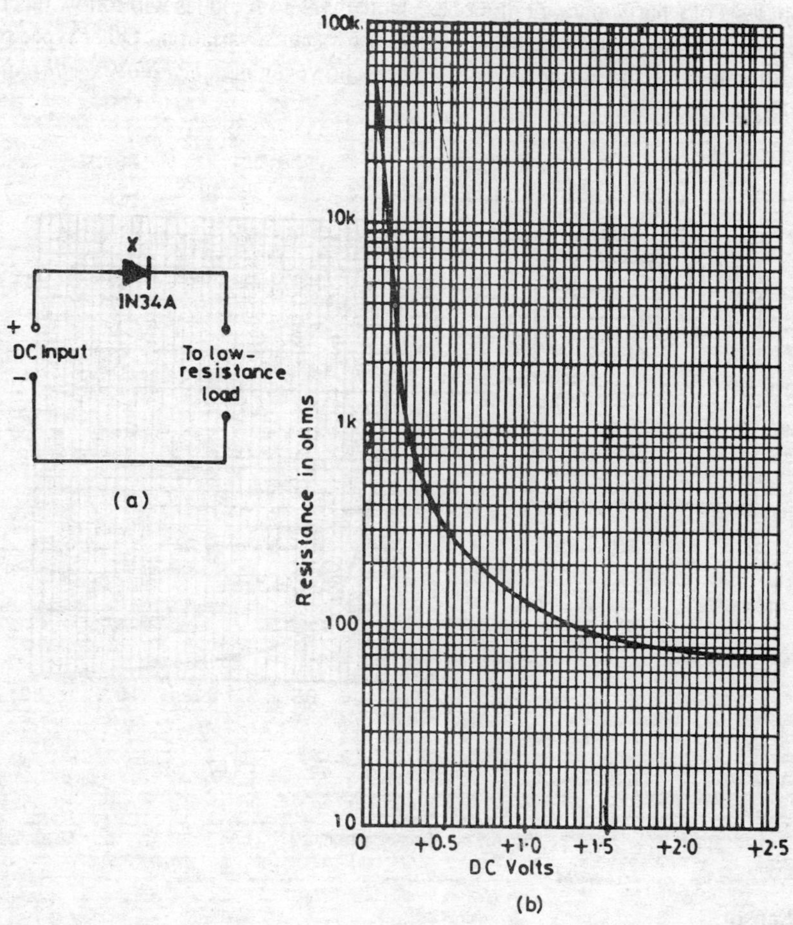


**Fig. 6.6** dc Varistor action of 1N 34 A

The resistance voltage characteristic is shown in Fig. 6.6(b). The curve was plotted from voltage/current data obtained from tests of a particular diode. Not all 1N 34 A diodes will follow this curve exactly, but all will exhibit resistance variation of the same order of magnitude. Note that the resistance of 1N 34 A changes most rapidly between +0.05 volts and +0.5 volts, this 10:1 voltage change producing a 151:1 resistance change.

An ac varistor may be made by connecting two diodes in parallel, *reversed with respect to each other*, as shown in Fig. 6.7(a). This is equivalent to a shunt arrangement of two dc varistors of the type shown in Fig. 6.6(a). One diode is forward biased by one half cycle of applied ac voltage and the other diode by the other halfcycle. Conduction during the reverse biasing half cycle is negligible consisting only of the diode leakage current.

Fig. 6.7(b) shows the performance. The resistance varies from 4,000 ohms at 0.025 V to 10.5 ohms at 2 V. All voltages are rms values. As with the dc varistor, the curve here was plotted from voltage/current data from tests of a particular pair of diodes. Not all 1N 34 A diodes will follow this curve exactly, but all will exhibit approximately the same degree of resistance variation. Other types of diodes may be used for different resistance values. With any type, however, *matched diodes will reduce distortion by keeping the current symmetrical.*

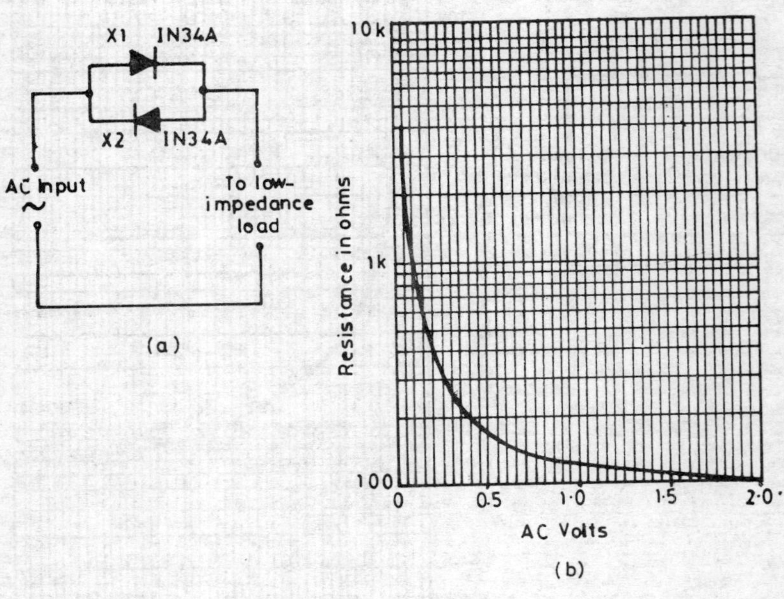


**Fig. 6.7** ac Varistor action of 1N 34 A

## 6.3 Photo Sensors

The photosensor, *a device that converts light to electricity*, is playing an increasingly important role in electronics. As a discrete unit, it is used as an exposure meter for cameras to measure brightness and also as a day light switch to distinguish day from night. These applications involve the detection of external light. However, together with the development of LEDs, IREDs, laser diodes and other light emitting devices, the application areas of photosensors are expanding rapidly.

Structurally there are two kinds of photosensors; *photo interrupters* and *photo-reflectors*. The photo interrupter is so structured that it has a photo-detector and a light emitting device facing each other so that it detects a change in *'through light'* when an object passes between them. The photo-reflector has a photo-detector and a light emitting device arranged in the same direction to detect a change in reflected light. Both the photo-interrupter and the photo detector consist of a light-emitting device, a photo-detector and a package.

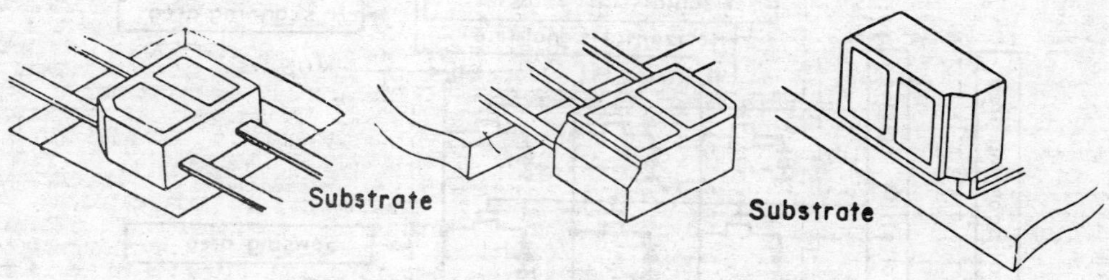


**Fig. 6.8** Photo-sensors—Mounting example of miniature photo-reflector.

Ga-As IREDs with relatively high output are mainly used as LEDs, while silicon photo-transistors, including photo-Darlington transistors, are used as photo-detectors in almost all cases. Photo detectors with signal processing circuits integrated on the same substrate have been recently developed.

There are two types of photo-reflectors. The first consists of two separately packaged devices that are contained in a package or holder to be made into a photo-reflector. The other is made as a photo reflector from the onset, with the two devices contained in the same package. The former is called the *assembled type* and permits a change in its size and shape as well as in the optical functions and is suitable for special or customised applications. The latter is the *integrated type*. The integrated type lends itself more easily to reduction and is more amenable to mass production because manufacturing is less complex. Therefore, the leading photo-reflector is the integrated type.

Photo-reflectors are used mainly for the detection of rotation systems in vcrs, audio-systems and other products.

### 6.4 Image Sensors

Solid-state image sensors can be classified into two categories; an MOS and a CCD type. An *MOS type* exhibits high picture quality and can be driven with a single low voltage. A *CCD type*, on the other hand, performs better under low light conditions due to its low noise operation. Camcorders (camera recorders) with both sensors are now available and they achieve high overall performance. The resolution of MOS sensors has improved significantly from 180,000 pixels (picture elements) to 300,000 pixels with sensitivity improvement of more than 20 dB. The progress in this aspect is attributed to the introduction of a fine lithography process and transverse signal line design (TSL).

Both MOS and CCD sensors employ essentially the same photo-diode structure to sense incident light and convert it into an electric-signal. The main difference is in the way they read out signals.

A CCD transfers electric signals into corresponding CCD lines-prepared on the same chip-in the first stage operation. Each signal is then moved along CCD lines and successively read out to the outside.

In an MOS device, no such CCD line is prepared. Rather, electric signals are read out directly through an array of MOS transistor switches. In essence, signals from each pixel are read out line sequentially under an x-y shift register arrangement in an MOS sensor.

The design of an MOS sensor, combined with a TSL structure has the following salient features:

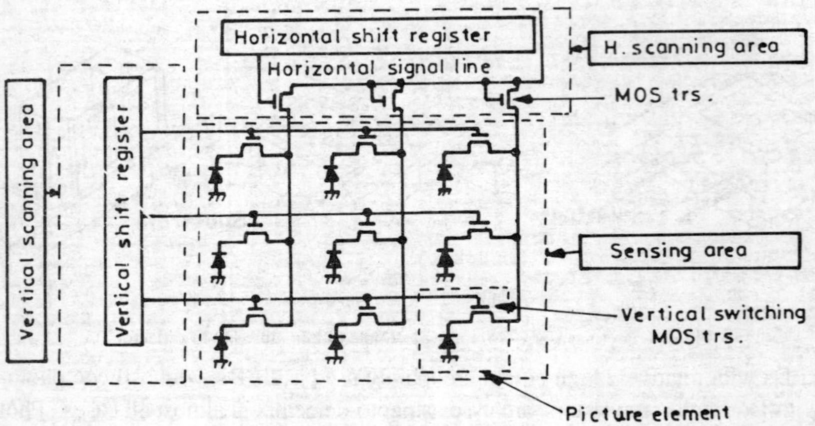


Fig. 6.9 A conventional circuit

1. *high picture quality*—due to independent readout of signal from each photodiode;
2. *high resolution*—a result of simple design without CCD;
3. *almost no smear*—due to TSL structure;
4. *versatile function*—because of an xy address configuration;
5. *a single low voltage driving*—due to an all MOS transistor structure; and
6. *low frequency drive*—due to zero phase and π phase readout within one clock cycle

A circuit for an MOS device without transverse signal line (TSL) is shown in Fig. 6.9. As can be seen in the figure, the device consists of three main areas: photosensing, vertical scan and horizontal scan circuits.

One *pixel* is composed of a photodiode and an MOS transistor. First, the vertical shift register selects a row of pixel data and feeds it into corresponding vertical signal lines. Then, the horizontal shift register circuit reads the information from each pixel sequentially from the left to the right.

This simple circuit was not sufficient after the performance of the circuit was much improved.

Demand for less smear noise became strong and TSL was finally incorporated in the imager to overcome the problem. The TSL is described in Fig. 6.10. In the TSL, a new MOS switch ($T_H$) is added to each pixel. It should be noted that the signal read out-line is now laid out horizontally (horizontal signal line).

The information stored in each photodiode is read out in the following manner; the vertical shift register turns on the tv and read out transistor Sv, and the horizontal shift register turns on $T_H$ by transferring information.

The essence of this structure is that electric charges from each pixel stay only a short time on the signal line. The *smear noise* arises from overcharging of the vertical signal lines in the conventional MOS sensor-during the horizontal scanning period (63.5 ms).

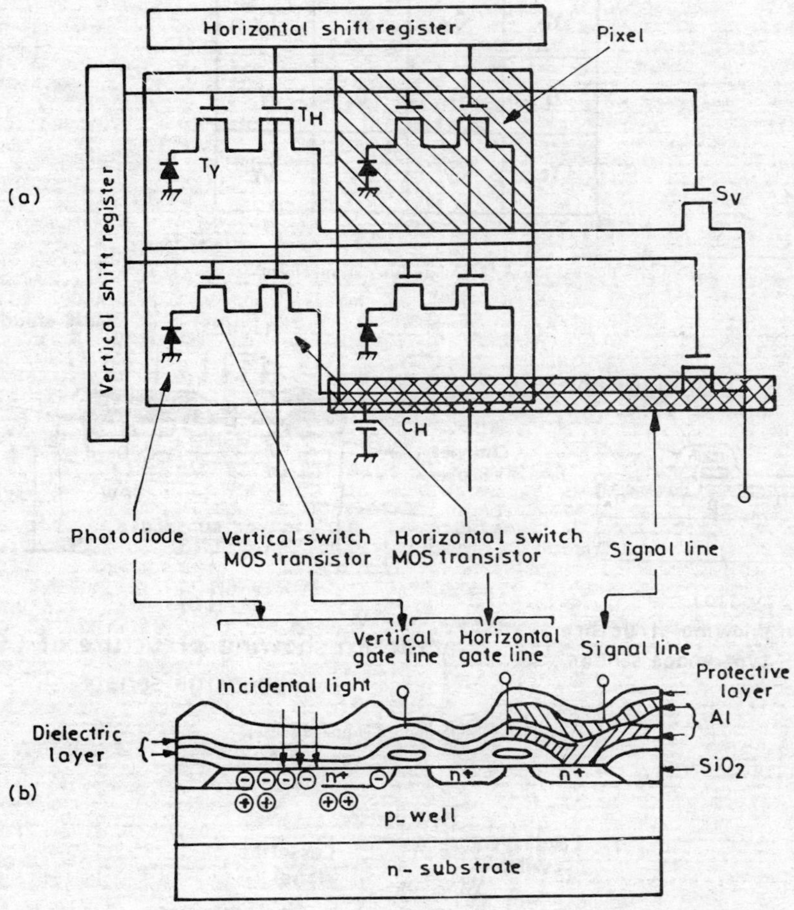


**Fig. 6.10** A TSL circuit and its cross section

The overcharge does not occur in the TSL. The TSL reads out the information charge, pixel by pixel, with an approximate readout time of 200 ns or less. A 320 fold improvement is expected from this change. In an actual measurement the smear noise was suppressed by more than 100 dB. The TSL structure smoothed out random and fixed pattern noises too, leading to drastic enhancement of MOS sensor.

To provide a colour image, it was necessary to add a colour filter process in the sensing chips. For MOS sensor, the monolithic colour filter is formed by gelatin coating followed by dying and patterning. White, cyan, yellow and green were selected and arranged as seen in Fig. 6.11.

The arrangement was designed to minimise undesired *moire pattern noise*. Complementary colouring was chosen by considering light utilising efficiency. Together with an independent readout system

| Ye | Cy | Ye | Cy |
|---|---|---|---|
| G | W | G | W |
| Ye | Cy | Ye | Cy |

**Fig. 6.11** Colour filter arrangement

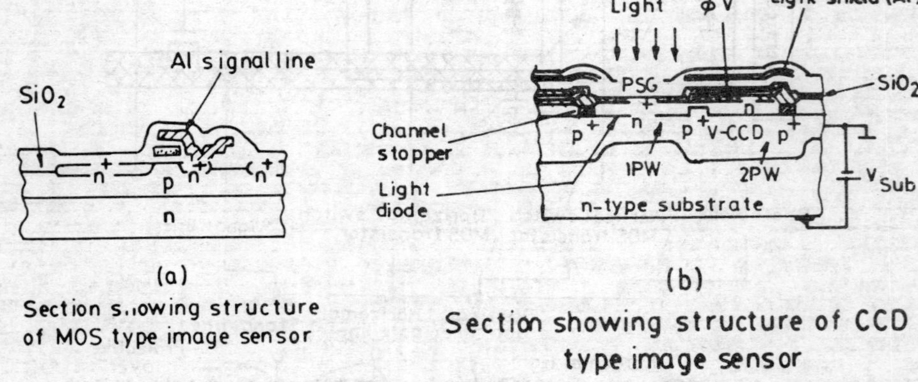


**Fig. 6.12** MOS and CCD image sensors

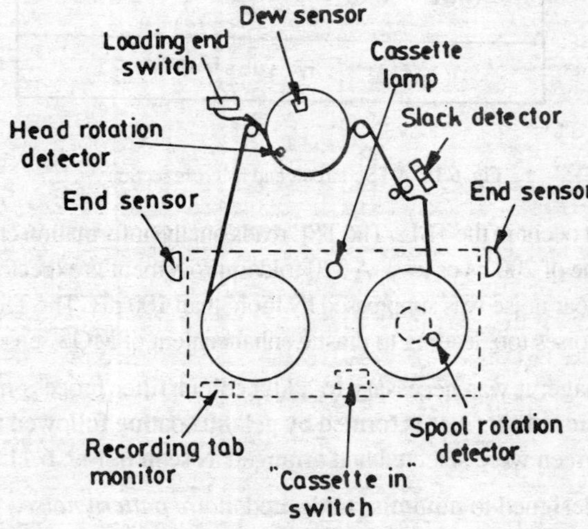


**Fig. 6.13**: Dew sensor on the deck of a video cassette recorder (vcr)

of an MOS sensor for colours, all improvements resulted in excellent reproduction of colour pictures with high fidelity. MOS image sensors are illustrated in Fig. 6.12.

### 6.5 Dew Sensor

Video cassette recorders are equipped with a heater which prevents condensation, or '*dew*', forming inside the unit. If excessive condensation is present the vcr cannot be operated.

If the tape itself is in good condition one of the most common causes of damaged creased tapes is condensation on the head drum when the tape adheres to it and is shredded. This is also prevented in the machine by the dew sensor on the drum assembly.

P-types dew sensor (Fig. 6.14) for *moisture detection* in miniature video recording equipment measure 5.5 mm × 3.3 mm × 0.64 mm. Resistance changes from <10 Ω to >100 kΩ in less than 20 seconds when subjected to condensation above the dew point. The sensors need a maximum operating voltage of 0.8 V d.c. and consume less than 2mW. They will operate over a temperature range of 0 to 60°C and from 0 to 100% relative humidity.

**Fig. 6.14:** Dew sensor

*Chapter 7*

# MICROPROCESSORS

---

*The past decade has witnessed a remarkable growth in integrated circuit (IC) technology. The development of large scale integrated circuits (LSI) and very large scale integrated circuits (VLSI) coincided with the design and the manufacture of highly sophisticated devices called microprocessors which form the subject of study in this chapter.*

---

## 7.1 Microprocessors

By the term *microprocessor* (MP) one understands a *central part of the computing system,* which consists of one or a few LSI circuits performing the functions of an operational arithmetic logic unit and a control unit, both fabricated on one chip or a few chips. A microprocessor sequentially executes instructions (microinstructions) specified by the program for realising a definite algorithm. In the general case, an MP can perform the following:

1. receive coded instructions;
2. collect, process, store and issue coded information;
3. issue signals controlling the operation of microprocessor and other circuits or specifying their state.

   An *algorithm* is a defined process or set of rules that leads and assures development of a desired output from a given input.

The microprocessor thus performs the same functions as the processor of a digital computer, but executes fewer commands, from 50 to 100 and deals with shorter words, commonly comprising 8 or 16 bits.

Microprocessors find use in controllers, microcomputers, and microprocessor systems. The *controller* is a digital automatic device which realises a fixed, or an adopted, algorithm. This control unit commonly controls peripherals, or non-interacting devices according to a specified algorithm. The *microcomputer* is a self contained unit built around a microprocessor set and complete with a software package and input-output (I/O) terminals. The *microprocessor system* is a system based on microprocessor sets intended to perform computing, monitoring and other functions.

At present, microprocessors make the building blocks of airborne and ground computers. They service telemetering stations, control manufacturing operations and telephone sets, act as coordinate converters in the peripheral equipment of computing systems, execute standard progress and subroutines, and also find use in optical recorders of medical equipment, commercial systems etc.

## 7.2 Microprocessor Organisation

The structure of a microprocessor must be *flexible*, ensure *high speed* and *economically efficient* realisation of specified tasks. Any microprocessor consists of an operational unit and a control unit.

The *operational unit* converts computer (machine) words. The *control unit* controls the course of a computing process. It ensures the retrieval of commands (*instructions*) from the storage, decoding of the command's sequential input of control signals (*microinstructions*), and execution of microinstructions in a definite sequence. Microprocessors can be of the fixed and extended word capacity types. The first microprocessor had a fixed word capacity (word size) and a fixed set of instructions.

Most microprocessors have microprogram control that imparts *flexibility* to these devices. The flexible structure of an MP enables the use of a large number of internal scratchpad registers and affords an advanced system of controlled interruptions.

The simplified microprocessor resembles a first generation computer which lacked microprogram control and scratchpad registers (storage cells and other elements play the role of registers). Generally, the registers that are used to store temporary intermediate results (data) in some type of random access memory (RAM) are referred to as *scratchpad registers.*

Fig. 7.1 illustrates the basic block diagram of such a microprocessor. The operating unit of the MP includes an *arithmetic logic unit* (ALU), *operand registers* A and B ($OR_A$ and $OR_B$), *instruction register* (IR), *general purpose registers* (GPRs) and *address register* (AR). The arithmetic block performs arithmetic and logic operations on one or two operands contained in registers A and B. The result of the operation is sent to a unibus combining all blocks. The instruction register serves to store the code of the current instruction. General registers are used to store on line information during computation. The program (instruction) counter (PC) determines and stores the address code of a random access memory (RAM) or that of an input-output (I/O) device (IOD).

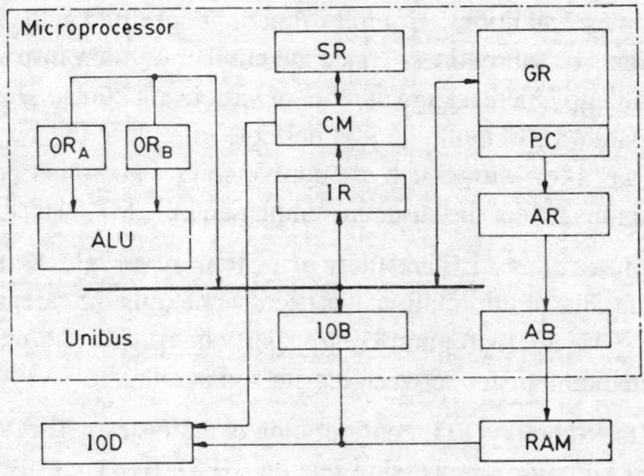


**Fig. 7.1:** Block diagram of fixed-word length microprocessor.

The *status register* (SR) fixes the state of the MP at each moment of program execution and ensures each transfer in the program according to the specified characters (flags), or conditions. The MP has a fixed length of data words, 4 or 8 bits. The set of instructions to be executed is determined by the structure of a control unit which is a *control memory* (CM) or micromemory. The MP operates together with I/O devices and RAM connected via I/O *buffer registers* (IOBs) or address buffer registers (ABs). The microprocessor usually has one common (shared) bus to transfer address, data, and instructions. The use of such a data path calls for time-interleaved multiplexing which decreases the speed of the MP. Microprocessors of this type are in essence specialised devices since they have a small word size and perform a limited, rigidly fixed, number of instructions (several tens of commands).

The *parameters* of a microprocessor are largely determined by the speed of micromemory (MM). For a microprocessor to be most efficient, the MM access time (cycle time) must be one-tenth of the RAM cycle time. The mean readout time for MM ranges between 30 and 120 ms. A micro-instruction controls the operation of ALUs which perform the required micro-operations and besides, specifies the address of the microinstruction stored in the RAM. A set of general registers as a rule includes an index register, stack pointer, program counter register, and other registers performing some specified functions. The width of GRs in the best models of MPs is 16. GRs are connected to other MP blocks with a bidirectional (two way) 8 or 16 bit bus. The register of a program counter is coupled with an address register connected to a 16 bit output bus. The general purpose registers (GRs) are always addressed directly from a microinstruction register.

The *arithmetic logic units* come in several types; binary, binary-decimal and symbolic. Most of the ALUs are of parallel action. An ALU generally consists of a binary ripple adder, scratchpad register, shift register and accumulator register.

For access to *external memory* microprocessors commonly employ indirect, indexed, and direct methods of addressing. In *indirect addressing* one of the GRs contains the memory address, the indication to which is given in the instruction. Using a 16-bit general register, a short instruction can address upto 4k words. In *indexed addressing*, the address contained in the instruction and complemented by the information from the index register serves as a memory reference address. The method of *direct addressing* uses instructions 3 and 4 bytes long.

The MP ALUs perform the simplest operations such as subtraction, transfer, logic AND and OR functions, shift, and modules 2 additions. Complex functions specified in the instruction are performed according to microprograms or subroutines. Since the number of microinstructions is commonly over 100, a microprocessor can implement a large number of various functions. Besides, the microprocessor is usually made as a separate LSI circuit, so it is possible to replace the set of micro-instructions. A microprocessor usually has a few *buses* such as address, control and other paths, which eliminate the need for multiplying and thus render the structure simpler and the MP faster.

The address and data buses of an ALU and those of a GR array are split, as it were, into equal sections (*binary layers*). To transfer bits of information, it is necessary to link the carry and shift circuits between corresponding sections. The microprogram control unit controls all the operations. The multiwire buses perform the information transfer between registers in parallel.

A microprocessor in the Schottky TTL configuration is the fastest. The MP consists of four chips, each accommodating a 4 bit control processing section (CPS), five control circuit chips and has two chips for micromemory. The MP has 16 GRs. The 4-bit unit on a single chip contains an ALU, general

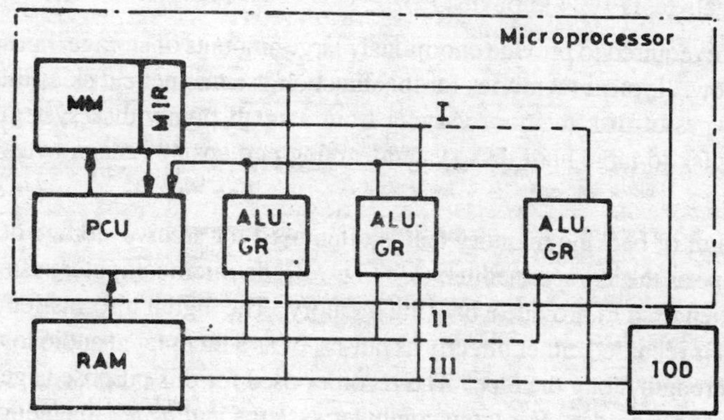


**Fig. 7.2:** Block diagram of extensible wordlength microprocessor with microprogram control, I-control bus, II-address bus, III-data bus.

register memory of 8 word capacity, two 4-bit registers, scaling and shift multiplexers (MUXs) that allow for a variety of interfaces, and, last, a programmable logic array (PLA) for 512, standard operations. The instructions programmed in the PLA during its fabrication ensure a high universality of the microprocessor. Such a large set of microinstructions enables emulating most diverse machines without the loss of *program compatibility* and without increase in the cost for software.

Longword microprocessors ensure high speed carriers for the entire adder and parallel reference to all control codes, data and input-output addresses.

For its operation a microprocessor requires a number of peripheral units, RAM, ROM, input-output (I/O) devices etc, designed with LSI. These LSI devices form a microprocessor set and number ten or more in the set depending on the organisation of a microprocessor and its hardware.

Input-output control circuits of MPs are very simple. Direct access channels used in some types of MPs enable a high speed exchange of a large body of information. The interrupt system is programmable and is responsive to external requests.

## 7.3 Memory

The memory system attached to a computer can be divided into two major sections, generally referred to as the '*main or working memory*' and the '*backing memory*'. The main memory is used to hold the program being executed and any data that it is currently using. The backing memory is used for achieving storage of data and programs.

The general requirements of the main working memory are that it should provide *rapid access to data* while allowing a reasonably *large storage capacity*. Many of the early designs of computers used a main memory system based on magnetic ring cores, but virtually all modern computers use solid-state electronic memory devices. The *capacity* of a computer memory is normally quoted in kilo bytes (K) where 1K = 1024 bytes. The main memory of a typical microcomputer generally has a capacity of

64-128K, whereas mini and mainframe computers may have memories with capacities upto a few 'megabytes' (a 'megabyte' is 1024 kilobytes.)

Backing stores are required to provide enormously large amounts of storage, but speed of data access may be relatively slow. Typical memories for this function use magnetic disk as the memory medium. There are various types of disk memory varying from a small floppy disk system with capacities of perhaps 250 K per disk to large hard disk systems giving perhaps 300 mega bytes from a single disk memory unit.

An alternative form of backing memory that is often used for archive storage of programs, or data, employ magnetic tape as the storage medium (see Fig. 7.3). In microcomputer systems standard audio-cassette tapes are often used to provide a backing memory. The digital information representing a program or a set of data is recorded either directly as pulses, or as a modulated audio tone signal by varying the magnetic field strength along the tape. The recorder used for this purpose is generally an ordinary domestic audio cassette recorder. For large computer systems *multitrack* magnetic tape machines are used to read or write nine parallel tracks of data on to 12.5mm (1/2inch) wide tapes. Because data have to be written or read *sequentially* along the tape the computer will take some time to search for and read a particular data item, and thus *access time* is generally much slower than that for a disk system.

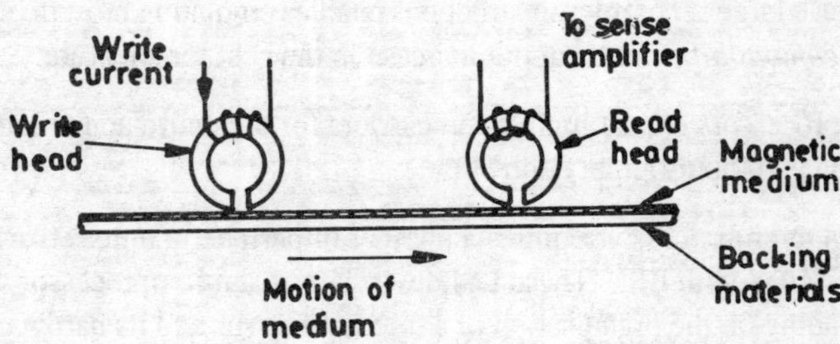


**Fig. 7.3:** Magnetic surface recording arrangement.

## 7.4 Classification of Memories

MSI and LSI circuits find wide application in computers, calculators, peripheral devices, control equipment, communication equipment etc. Recent advances in semiconductor technology have provided a number of reliable and economical memory circuits. The typical semiconductor memory consists of a rectangular array of *memory cells* fabricated on a silicon wafer, and housed in a convenient package, such as DIP. *The basic memory cell, typically a transistor flip-flop, or a circuit capable of storing charge, is used to store 1 bit of information.* LSI memories, or storages, fall into a number of categories depending on the criterion for classification; static, dynamic, bipolar (BJT based), MOS on-line (main), permanent, random access, serial access and other types.

The method of storing information whereby a loss of power will result in loss of data stored is referred to as *volatile storage*. On the other hand, the method of storing information whereby a loss of power will not result in loss of stored data is referred to as *non-volatile storage*.

*Static Memory* is a matrix of elements each of which can be driven in one of the two states that remain *stable* in the given condition of power supply. The basic element of this storage is a flip-flop, or a gate. In its operating condition, such a matrix obviously consumes power uninterruptedly.

*Dynamic Memory* is a matrix of elements which requires a *periodic restoration* (regeneration) of stored data. This is a substantial drawback of this storage, but then it is fast and consumes a small amount of power (only during reading, recording, and refreshing).

*Bipolar LSI Memories* are much *faster* than *MOS LSI memories,* but have a *lower packing density*. In wide use today are memories designed with TTL and ECL. $I^2L$ (Integrated Injection Logic), which is a new LSI technique using BJTs, affords a considerable increase in the packing density of microcircuits. $I^2L$ circuits are most suitable for use in high speed buffer storages of large systems. *MOS Memories* are largely built around p-channel and n-channel MOS LSI circuits. *The highest achievement of MOS technology is the creating of charge-coupled devices and memories based on these devices.* LSI CCDs are adaptable to industrial production and are used for a *very high packing density.*

*Random Access Memories (RAM)* feature the same rate of writing (or reading) of any bit of the word in the matrix. It is expedient that the main memories of computers should allow random access. RAMs use both bipolar and MOS structures.

*Serial Access Memories (SAMs)* allow access to data records in a certain sequence starting at a definite address. They serve as peripheral units of computers. Until recently they were built only with MOS structures.

*Permanent Memories* are of three modifications; *read only memories or storages* (ROMs, ROSs) where information cannot be rewritten, *programmable ROMs* (PROMs) where information can be rewritten once, and *electronically programmable ROMs* (EPROMs) allowing information to be rewritten several times.

In semiconductor ROMs, the storage unit is a matrix of diodes, or BJTs, or MOSFETs with a certain interconnection pattern. Programmable or reprogrammable ROMs can be built of metal-nitride-oxide silicon (MNOS) devices. The information in such ROMs is *erasable* by ultra-violet rays. ROMs serve as sequential control blocks, code converters, symbol generators, display character generators, and as program storages.

## 7.5 Parameters of Memories

The main parameters of the LSI memories are the following.

**1. Memory Capacity:** This is the maximum volume of data, expressed in bits, that the memory can store. The *number of words* and the *word length* characterise the organisation of a memory.

**2. Static Parameters:** These are the address input current and logic-0 output voltage, both specified at limiting values of temperature and supply voltage. To these parameters also belong static noise voltages $V_n^1$ and $V_n^0$ determined in the same way as for logic ICs. They define the static noise immunity of LSI circuits.

**3. Power Consumption:** This is the power consumed by a microcircuit during data storage and access at a definite voltage and current.

**4. Dynamic Parameters:** These are time parameters determined from timing diagrams of read and write cycles and counted off at a 50% level of the output signal.

For the *read cycle* the time parameters are:

(a) minimum cycle time for read, (b) required access time and (c) time during which the old data are valid after the address changes.

For the *write cycle* the time parameters are:

(a) minimum write pulse width, (b) minimum cycle time for write, (c) minimum data set up time, (d) minimum data hold time, (e) minimum address set-up time, and (f) minimum address hold time.

In practice, technical manuals and specifications commonly give three main parameters; access time, read cycle time and write cycle time.

The *access time* is the time interval between the moment of applying a new address for data transfer into the memory, or from the memory, and the beginning of data transfer. This parameter characterises the speed with which a word can be read out or written, into the storage.

The *read cycle time* is the total time from the moment of arrival of an address signal, or sense signal, to the moment when the circuit completes the operation and can begin to carry out a new operation.

The *write cycle* time is the total time from the moment of arrival of a write signal to the moment when the circuit is ready to accept the next write signal.

## 7.6 Random Access Memory (RAM)

Semiconductor memories have the same organisation as magnetic memories. The array of storage elements is made in the form of a square or rectangular matrix, with the location of each element by x and y coordinates. Most of the available LSI random access memories used for data recording, storage and retrieval, contain on one chip 1024 words by one bit or more.

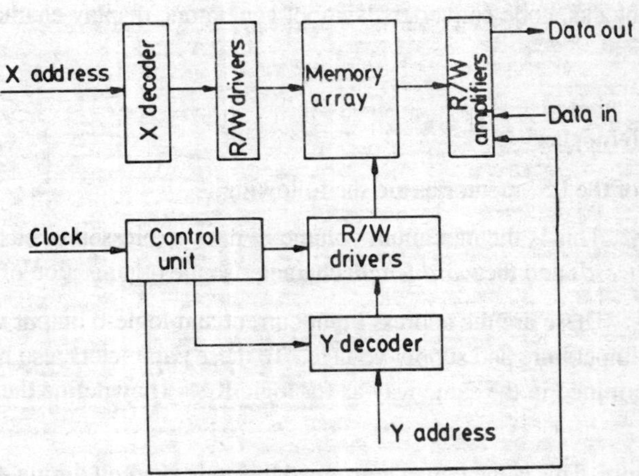


**Fig. 7.4:** Block diagram of random access memory.

Fig. 7.4 illustrates the block diagram of a RAM unit. Apart from the memory array, the RAM includes pulse decoders and driver to access, write, and read data along the x and y coordinates, and also read/write amplifiers, control unit, and buffers. RAM systems can be of the bit organised, word-organised and composite types. A *bit organised memory* allows access to each storage element, independent of the other elements arranged on the output at intersections of corresponding x and y access lines. A *word organised memory* allows a simultaneous access to a group of locations for one word. The first type of memory organisation, i.e. LSI form, is preferable since it requires fewer amplifiers and a much smaller number of outputs. In the *composite type of memory* organisation, access is made to a row of elements connected to the x access line, with the sense amplifiers, or write drivers brought into action to access the storage element of the corresponding x coordinate.

With the 'black box' concept of a RAM system, we can single out the following data streams moving to and from the peripheral units:

- input data to be written
- output data to identify the location to be accessed
- output data to be read out
- control signal to enable the memory to operate in the write, read, or store mode.

## 7.7 Serial Access Memory (SAM)

As an example of serial access memory devices consider shift registers. By the methods of writing and reading, shift registers fall into serial, parallel, and parallel-serial types. The shift register proper can be thought of as an *array of interconnected storage elements* called register storages. Links between the stages enable a sequential transfer of information from one element to the next. In the *serial shift register* of Fig. 7.5, only the first and the last elements are accessible for direct recording and reading of the data respectively. For data to be written into or read from the elements, bits should be sequentialy shifted left or right from one element to the other. In the *parallel shift register* data can be written in, or taken from, any stage in parallel form because of the limits on the number of output leads and high requirements on the packing density. LSI memories mostly use shift registers of the serial type. For a simple non-destructive storage to be built using a serial shift register, there is a need for a feedback loop to connect the output to the input as illustrated in Fig. 7.5.

This ring type configuration enables non-destructive read out by shifting bits along the loop into free locations. In order to access, say, the nth location, its contents must be rewritten into the nth location, thereby routing the shifted bits along the loop to occupy free locations 1st, (n – 1)th.

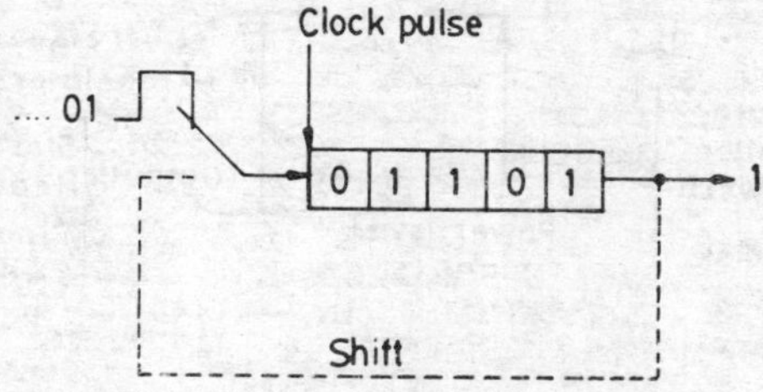


**Fig. 7.5:** Principle of action of serial shift register.

The input data can flow into a register along one or two channels ( a single-phase or a two-phase transfer method). Each data bit shifts from one location to the next as a clock pulse arrives at the register circuit. By the principle of action, memory circuits that form the basis of shift registers may be classified into static, quasi static, and dynamic types.

## 7.8 Micro-computer

Micro-computer is a general term referring to a complete tiny computing system, consisting of *hardware and software*, whose main processing blocks are made of semiconductor integrated circuits. In function and structure, it is somewhat similar to a minicomputer, with the main difference being price, size, speed of execution, and computing power. The hardware of a microcomputer consists of the microprocessing unit (MPU) which is usually assembled on a printed circuit board with memory and auxiliary circuits. Power supply, control console and cabinet are separate.

The basic inter-relationships between the principal parts of the microsystem and the methods (or paths) by which data can be made to flow within a system is referred to as *microcomputer architecture*. Architecture includes number of registers available, how registers are used, instruction sets, input-out-

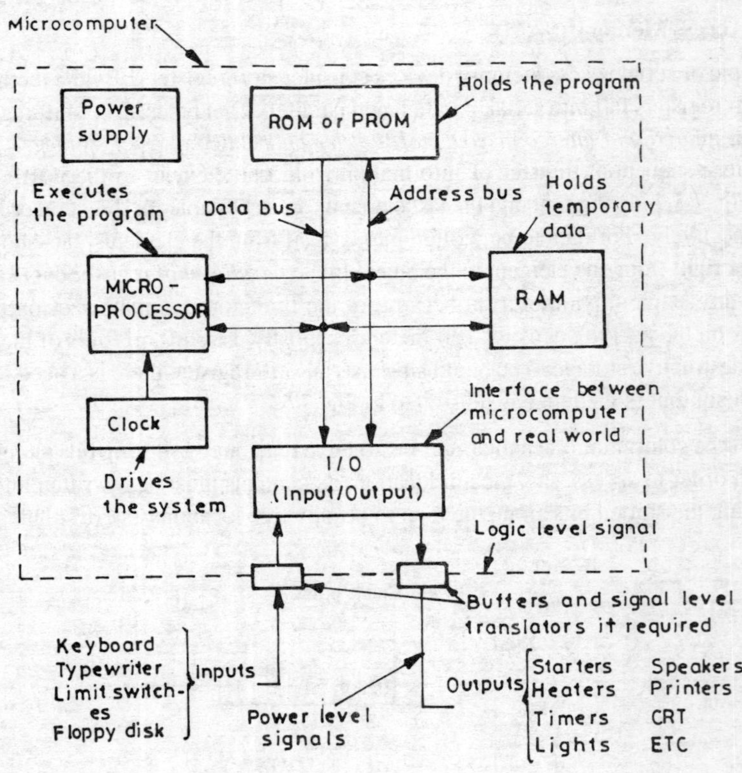


**Fig. 7.6:** Micro-computer components.

put operations, number of stacks available, how stacks are used, hardware interrupt structure, data paths into memory (number and types), data paths into CPU (number and types), and others. Most architecture are classified into either CPU oriented, memory oriented, or bit-store.

Microcomputers are tiny digital computers developed from a few large scale integrated circuit (LSI) chips. A central processing unit (CPU) may be on a single LSI chip (microprocessor chip) or several such chips. A microcomputer system also requires ROM storage for programs and data, clock circuits, input/output interfaces, shift registers, and control circuits (see Fig. 7.6).

Various programming aids that are frequently supplied by the manufacturers to facilitate in the purchaser's efficient operation of the equipment are referred to as *software*. Such software items include various assemblers, generators, subroutine libraries, compilers, operating systems and industry application programs.

The electric, electronic and mechanical equipment used for processing data consisting of cabinets, racks, tubes, transistors, wires, motors etc. are referred to as *hardware*.

# PART 2
# SOLID-STATE CIRCUITS

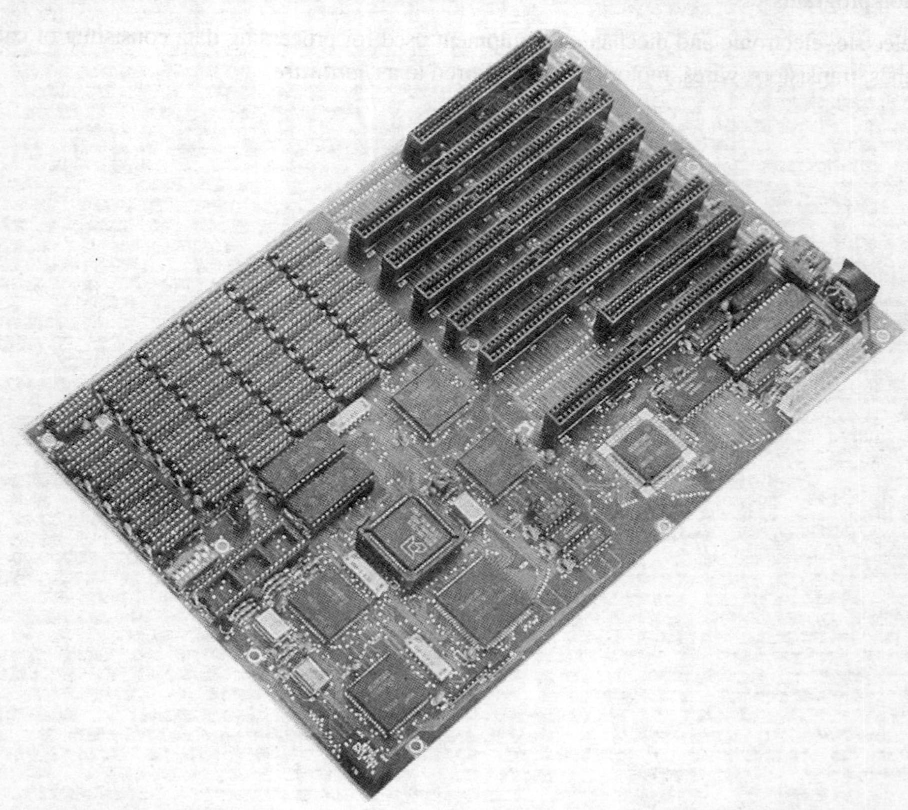

1. POWER SUPPLIES
2. ANALOG CIRCUITS
3. DIGITAL CIRCUITS
4. CONTROL CIRCUITS

*CHAPTER 1*

# POWER SUPPLIES

*Primary batteries are often used as the collector supply for transistor circuits in portable equipment and low level circuits. In the majority of commercial transistor equipment, however, mains operation is required. The production of rectified supplies from ac mains, in general, follows the established practice, using step-down power transformers and silicon rectifiers. We study these in this chapter.*

## 1.1 Half-Wave Rectifier

If an alternating source of EMF is connected across a p-n junction, it will be forward biased during the positive half-cycle and current will flow through the junction in the forward direction. During the negative half-cycle the p-n junction will be reverse biased and almost no current will flow through the junction in the reverse direction. In other words, *the p-n junction operates as a switch which makes and breaks at the frequency of the applied ac voltage*. This is called half-wave rectification. Silicon rectifiers for power supplies are shown in Fig. 1.1

In Fig. 1.2(a) the p-n junction operates in the forward direction and positive going output is obtained. In Fig. 1.2(b), the p-n junction operates in the reverse direction and no output is obtained.

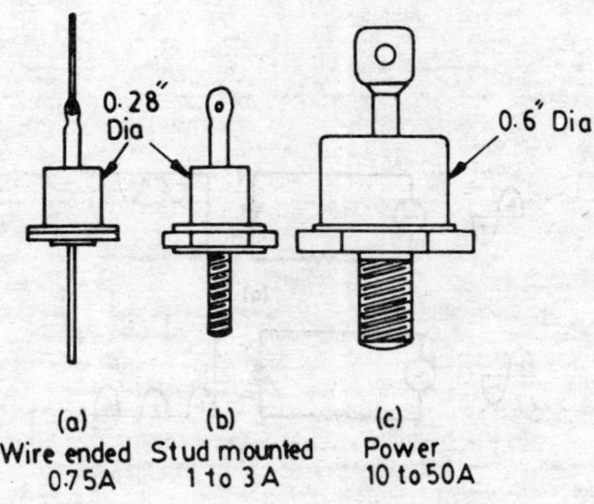


**Fig. 1.1** Silicon rectifiers for power supplies

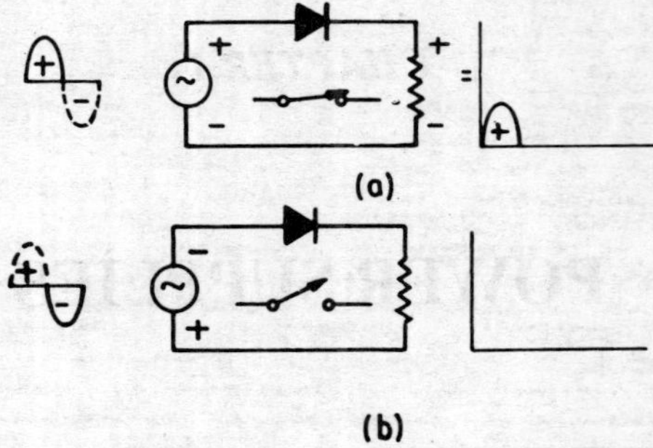

Fig. 1.2 Half-wave rectification

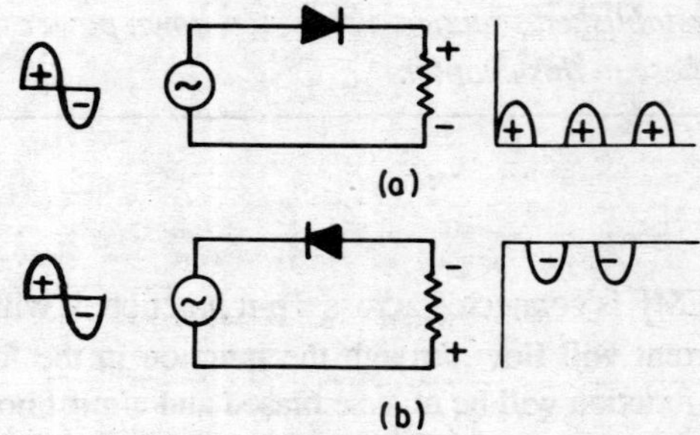

Fig. 1.3 Positive and negative going half-wave rectified pulses

Fig. 1.3(b) illustrates how negative going output is obtained by inverting the rectifier in Fig. 1.3(a).

The output of a half-wave rectifier can also be taken from across the diode, instead of the load. This is illustrated in Fig. 1.4.

In any rectifying action, negative dc output is obtained from the anode of the rectifier, whereas *positive dc output is obtained from the cathode of the rectifier.*

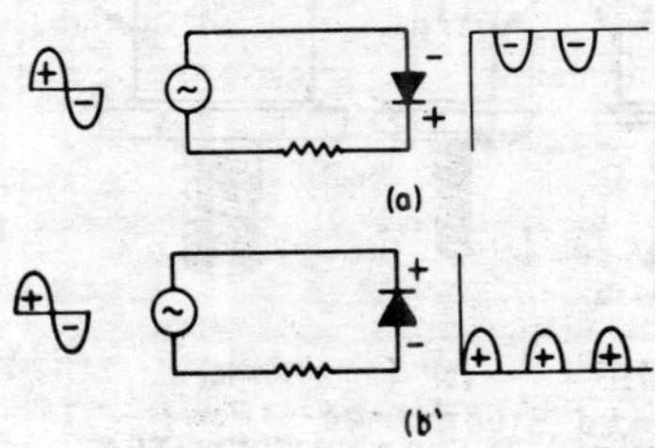

Fig. 1.4 Output waveform of a half-wave rectifier

## 1.2 Full-Wave Rectifier

The *efficiency* of a half-wave rectifier is low. Two half-wave rectifiers can be combined to provide full-wave rectification. This is illustrated in Fig. 1.5. Each rectifier deals with one half-cycle of input waveform and the combination of two rectifiers provides full-wave rectification.

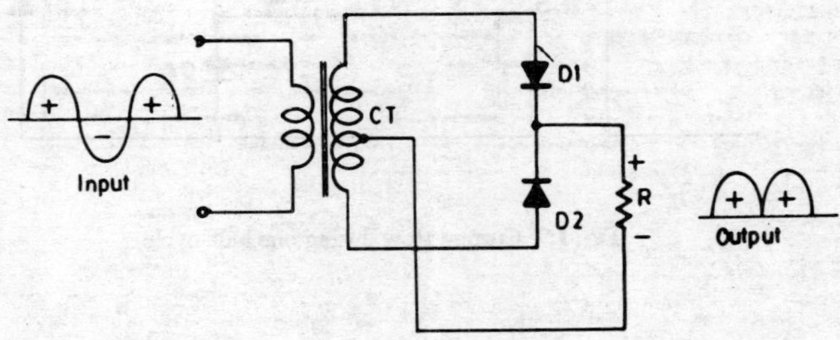


**Fig. 1.5** Full-wave rectification

*For the load current to always flow in the same direction, a power transformer with a centre tapped secondary winding is required.*

The efficiency of a half-wave rectifier is 31.8%, that of a full-wave rectifier is, 2 × 31.8 or 63.6%.

## 1.3 Bridge Rectifier

Four diodes connected to form a bridge, Fig. 1.6, can also be used for full-wave rectification. In this particular case, *a power transformer with a centre tapped secondary winding is not required.*

The two diodes, connected in the opposite arms of the bridge, are connected in series for each half-cycle of input waveform.

Diodes D1 and D3 work in forward direction for one half-cycle of input waveform and deliver output to the load, $R_L$. This is shown in Fig. 1.7.

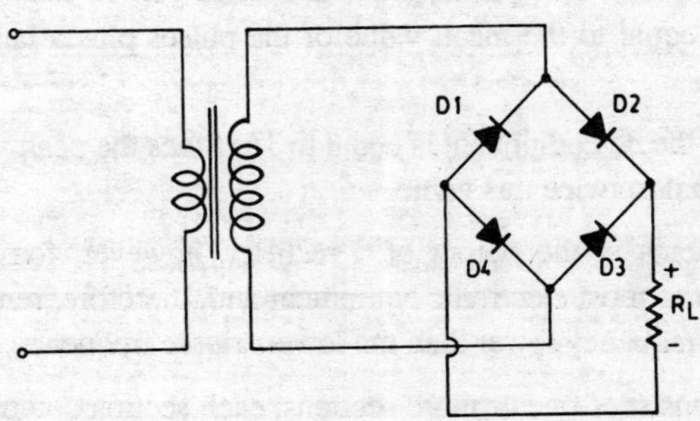


**Fig. 1.6** Bridge rectifier

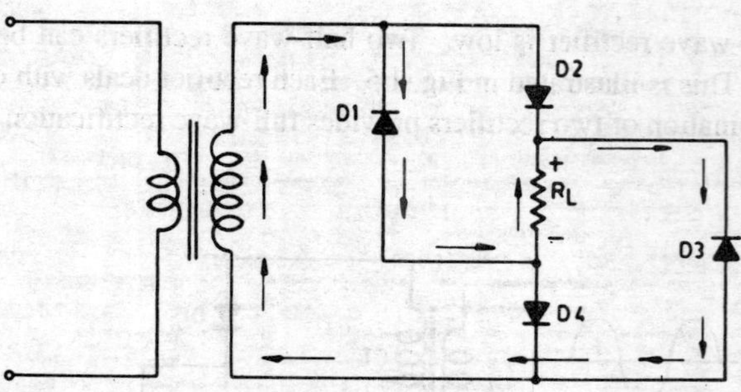


**Fig. 1.7** Current flow during one half-cycle

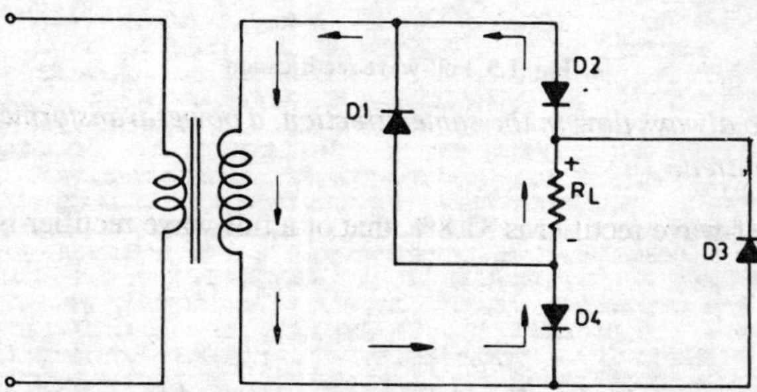


**Fig. 1.8** Current flow during second half-cycle

During the second half-cycle, diodes D2 and D4 work in forward direction and deliver current to load, $R_L$. This is shown in Fig. 1.8. The current flowing through the load during each half cycle is in the same direction.

## 1.4 Smoothing Circuits

The output from a rectifier consists of pulses of dc and these can be shown, by Fourier analysis, to consist of a *dc component* equal to the mean value of the pulses plus a large number of *alternating components*.

For half-wave rectifiers, the dc component is equal to $1/\pi$ times the peak value of the pulses and for full-wave rectifiers, it is equal to twice this value.

The alternating components in the output of a rectifier, however, form a *ripple* that would be detrimental to the operation of most electronic equipment and, therefore, removed by means of a *low-pass filter* having a cut-off frequency lower than the lowest ripple frequency.

These filters normally consist of one or more sections, each section comprising a *choke* or inductor in series with the rectifier output and a *capacitor* in parallel with, or across the output, as shown in Fig. 1.12.

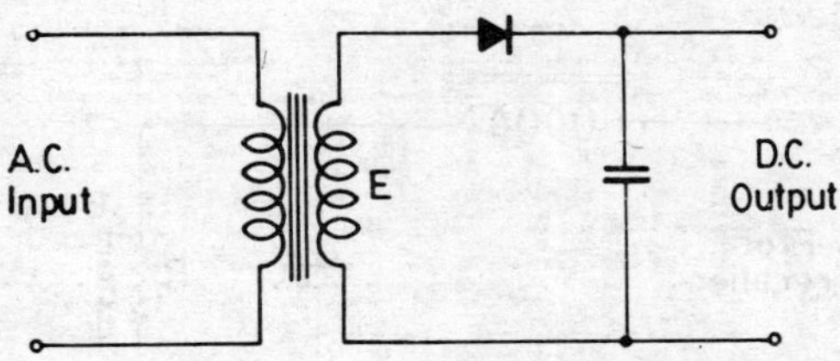


**Fig. 1.9** Smoothing the output of a half-wave rectifier

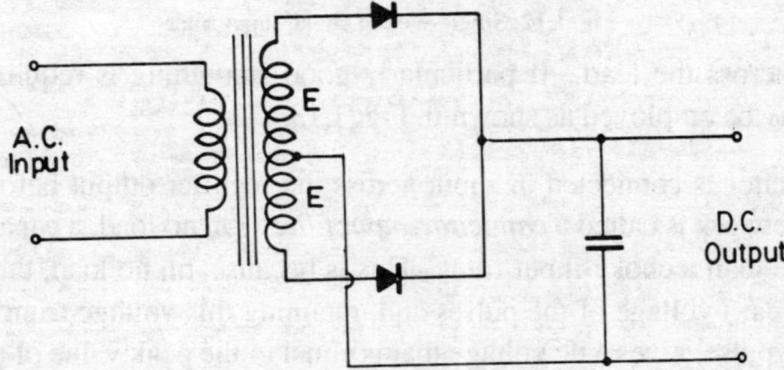


**Fig. 1.10** Smoothing the output of a full-wave rectifier

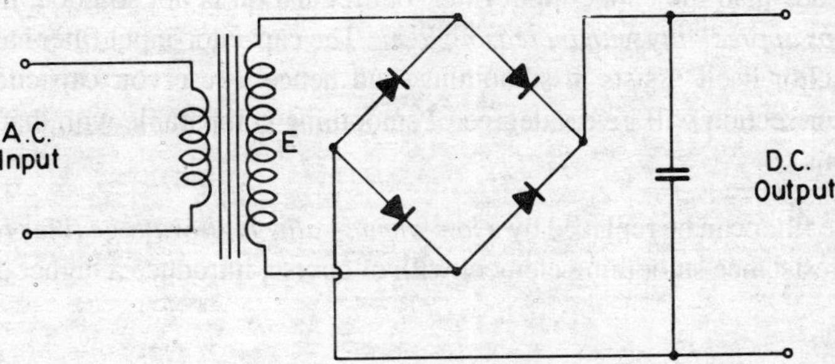


**Fig. 1.11** Smoothing the output of a bridge rectifier

The operation of the smoothing filter may be understood by considering the choke and the capacitor to form a *potentiometer* across the rectifier output. At *zero frequency* (i.e., for dc) the inductance offers a low impedance (merely that of its dc resistance) while the capacitor offers an infinite impedance (merely that of its reactance for dc). The whole of the dc voltage developed by the rectifier is, therefore, applied to the load, and this voltage is equal to the mean value of the rectifier output. *At the frequencies of the various ripple components, however, the inductance offers a high impedance, whereas the capacitor offers a low impedance.* Only a fraction of the ripple voltage thus appears across the

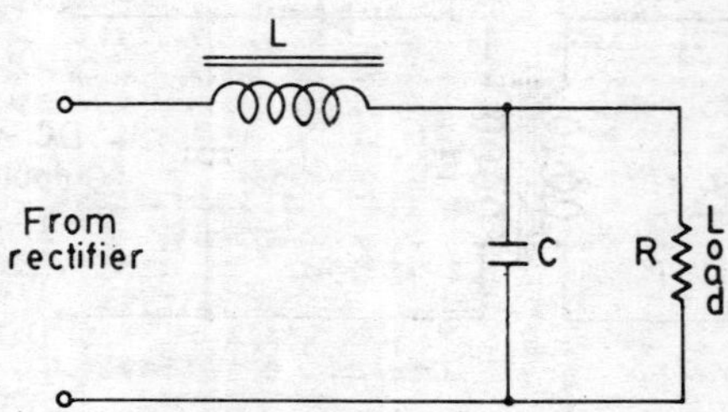


Fig. 1.12 Single section choke-input filter

capacitor and hence across the load. If particularly good smoothing is required, two or more such sections in tandem may be employed as shown in Fig. 1.13.

If a reservoir capacitor is connected in shunt across the rectifier output before the first choke, the resulting smoothing network is called a *capacitor-input filter*. On no load, a capacitor-input filter gives a larger output voltage than a choke-input filter. This is because, on no load, the capacitor charges up to the *peak* (not the mean) voltage of the pulses and, retaining this voltage from the peak of one pulse to the peak of the next pulse, gives a dc voltage that is equal to the peak value of the alternating voltage applied to the rectifier.

On load, the capacitor *partially discharges* through the load during the periods between pulses (see Fig. 1.14), and the mean dc output voltage drops. Thus a capacitor-input filter gives a higher output voltage on light loads than the choke-input filter, but regulation is not so good, that is to say, *the dc output voltage drops appreciably with increasing load.* The capacitor-input filter has the advantage that the reservoir capacitor itself assists in smoothing, and hence a reservoir capacitor followed by one choke-and-capacitor section will give a degree of smoothing comparable with that obtained from two choke-input sections.

The choke in the filter can be replaced by a *low-ohmic-value high-wattage filter resistance*, as shown in Fig. 1.16. The resistance smoothing element will, of course, introduce a higher drop in voltage than

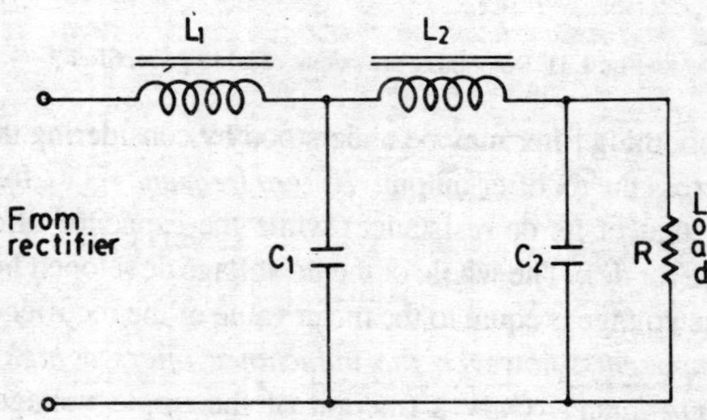


Fig. 1.13 Double-section choke-input filter

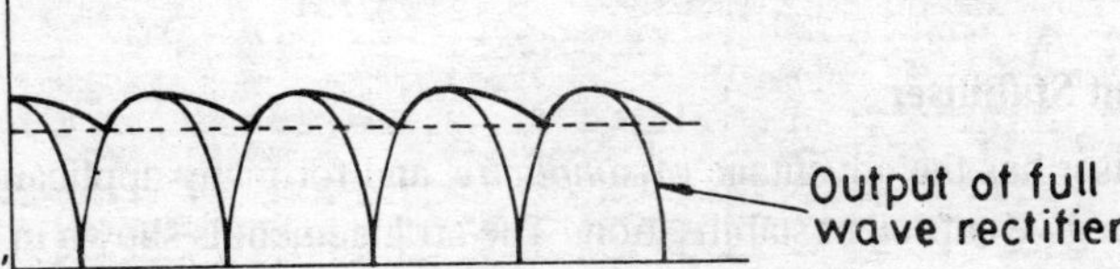


Fig. 1.14 Output of full-wave rectifier with reservoir capacitor

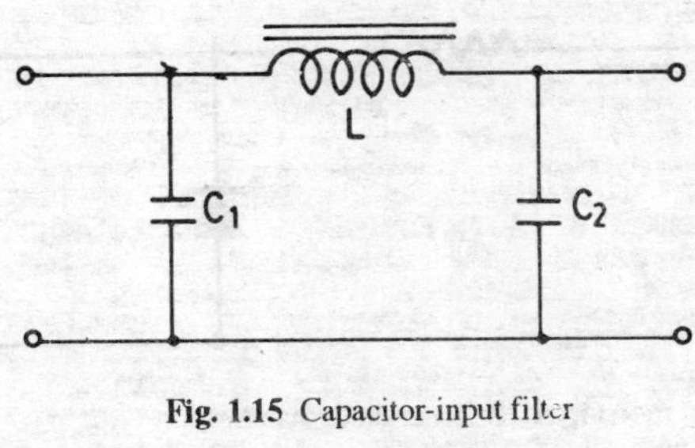


Fig. 1.15 Capacitor-input filter

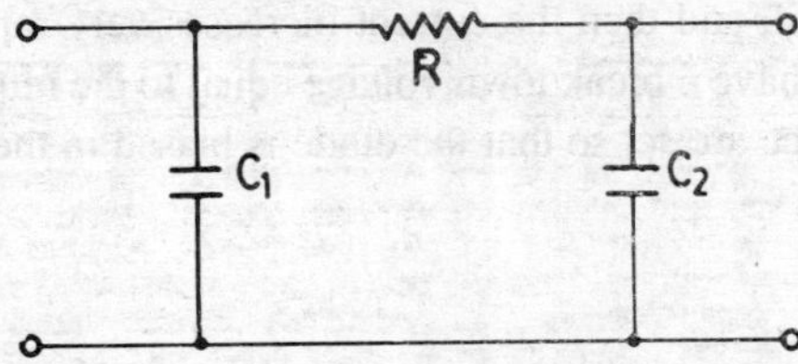


Fig. 1.16 R-C filter for low current, constant load applications

the filter choke, and will result in more severe regulation. This type of filter is normally used for *low current constant load applications.*

## 1.5 Voltage Stabilisation

Since transistors usually operate with relatively low collector voltage, medium and high power operation demands quite high current swings, and so the subject of output voltage *regulation* must receive special attention. Considerable developments have taken place in the design of voltage stabilising circuits, and it is significant that these stabilising circuits form an important application of transistors, not only for stabilised supplies for transistor circuits but also to provide *constant and reference voltages* for many other items of electronic equipment.

As the load current taken from a normal rectifier circuit is increased, the output voltage will fall, partly due to the ohmic resistance of the transformer windings, the rectifier diodes and the smoothing elements, and partly due to the lower effective storage efficiency of the reservoir capacitor with the lower load resistance. In many applications, the load current will vary, as for example, in class B and switching circuits, and it is then desirable to include some form of stabilisation to maintain the output voltage constant. Various arrangements can be used depending upon the degree of stabilisation required.

## 1.6 Zener Diode Shunt Stabiliser

The zener diode stabiliser has the advantage of *simplicity*, and for many applications, particularly for low current loads, it can give sufficient stabilisation. The arrangement is shown in Fig. 1.17. The zener diode is connected in the reverse direction across the load and is fed from the unstabilised voltage through a series resistance, $R_S$.

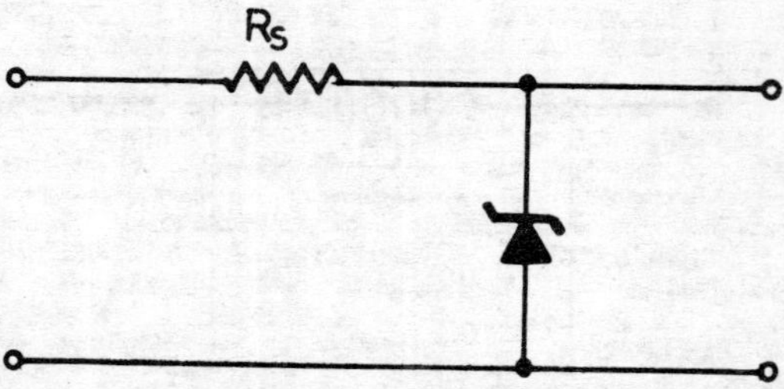


**Fig. 1.17** Zener diode stabilisation

The characteristic of a zener diode is illustrated in Fig. 1.18. In the reverse direction the current is small until breakdown is reached, and then the current increases very rapidly for a small increase of voltage. The diode is chosen to have a breakdown voltage equal to the required output voltage and the input voltage and series resistance are set so that the diode is biased in the breakdown current such as the value I in Fig. 1.18.

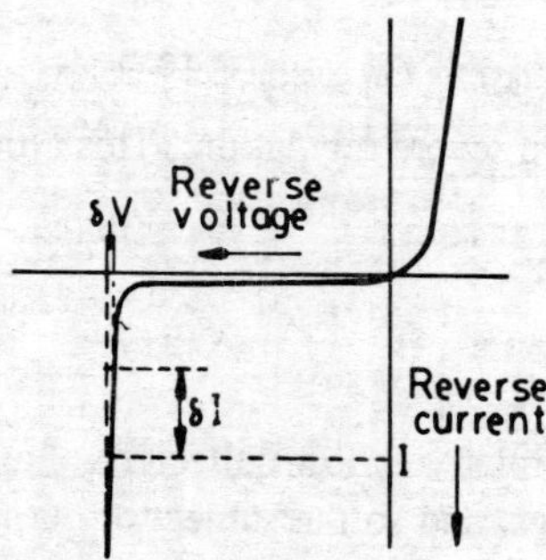


**Fig. 1.18** Characteristic of a zener diode

Now, if the load current increases, the output voltage will tend to fall, but this will result in a decrease in the diode current such that *the sum of the load plus diode currents remains almost constant*. Thus the output voltage will remain almost constant, the only voltage change being δ V due to the change in the diode current δ I.

Similarly, if the input supply voltage increases, the voltage across the zener diode will tend to increase, causing an increased current through the zener diode. This extra current will produce a voltage

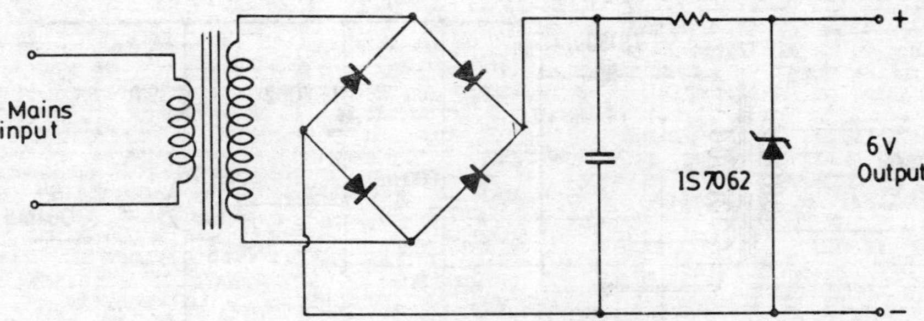


**Fig. 1.19** Zener diode shunt stabiliser

drop across the series resistance, $R_S$, almost equal to the increase of input voltage, thus keeping the output voltage almost constant.

At minimum load current and minimum input voltage there must still be sufficient current flowing through the zener diode to *hold it in the breakdown region,* at least 2 or 3mA. If this value is $I_M$ and the maximum load current $I_L$, the total current flowing through the series resistance $R_S$ will be $(I_M + I_L)$. If the minimum input voltage is $V_M$ and the stabilised output voltage $V_O$, the value of $R_S$ must be equal to $(V_M - V_O)/(I_M + I_L)$. For good stabilisation, the value of $V_M$ should preferably be at least 1.5 times the value of $V_O$

A practical example of a zener diode shunt stabiliser circuit is shown in Fig. 1.19. The circuit gives an output voltage of 6 volts. It allows an input voltage variation of ±15% and a load current variation of 5 to 50 mA with a maximum output voltage change of only 0.1 volt.

For higher output voltages, diodes with correspondingly high breakdown voltage can be used, but the value of the diode slope resistance in the breakdown region increases faster than the value of breakdown voltage, and better performance can be obtained by two or more *low voltage* (5-6 volts) zener diodes in series to obtain the required output voltage. The series arrangement also gives the optimum performance from the viewpoint of stability of the breakdown voltage with temperature changes as the change is nearly zero with the diodes in the 5-6 volts region. The choice of the method is entirely dependent upon the degree of stabilisation required and, of course, upon economic considerations.

## 1.7 Transistor Shunt Stabiliser

An improvement over the diode shunt stabiliser can be obtained with a transistor circuit as shown in Fig. 1.20. A transistor is connected across the supply with its base electrode fed through a zener diode. The input voltage and base resistance are again set so that the zener diode is biased into the breakdown region. If the input voltage increases, the zener diode passes more current, which increases the base current of the transistor and causes a much larger current to flow down through the transistor. This increase of transistor current results in an increased voltage drop across the series resistance $R_S$, so that the output voltage is maintained almost at its original value.

The circuit shown in Fig. 1.20 was designed to give a stable 9 volts supply from the ac mains to repalce the battery in portable transistor radio receivers. A mains voltage variation from 190 to 250 volts

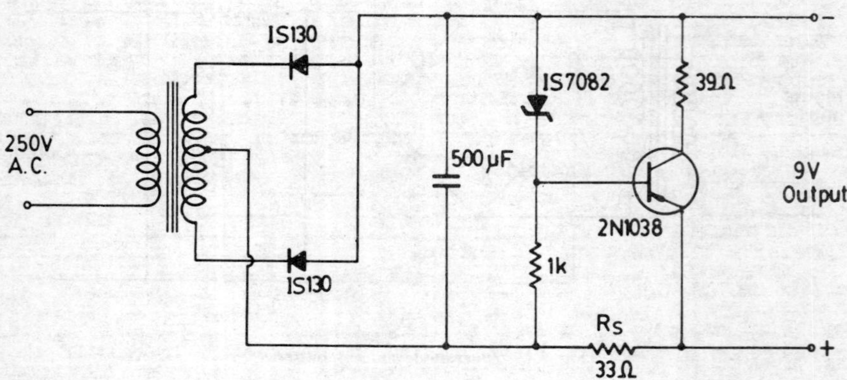


**Fig. 1.20** Transistor shunt stabliliser

will result in an output voltage change of only 0.2 volt and a load current variation from zero upto 125mA gives an output voltage change of 0.25 volt.

## 1.8 Transistor Series Stabiliser

In a series stabilising circuit, it is arranged that the voltage change of an unstabilised supply appears across a *series control transistor* so that the output from the transistor remains constant. The simplest arrangement of this type of stabiliser is the *emitter-follower* circuit shown in Fig. 1.21.

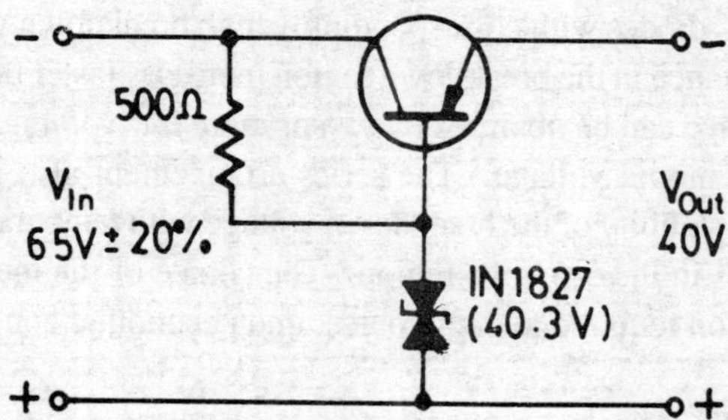


**Fig. 1.21** Emitter-follower stabiliser

The base electrode of the transistor is held at a constant potential by means of the zener diode. The output voltage at the emitter of the transistor can only differ from this potential by an amount equal to the base-emitter voltage of the transistor, which is only a small fraction of 1 volt for germanium transistors, even over the full range of emitter current.

*The excess voltage between the variable input voltage and the constant output voltage will appear across the collector and emitter of the transistor, and so it must be chosen to have a sufficiently high collector voltage rating.* Moreover, if the output emitter terminal is inadvertently shorted to earth, the full input voltage will appear across the transistor and if the collector voltage rating is not high enough to withstand this eventuality, some special *protection circuit* will have to be included.

## 1.9 Regulated Power Supply

The circuit of a regulated power supply is given in Fig. 1.22. A pnp germanimum low power transistor, AC 188, is used as the series control element and the silicon pnp lockfit type BC 157, or metal type BC 177, is used as the dc amplifier with the base-to-emitter voltage drop providing the reference voltage. In this circuit, a fairly good regulation of about 2% is obtained for a load variation of 230 mA with the AC 188 as series regulator.

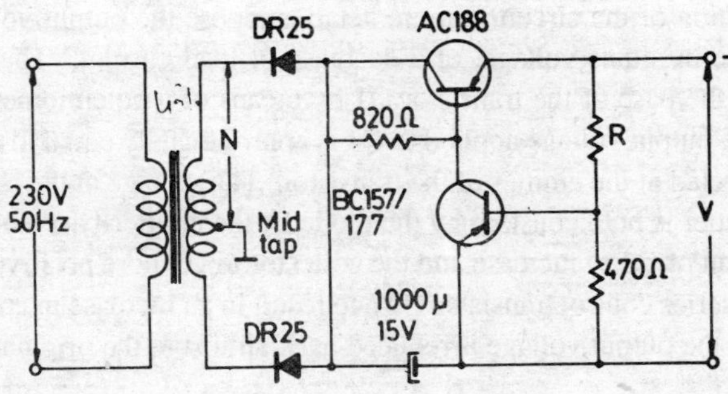


**Fig. 1.22** Regulated power supply

**Table 1.1**

**Performance Specifications**

| Resistance value kΩ | No. of secondary turns of transformers N | DC input voltage at $I_L$ = 20 mA Volts | AC Input required at $I_L$ = 20 mA Volts | Peak ripples content at $I_L$ = 240 mA mv | Output voltage at $I_L$ = 250 mA Volts | Output voltage fluctuations at $I_L$ = 150 mA for ± 10% mains supply variations Volts |
|---|---|---|---|---|---|---|
| 2.7 | 79 | 4.5 | 6.4-0-6.4 | 34 | 4.4 | ± 0.1 |
| 3.9 | 94 | 6.0 | 7.6-0-7.6 | 41 | 5.8 | ± 0.1 |
| 5.6 | 116 | 9.0 | 9.4-0-9.4 | 64 | 8.8 | ± 0.2 |

## 1.10 Feedback Series Stabilisers

The performance of the simple emitter-follower circuit can be improved by including *amplified feedback control* as shown in the basic circuit of Fig. 1.23. A zener diode D is used as a constant voltage reference, being biased into the breakdown region through the resistance R1. The diode holds the emitter of transistor Q1 at the constant reference voltage. Any variation on the output voltage appears at the

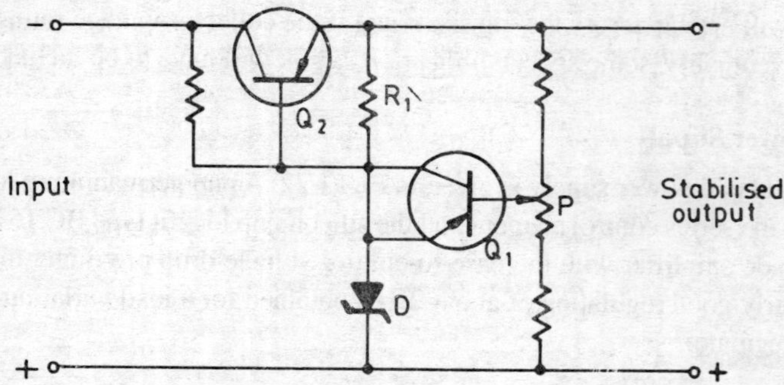


**Fig. 1.23** Basic series stabiliser circuit

base of Q1, is amplified, and fed to the base of the series control transistor Q2, resulting in the collector-to-emitter voltage of Q1 changing so as to oppose the initial output variation.

Looking at the operation of the circuit in more detail, suppose the output voltage tends to *increase* either due to an increase in mains voltage, or a decrease of load current. The percentage change in output voltage is fed to the base of the transistor Q1 by means of a potentiometer P across the output voltage. The level of the output voltage applied to Q1 is approximately equal to the breakdown voltage of the zener diode connected at the emitter of the transistor. The voltage at the base of Q1 will go more negative, and as the emitter is held constant by the diode D, the emitter-base voltage will increase and cause the collector current of Q1 to increase and the collector to go more positive. This will reduce the emitter-base bias of the series control transistor Q2 and result in an increase in collector-emitter voltage of that transistor, so that the output voltage is reduced back almost to the original level.

## 1.11 Switching Regulator

The switching regulator adjusts voltage by controlling the 'on-off' time of a transistor used as a switch. Fig. 1.24 shows the basic composition of a PWM (pluse-width modulation) switching power supply in common use.

The most important part is the dc-to-dc converter, which consists of a semiconductor switch, such as a transistor, diode, a *transformer*, a reactor and a smoothing capacitor. The transformer plays an important role among the parts that are used to form a switching power supply. There are two circuit configurations for transformers.

One is the *'on-on'* configuration, in which when the switching transistor is in the 'on' position, the secondary side is also in the on state. Among the products using this system is the *forward converter* shown in Fig. 1.26.

The other is the 'on-off' configuration, in which the switching transistor is put in the 'off' position and, the secondary side is in the 'on' position. The *flyback system* is one of these.

Fig. 1.26 illustrates the basic circuit configuration of a one transistor forward converter. When bias (pulse) in the forward position is impressed on the base of the transistor Q1, the voltage $e_1$ is impressed on the primary coil $n_1$ of the transformer $T_1$, and a voltage $e_2$ is generated in the secondary coil $n_2$, where $e_2 = e_1\, n_2/n_1$ and

$e_1$ = input voltage of the transformer

$e_2$ = output voltage of the transformer

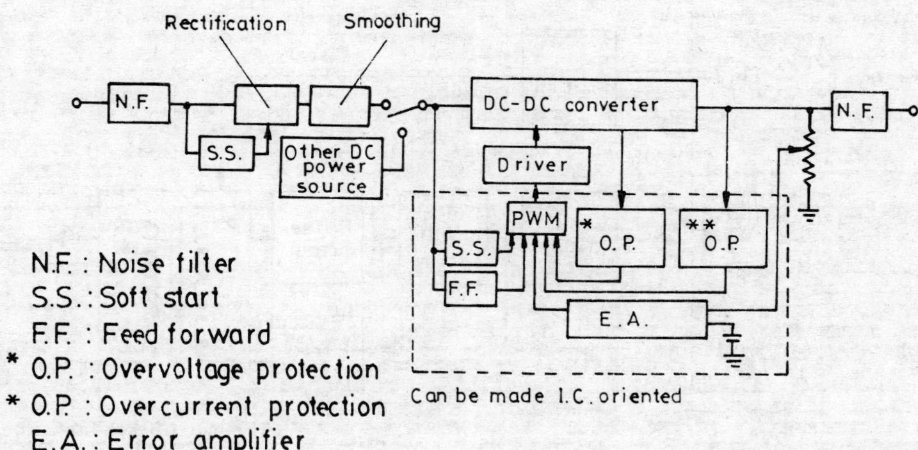


**Fig. 1.24** Switching regulator

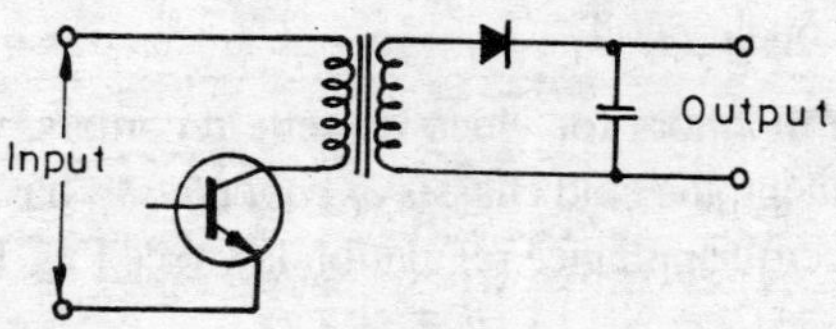


**Fig. 1.25** dc-to-dc Converter and transformer

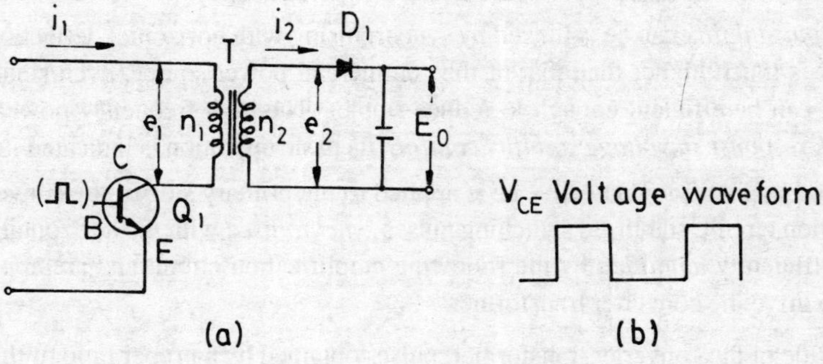


**Fig 26** The 'on-on' system

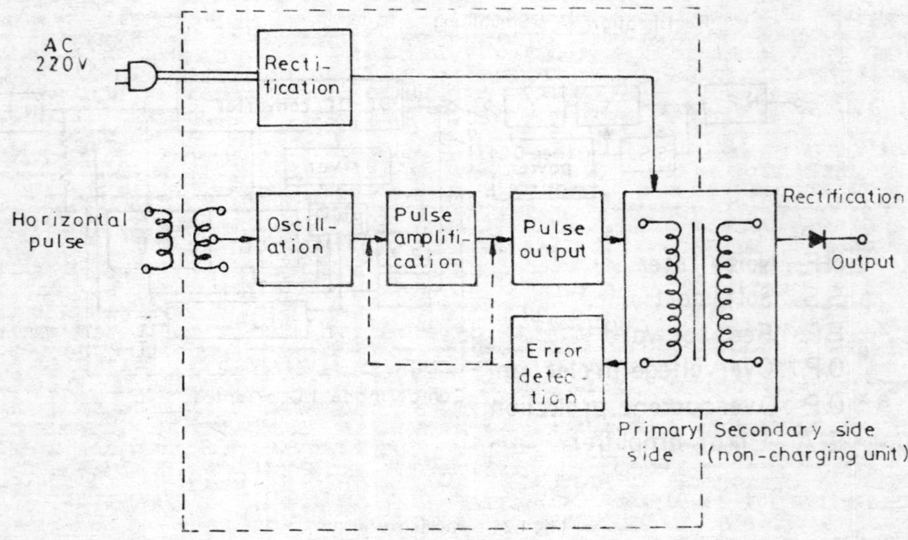


**Fig. 1.27** Basic power supply circuit for AV, tv

$n_1$ = number of turns of the primary coil

$n_2$ = number of turns of the secondary coil

In colour tv receivers used as monitors for video cassette recorders, video disc players and other audio, video and computer equipment, the cold chassis or *cool chassis* has become essential because of the limitations imposed by the electric appliance regulation law (see Fig. 1.27).

For insulating 220 V ac, a separate transformer for signals is used just before the terminal connected to the other equipment, to make a part of the connection terminal a cold chassis.

As the number of input-output terminals is increased as in recent A-V tv receivers, circuit configurations become more complex, and efficient transformers, such as flyback transformers are increasingly used. *The high frequency power supply system has become more common.*

In tv receivers, *output stability* can be achieved by synchronising with horizontal deflection frequency. As the frequency is much higher than that of the commercial power source, even small capacity smoothing capacitors can be efficient enough to reduce ripple. The high frequency power supply is characterised by *quick response in voltage stability control.* Its basic operation is indicated in Fig. 1.27.

DC voltage obtained by rectifying ac 220 V ac is applied to the primary side of the converter transformer. In the oscillation circuit, stabilised switching pulses *synchronised* with the horizontal pulses are generated, and are sufficiently *amplified* by the following amplification circuit and are applied to the pulse output circuit to drive the converter transformer.

On the secondary side of the converter transformer, pulses obtained by a proper ratio of the numbers of coil turns are rectified and supplied to various circuits as the B+ power supply of the cold chassis.

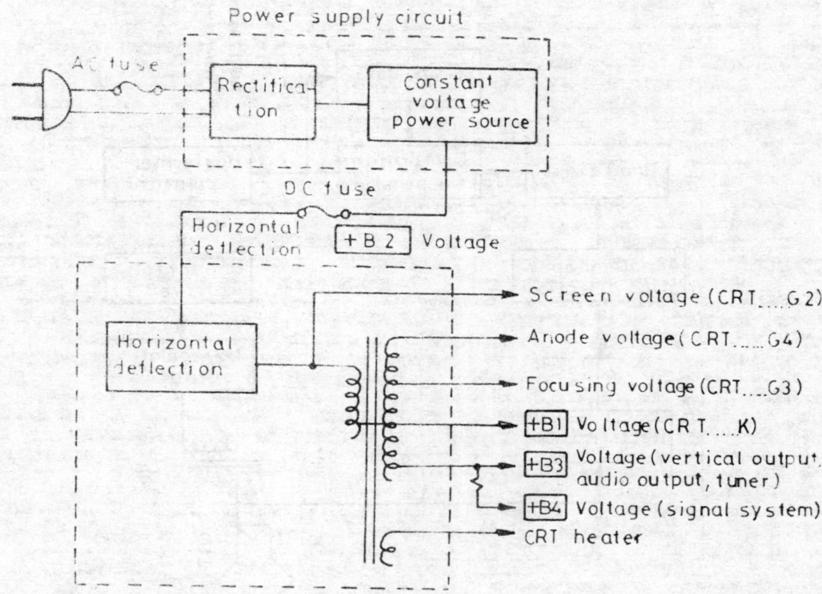


**Fig. 1.28** Power supply system

Any change in the load voltage of the secondary side, brings a corresponding change in the pulse of the primary side. This change is used as a *control voltage* by the error detection circuit to control voltage.

The *hot chassis* is insulated from the *cold chassis* after the secondary side of the converter transformer. In voltage measurement, as in repairing or inspection, *it is necessary to distinguish a cold chassis from a hot chassis, and not to make any error in using the grounding points.*

The power supply in AV, tv receivers is shown in Fig. 1.28. The commercial 220 V power source is rectified and +B voltage is generated by a constant voltage circuit.

The horizontal deflection circuit is operated by this $+B_2$ voltage which is supplied to various circuits like $+B_1$, $+B_3$ and $+B_4$. In this way, the pulse lighting system is used as the heater power source of the crt in recent tv receivers.

This pulse lighting system has an *overvoltage protection* circuit, which prevents the deterioration of the heater due to abnormal oscillation and protects circuits.

As shown in Fig. 1.29, this circuit supplies the horizontal pulse, which is the heater voltage of the crt to the rectification circuit of the overvoltage protection circuit at the same time.

In normal operation, the rectified and smoothed voltage is not sufficient to put the zener diode in the 'on' position so that no voltage is applied to the *abnormal operation detection circuit*. The following circuits are in the 'off' state.

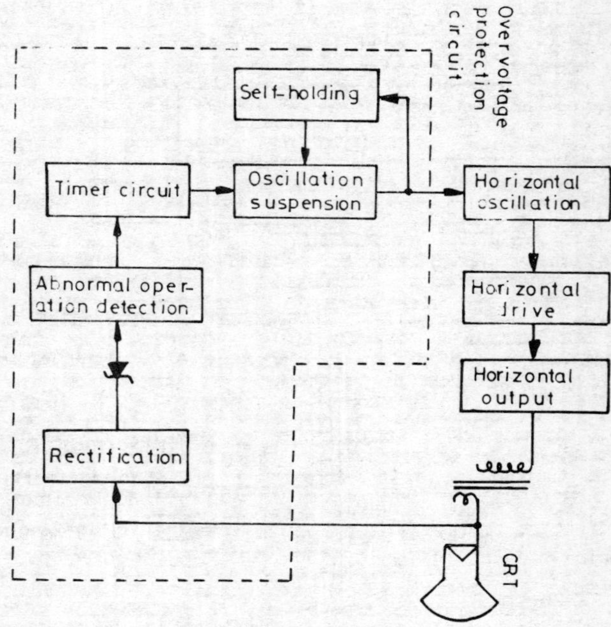


**Fig. 1.29** Overvoltage protection circuit

When the zener diode is put in the 'on' position due to some abnormality in horizontal oscillation, voltage is applied to the abnormal operation detection circuit, and the circuit is put in the 'on' position and the *oscillation suspension circuits* driven by the timer circuit.

Horizontal oscillation is stopped by the operation of the oscillation suspension circuit, and the horizontal power is not generated.

As the oscillation is stopped, input voltage to the abnormal operation detection circuit is lost and no operating voltage from there is also lost. As the self-holding function of the oscillation suspension circuit operates, this circuit continues its 'on' operation.

The *timer circuit* is intended to prevent operating error caused by a momentary pulse, and issues an order for oscillation suspension when abnormal pulses continue.

## *Chapter 2*

# ANALOG CIRCUITS

*The word ANALOG refers to the representation of numerical quantities by means of physical variables such as translation, rotation, voltage or resistance. An analog device is a mechanism that represents numbers by physical quantities e.g., by lengths as in a slide rule, or by voltages or currents as in a differential analyser or a computer of the analog type. Analog circuits refer to the class of circuits that are linear, or non-digital. We study these circuits in this chapter.*

### 2.1 Limiters

A limiter circuit is *one which limits the travel of a waveform in a certain direction.* For example, in Fig. 2.1 the circuit has clipped a portion of the positive half-cycle of the sine wave. For this reason, the limiter is sometimes called a *clipper.*

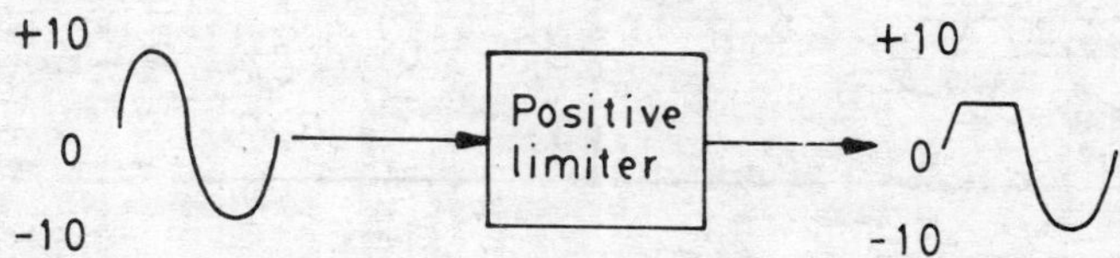


**Fig. 2.1** Limiting action

#### *Series Limiter*

The *series limiter* is used in communication and other electronic equipment where amplitude limiting is desired. It may be used to limit *either the positive or negative excursion* of the waveform. A typical circuit used to limit the positive half-cycle of the waveform is shown in Fig. 2.2.

Diode $D_1$ is connected in series between the input and output, with $R_1$ serving as the load resistor. When a positive input is applied between $D_1$ cathode and ground, the cathode is made more positive than the anode and the diode does not conduct. For this condition only, the reverse resistance of the diode will not allow any signal to pass to the load. When the input signal is negatively polarised, however, the cathode is made more negative than the anode and the diode conducts. When it conducts, *electrons flow*

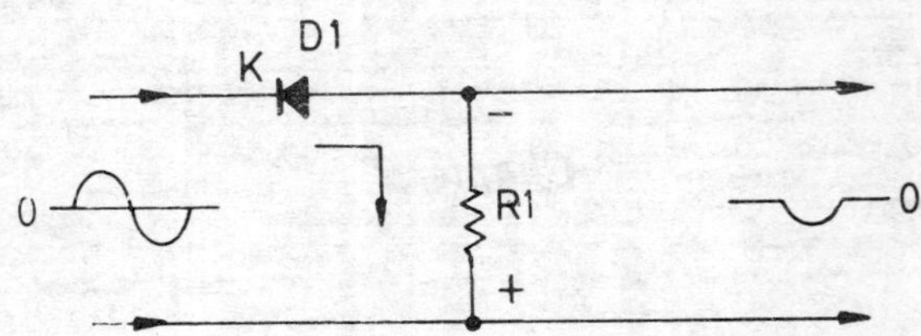


**Fig. 2.2** Series diode, positive alternation limiter

*opposite the di:ection of the arrow* through $R_1$ to ground, and the polarity of the voltage developed across $R_1$ is negative with respect to ground.

The amount of current flow and the resistance value determine the output voltage produced. Since the forward resistance of the diode is in series with $R_1$ to ground, together they form a *voltage divider* and the output voltage taken across $R_1$ is always less than the input voltage. Also, since the load resistance is not frequency selective, the waveform of that portion of the signal produced by forward current flow through the resistor is the *same* as the waveform of the original signal and has the same phase.

The amount of clipping of the waveform can be *selected* by using a diode which has the proper value of reverse resistance or by placing a negative potential in series with $R_1$, as shown in Fig. 2.3. In this case, the diode will not conduct until the input signal is more negative than the *applied bias*. The reverse resistance loading of the diode on the input circuit is effectively nullified. The complete positive lobe is eliminated. Note, how even the voltage divider action produced by the forward resistance and $R_1$ still remains. The total negative signal amplitude is reduced by the amount of forward voltage drop and the negative bias.

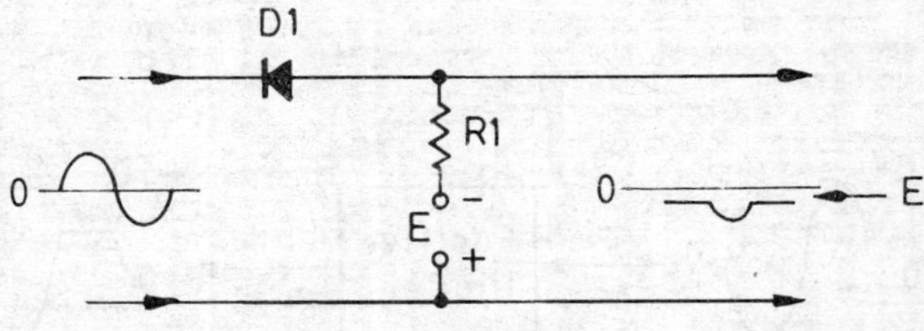


**Fig. 2.3** Series, biased limiter

The same circuit can also be used for negative *clipping*, by *reversing* the diode so that only positive pulses reach the load. If a bias voltage is applied, it will also be *reversed*.

### *Parallel Limiter*

Parallel limiters may also be used to limit either the negative or positive half-cycle of the waveform.

A parallel, positive diode limiter is shown in Fig. 2.4. In this circuit, diode $D_1$ conducts only during the positive portion of the input signal. Since the value of $R_1$ is very large compared to the conducting resistance of the diode, practically the entire value of the input voltage drops across $R_1$, while only a very small voltage drop appears across $D_1$. When the input signal goes negative, the diode does not

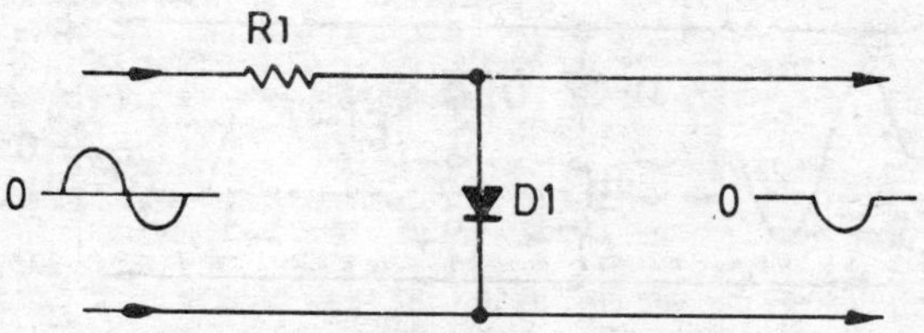


**Fig. 2.4** Parallel limiter

conduct and current flow through $R_1$ almost ceases. A small reverse current still flows through $D_1$ (because of the reverse resistance of the diode), and a small portion of the input voltage is dropped across $R_1$. The amount of reverse resistance of the diode depends upon the characteristics of the diode selected.

A parallel, positive limiter may also be used to limit only the peaks of the positive waveform while allowing a given value of the positive signal to pass through the circuit to the output. This may be accomplished by applying a biasing voltage, having a value equal to the value of the positive signal to be passed by the circuit, to the cathode of the diode as shown in Fig. 2.5.

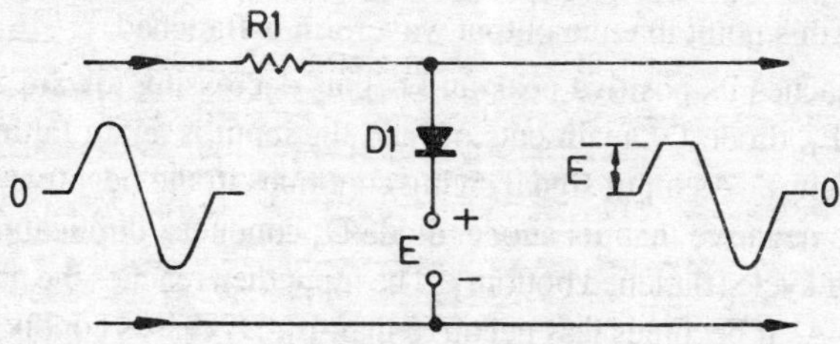


**Fig. 2.5** Parallel, biased positive peak limiter

When connected in this manner, with the cathode of $D_1$ connected to the positive terminal of the dc source, the cathode of the diode is held more positive than the anode by the value of E. As long as the positive half-cycles of the input voltage remain less positive than E, the diode remains cut off because its cathode is positive with respect to the anode and the output voltage is equal to the input voltage minus the voltage developed by the reverse resistance of the diode. Since all of the negative half-cycles of the input voltage are less positive than E, these also cause the diode to remain cut off. When the input signal increases to a value which exceeds E, the anode becomes positive with respect to the cathode, the diode conducts and continues conducting as long as the input remains more positive than E.

***Double Diode Limiter***

The parallel, double diode limiter is used when it is necessary to limit a portion of both the positive and the negative half-cycles of the waveform, and allow the remainder of the signal to pass through without modification. This limiter is used to square off the peaks of an applied signal, to obtain a rectangular waveform from a sine wave signal, or to eliminate the negative or positive portion of a waveform and clip the other portion.

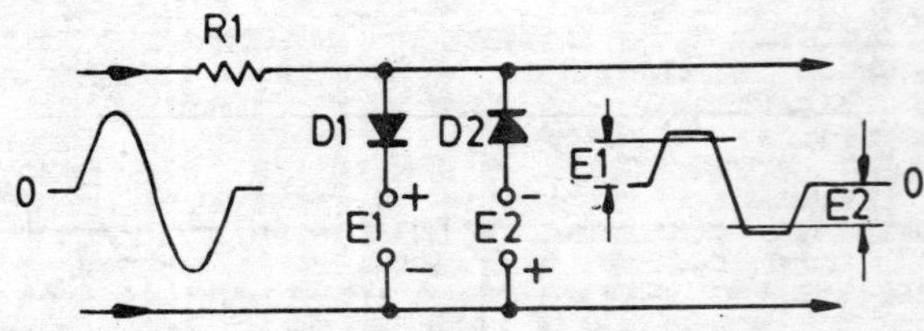


**Fig. 2.6** Biased, double-diode positive and negative peak limiter

In Fig. 2.6, diode $D_1$ limits the positive half cycle of the input, and $D_2$ limits the negative half. Separate supplies $E_1$ and $E_2$ bias their respective diodes. Resistor $R_1$ is the input load. The $E_1$ and $E_2$ voltages reverse bias the diodes, that is, the cathode of $D_1$ is positive with respect to its anode, and the anode of $D_2$ is negative with respect to its cathode. During no signal, therefore, they do not conduct.

As the sinewave signal is applied at the input and begins increasing in a positive direction, both diodes remain cut off due to the bias, and the beginning positive half of the reproduced positive going ramp appears at the output. The output continues to follow the input signal until a point is reached where the signal becomes more positive than the positive bias applied to the cathode of $D_1$. At this point, the $D_1$ anode becomes more positive than the cathode, and $D_1$ conducts. When $D_1$ conducts, it provides a low resistance path for forward current to ground, and shunts the output through the diode instead of the external load. It is at this point that the output waveform is flattened.

As the input signal reaches its positive peak and begins decreasing toward zero, it again reaches a level which is less than $E_1$, diode $D_1$ again cuts off and the input is again faithfully reproduced at the output (negative going ramp). As input signal swing continues in the negative direction, and when its amplitude becomes more negative than its anode, diode $D_2$ conducts, duplicating the action which occurred on the positive half cycle (flattened bottom). The input then reaches the negative peak and begins decreasing toward zero. As it becomes less negative than $E_2$, $D_2$ ceases conducting and the remainder of the input signal is reproduced at the output.

## 2.2 Clampers

Clampers, also referred to as *dc restorers* and base line stabilisers, are used in electronic circuits to *hold* either amplitude extreme of a waveform to a given reference level. The diode clamper circuit can be arranged to clamp either the positive or negative amplitude extreme and thereby permit the waveform to extend in *only one direction* from the reference level. In other words, the circuit can be used to hold either the positive extreme or the negative extreme of a waveform to a desired reference voltage level.

Diode clampers which hold the positive extreme of the waveform to a desired reference level are called *negative clampers* because the entire waveform is shifted negatively with respect to the reference level. Those which hold the negative extreme to a desired reference level are called *positive clampers* because the entire waveform is shifted positively with respect to the reference level. Fig. 2.7 shows the variation of signal voltage with respect to a reference level without clamping and when negative and positive clamping are used.

### *Diode Clampers*

To effect clamping, the semiconductor diode is operated as a simple *switch* controlled by the polarity of the input waveform. In the direction of forward conduction it passes the signal, but in the direction of

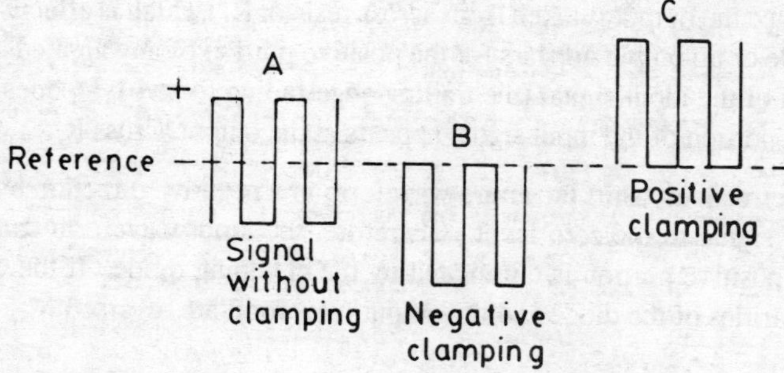


**Fig. 2.7** Clamping action

reverse conduction, it is in effect an open circuit and blocks the signal. Although the forward resistance of the diode is low, the reverse resistance is not infinite like that of an open switch. In addition to the inherent disadvantage of a *relatively low reverse resistance*, the semiconductor diode also has *capacitive effect* which varies with applied voltage. Usually, the capacitive effect is small enough to compare favorably with that of the vacuum diode. However, when the semiconductor diode is used as a dc restorer at video and higher frequencies, the shunting capacitance may affect the waveform of the signal.

Since the diode may be connected to operate on positive, or negative, signals and may also be biased positively, or negatively, four basic variations of the diode clamp exist. Operation of these four basic configurations is practically identical, the major differences being in the *polarities* of the diodes and the bias supplies.

The *unbiased diode clamp* is usually employed as a shunt across the resistor portion of an RC coupling circuit. By providing a low resistance path during conduction periods and a high resistance path during non-conduction periods, the diode provides different charge and discharge times for the coupling capacitor. When the positive portion of the input waveform causes the diode to conduct, clamping is produced.

### Negative Clamp

A basic unbiased *negative clamp* is shown in Fig. 2.8. $C_1$ is the coupling capacitor of an RC coupling network. Resistor $R_1$ is the input resistor of the network and determines the *long time constant* of the circuit clamping diode. $D_1$ is connected in shunt and determines the *short time constant* of the circuit.

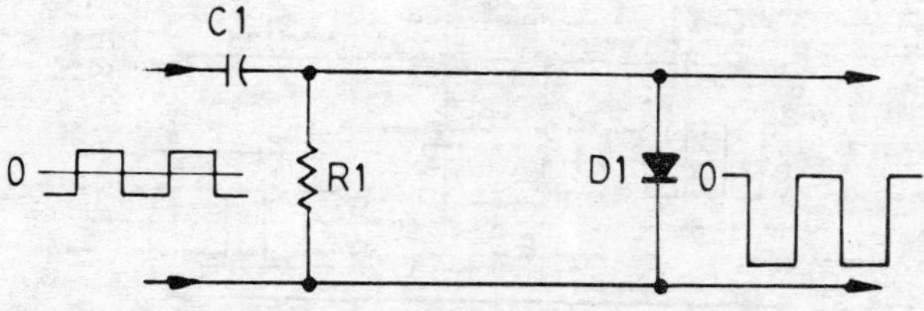


**Fig. 2.8** Negative clamper

When a positive going input signal is applied, it causes $D_1$ to conduct and C1 is quickly charged to the input potential. Since the output is taken from across resistor $R_1$ (which is effectively shorted by the conducting diode), little or no output appears for the positive portion of any applied signal. During the negative going portion of the input signal (the trailing edge falling to zero), $D_1$ does not conduct; consequently the negative portion of the input signal appears at the output across $R_1$.

This circuit acts to effectively shift the entire waveform in a negative direction by holding the positive peak of the input signal to the zero level. Therefore, the input waveform can only appear as a negative output. The positive portion is eliminated by the clamping diode. If the circuit is used as a positive clamp the polarities of the diode and the output waveform are reversed.

### *Biased Clampers*

In Fig. 2.9, $C_1$ and $R_1$ form an RC coupling network and determine the long time constant associated with the circuit. Diode $D_1$, during the time of conduction, together with $C_1$, determines the short time constant. *The bias supply alters the reference level from zero to a level equal to the bias.*

When the circuit is initially energised, and with no signal applied to the input, the diode begins conducting because of the negative potential applied to the cathode. As $D_1$ conducts, $C_1$ begins to charge, and when the charge is equal to the bias supply, the diode cuts off since its anode and cathode potentials are now equal. The voltage at the output is at this time equal to the bias voltage E.

When a signal is applied the voltage increases almost instantly, capacitor $C_1$ cannot change its charge immediately and the anode of the diode $D_1$ suddenly becomes more positive than its cathode and $D_1$ begins conducting. Because $C_1$ cannot immediately change its charge, the entire input voltage is developed across the diode. The output, taken from across the diode, increases in a positive direction. The conducting state of $D_1$ provides a very short time constant for the capacitor, however, and $C_1$ rapidly charges to the new voltage. As $C_1$ charges, the voltage drop across $D_1$ decreases, and reaches the reference level when $C_1$ is fully charged.

The output *remains* at this voltage, until the input swings negative. Again $C_1$ cannot immediately change its charge, but this time the diode cannot conduct, because its anode is negative with respect to its cathode. The entire input voltage, therefore, is developed across $R_1$, and the output voltage swings in a negative direction from reference level. Because the diode is not conducting $R_1$ provides a long time constant for $C_1$ and the capacitor begins charging very slowly to the input signal. At this time, the input again swings positive, bringing $D_1$ into conduction. This sudden rise also produces an increase in

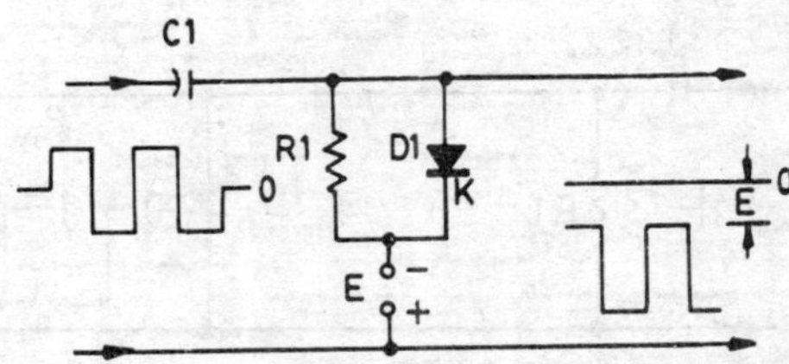


Fig. 2.9 Negatively biased negative clamper

the output. Because of the short RC time constant provided by $D_1$, the capacitor quickly charges again to the reference level and remains there until the input swings negative, when the cycle repeats.

By reversing the diode, the circuit can be converted into a negatively biased positive diode clamper. The difference at the output then will be that the entire output waveform will be clamped *above* the negative bias voltage level instead of below it as in the negative clamper in Fig. 2.9.

## 2.3 Transistor Linear Amplifiers

The basis of most linear integrated circuits is the linear transistor amplifier. The following generalisations apply to a transistor circuit that is operated as class A.

1. The *first letter* of the transistor type (npn or pnp) indicates the polarity of the emitter voltage with respect to the base.
2. The *second letter* of the transistor type indicates the polarity of the collector voltage with respect to the base.
3. The electron current direction is always *against* the direction of the arrow on the emitter. If the electrons flow *into* the emitter, they flow *out* of the collector, and if they flow out of the emitter, they flow into the collector.
4. The emitter and collector currents *increase* with an input voltage that increases the forward bias. Likewise, the emitter and collector currents will *decrease* with an input voltage that opposes or decreases, the forward bias.

## 2.4 Basic Circuit Configurations

The transistor can be connected in three circuit configurations, common-base, common-emitter and common-collector. These configurations are also referred to as grounded-base, grounded-emitter and grounded-collector. *The term common or grounded refers to the terminal that is common to the input and output circuits.*

Fig. 2.10 shows a transistor connected in common-base configuration. The input is applied between the emitter and base and the output appears between the collector and base. The salient features of this connection are: *no phase shift,* medium voltage gain (100 to 200), current gain less than unity, medium power gain, very low input resistance and very high output resistance.

Fig. 2.11 shows a transistor connected in common-emitter configuration. The input is applied between base and emitter and the output appears between collector and emitter. The salient features of this

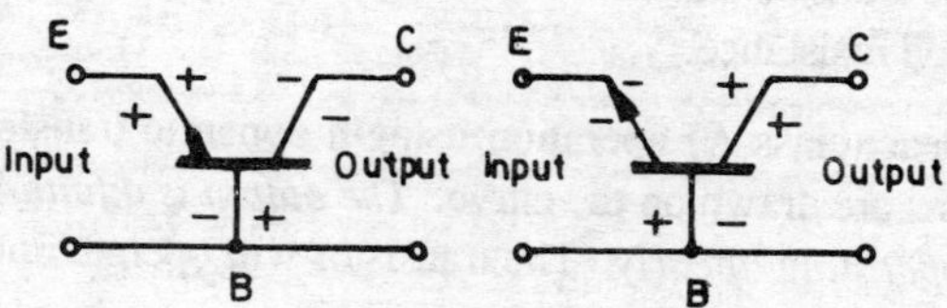


**Fig. 2.10** Common-base (CB) configuration

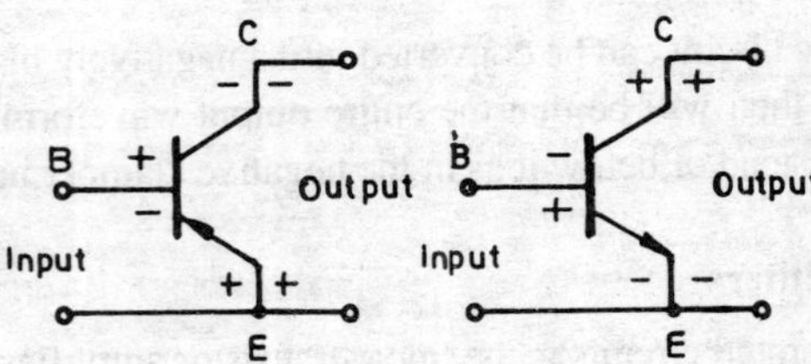


**Fig. 2.11** Common-emitter (CE) configuration

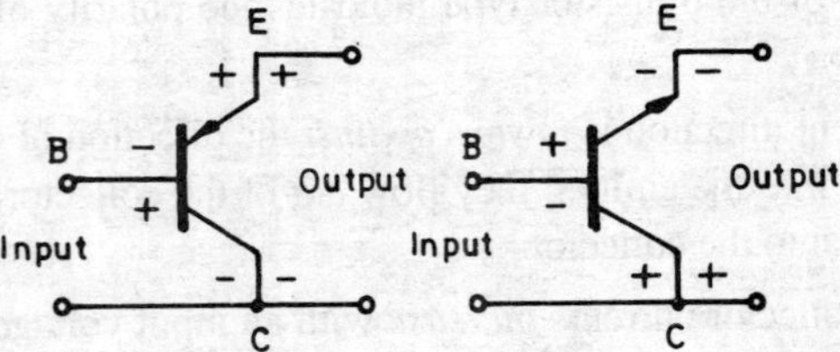


**Fig. 2.12** Common-collector (CC) configuration

connection are: *180° phase shift*, high voltage gain (300 to 600), medium current gain (20 to 200), high power gain, low input resistance and medium output resistance.

Fig. 2.12 shows a transistor connected in common-collector configuration. The input is applied between base and collector and the output appears between emitter and collector. The salient features of this connection are: *no phase shift*, voltage gain less than unity, medium current gain (20 to 200), low power gain, very high input resistance, and very low output resistance. The common-collector configuration is frequently referred to as the *emitter-follower circuit*.

In each of the above three cases, *the base forms one terminal of the input circuit and the collector forms one terminal of the output circuit.*

## 2.5 Class A Transistor Amplifier

The proper bias for a class A transistor amplifier consists of forward bias on the emitter-base circuit and reverse bias on the collector-base circuit. In order to study the dynamic behavior of an amplifier, it is helpful to plot the dynamic *transfer characteristic* which is simply a graph of collector current versus base current for a given load resistance.

Fig. 2.13 illustrates linear (class A) operation using a dynamic transfer characteristic curve. The input and output waveforms are drawn on the curve. *The output is a faithful reproduction of the input, because the transistor is operating linearly.* The transistor will operate linearly when:

1. the proper *operating point* is established by the right value of base bias current; and
2. the *excursion* of base current is within the linear portion of the dynamic characteristic curve.

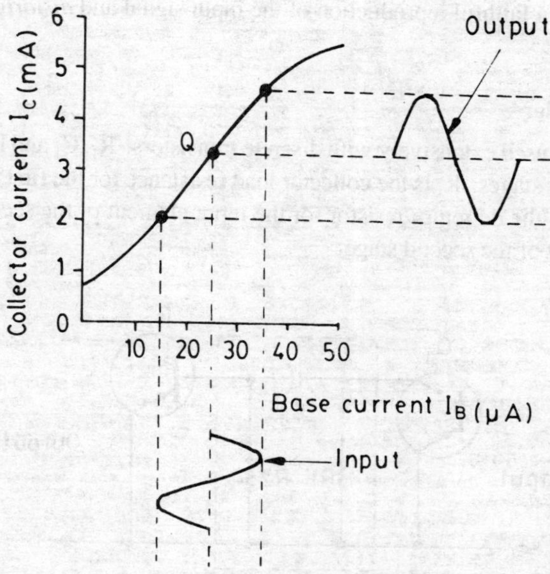


**Fig. 2.13** Linear class A operation

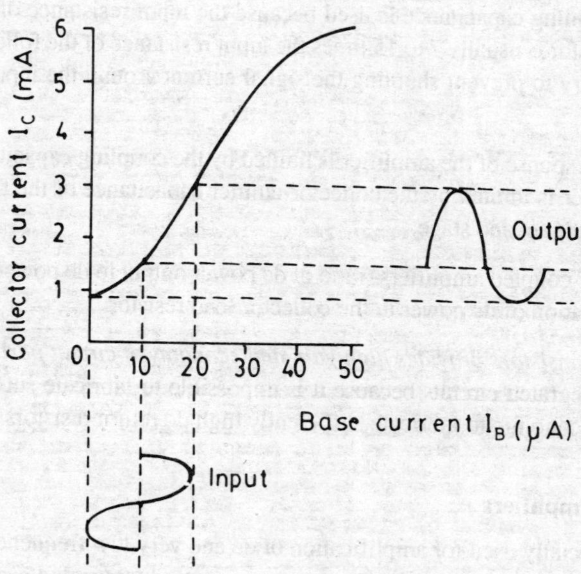


**Fig. 2.14** Non-linear operation

If the operating point is either too low or too high, the change in base current will go beyond the linear portion of the dynamic characteristic curve. When the operating point is *too-low* the output signal is distorted during the negative half-cycle of collector current. When the operating point is *too-high* the output signal is distorted during the positive half-cycle of collector current. In both of these cases the output signal is no longer a faithful reproduction of the input signal and *distortion* is introduced by the amplifier.

## 2.6 RC Coupled Amplifier

RC coupling, Fig. 2.15, is used extensively with discrete transistors. $R_1$, $C_1$ and $R_2$ form the RC *network* between the two transistor stages. $R_1$ is the collector load resistance for the first stage. $C_1$ blocks dc and couples the ac signal. $R_2$, the dc return resistor for the input element of the second stage, develops the signal applied to the input of the second stage.

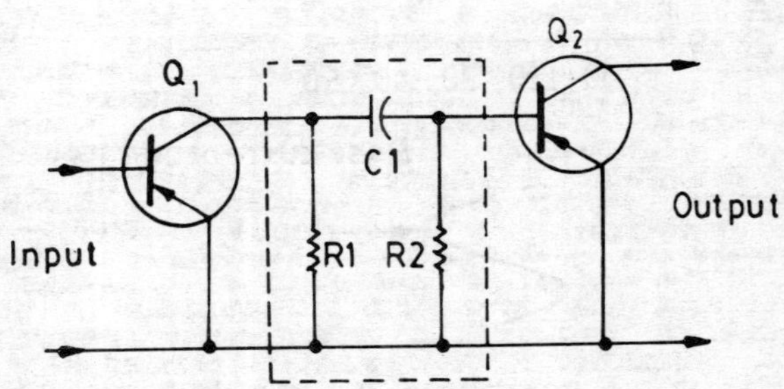


**Fig. 2.15** RC coupling

A high value of coupling capacitance is used because the input resistance of the following stage is low. The dc return resistor is usually 7 to 15 times the input resistance of the following stage. This ratio of resistance is necessary to prevent shunting the signal current around the input circuit of the second stage.

The *low frequency* response of the amplifier is limited by the coupling capacitor. *The high frequency response* of the amplifier is limited by the collector-emitter capacitance of the first stage and the base-emitter capacitance of the second stage.

The *efficiency* of RC coupled amplifiers (ratio of dc power output to dc power delivered to the stage) is low due to the dissipation of dc power in the collector load resistor.

*RC coupling with transistors provides high gain and economy of circuit parts.* This method of coupling is not used in integrated circuits because it is impossible to fabricate sufficiently large coupling capacitors and it is uneconomic to fabricate sufficiently high dc return resistors using IC technology.

## 2.7 Direct Coupled Amplifiers

Direct coupling is especially used for amplification of dc and very low frequencies. *It has an economy of parts and lends itself to use in complementary symmetry and integrated circuitry.* However, when more than a few such stages are connected in cascade, noise becomes a problem. Also direct coupled stages have a rather low efficiency. These disadvantages are outweighed in microelectronics by the relative ease with which such amplifiers can be realised in monolithic ICs. *The relative inefficiency of the direct coupled amplifiers is less important in IC applications than in high power applications.*

Fig. 2.16 shows direct coupling between two stages of audio amplification. The first stage is biased so as to provide a quiescent (no signal) value of 7 V on the collector of $Q_1$. Resistor $R_1$ is the base biasing

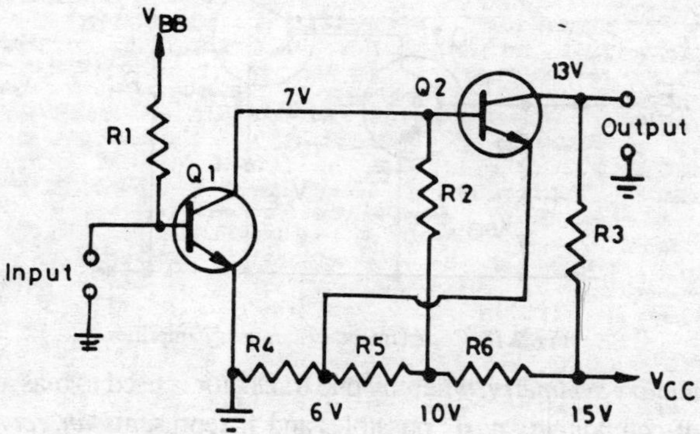


**Fig. 2.16** Direct coupling

resistor. Resistor $R_2$ serves both as collector load for $Q_1$ and also as the base input resistor for $Q_2$. Resistor R3 is the collector load for $Q_2$, Resistors R4, R5 and R6 form a *voltage divider network* to provide the necessary circuit voltage requirements. The 7 V collector voltage of $Q_1$ is also the base voltage of $Q_2$. The emitter of $Q_2$ is at a potential of 6 V, which is developed by the voltage divider network. This forward biases $Q_2$, and determines the *quiescent operating point*.

Fig. 2.17 shows another basic direct coupled amplifier connection. The common-base circuit of $Q_1$ is directly connected to the common-emitter circuit of $Q_2$. Thus, the input circuit of $Q_2$ is the load for $Q_1$, and collector bias for $Q_1$ is obtained through the collector-base junction of $Q_2$. Since $Q_2$ biases $Q_1$, *only one power source is required.*

A complementary symmetry direct coupled amplifier is shown in Fig. 2.18. In this circuit, the different natures of the npn and pnp transistors *complement* one another. The collector bias for $Q_1$ is obtained form the base-collector junction of $Q_2$.

If another transistor were added to the circuit of Fig. 2.18, an additional and larger collector supply would be required to maintain the negative collector-base potential for each stage. It is evident that a *shift* of the dc bias potential would be amplified and passed along to the second amplifier whereas in an ac coupled amplifier, such a dc shift would be blocked by the coupling capacitor.

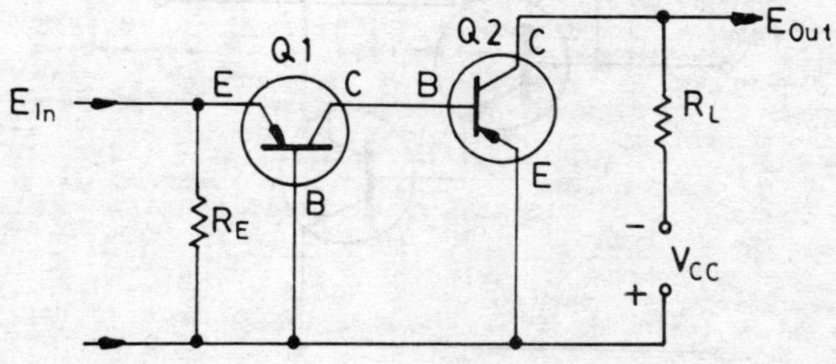


**Fig. 2.17** Direct coupling between common-base and common-emitter amplifiers

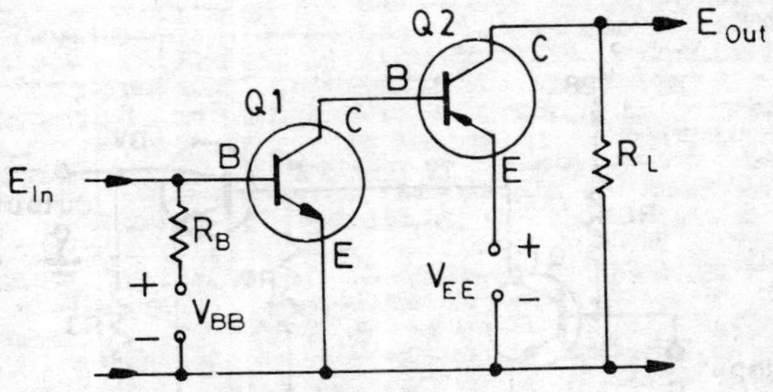


**Fig. 2.18** Complementary symmetry amplifier

The use of *complementary symmetry*, whereby one transistor is used to bias another with dc coupling, affords the *minimum of components parts* possible, and it represents *an economic advantage* that is possible only with transistors.

## 2.8 Darlington Amplifiers

A very convenient method of arranging a two stage direct coupled amplifier is shown in Fig. 2.19. This arrangement is known by a variety of names — *Darlington pair* (after S. Darlington who proposed it in 1953), compound connected pair, double emitter follower, super alpha transistor and so on.

Both the transistors are operated in the common-emitter mode. Signal current is fed to the base of the first transistor, and the resulting amplified emitter current is fed directly into the base of the second transistor. The overall current gain is equal to the *product* of the common-emitter current gains (at the operating currents) and the input impedance is very high-higher than that of a single transistor by a factor roughly equal to the current gain of the first transistor.

The combination is very convenient *for low noise input circuits*. For such use, the first stage must be operated at a very low collector current. The circuit is also very useful at *higher current levels* for the output stages of an amplifier, providing a very economic arrangement for the driver and output stages.

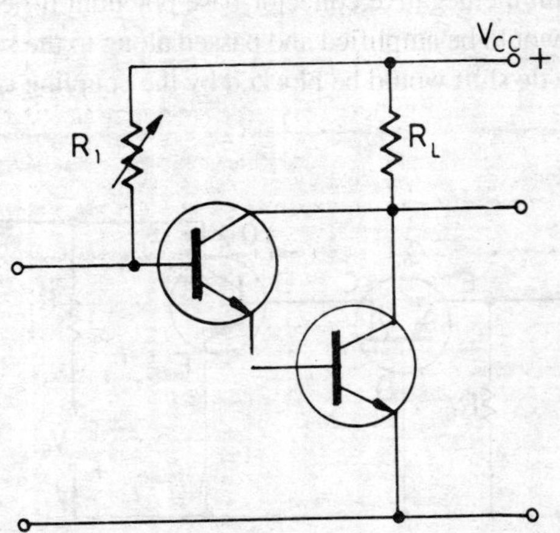


**Fig. 2.19** Darlington. direct coupled circuit

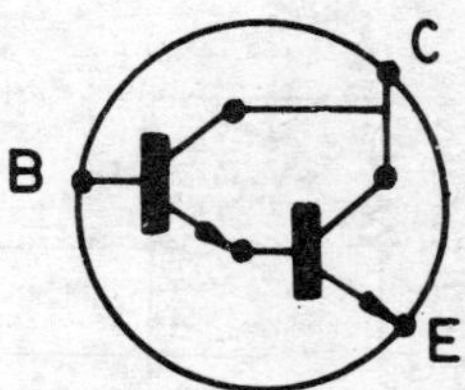


**Fig. 2.20** Composite, direct coupled transistor

The circuit can be used in the common-collector mode with a load resistance connected in the emitter circuit of the second transistor to give a very high value of input resistance. A limit is imposed, however, by the *shunting effect* of the collector resistance of the first transistor. Input resistances of the order of 1 Megohm can be obtained.

The arrangement can be treated as a single device as shown in Fig. 2.20. Some manufacturers supply units consisting of two transistor wafers assembled into a single unit. Advantage can be taken of the fact that by using two very ordinary low gain transistor wafers a very useful high gain unit can be made.

Compound connected transistors have at least two active elements. A great number of transistors may be used in the compound connection. Although two transistors are used, effectively *only three leads are used*: the base lead of one transistor, the emitter lead of the second transistor, and the collector leads of both transistors are connected to a common point.

## 2.9 The Differential Amplifier

The differential amplifier (diff-amp) has its greatest application in direct-coupled linear integrated circuit amplifiers such as the operational (op-amp) amplifier. Its design is, therefore, normally related to IC fabrication techniques, although it is also used in some circuits employing discrete components.

Fig. 2.21 illustrates the schematic diagram of the basic differential amplifier with two inputs and a single output. The circuit is *symmetrical*, that is, the two transistors $Q_1$ and $Q_2$ have identical characteristics. The emitter resistance $R_E$ is common to both transistors. Collector load resistor $R_{L1} = R_{L2}$. The two input circuits are also identical; that is $V_1 = V_2$ and $R_1 = R_2$.

*The output signal is proportional to the difference between the two input signals.* i.e.

$$v_{out} = A(v_1 - v_2) \tag{2.1}$$

where A = the gain of each transistor

$v_1$ and $v_2$ = the signal voltages at the base of each transistor relative to ground.

When the phase of the two input signals is the same, and their amplitudes are equal, the condition for *common mode operation*,

$$v_1 - v_2 = 0 \text{ and } v_{out} = A(0) = 0 \tag{2.2}$$

In the common mode, a diff-amp rejects the common mode signal, and its output for the common mode signal is zero. In practice, the two halves of the diff-amp are *never completely balanced*, and there is a *very low output* for the common mode signal.

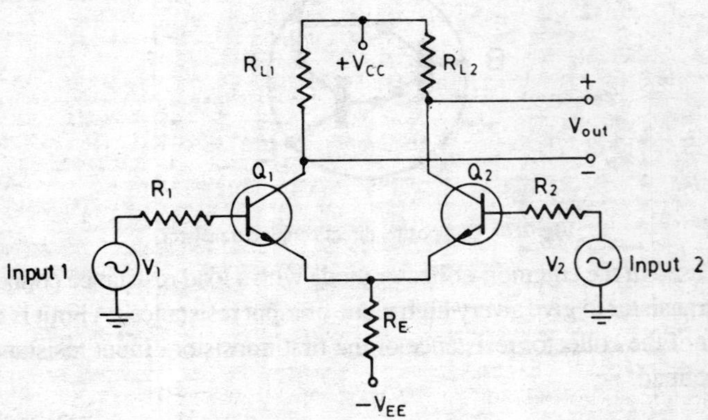


**Fig. 2.21** Differential amplifier

When the two input signals are equal in amplitude but 180° out of phase, the condition for *differential mode operation*,

$$v_1 = -v_2 \text{ or } v_2 = -v_1 \text{ and } v_{out} = A[v_1 - (-v_1)] = A(2v_1) \tag{2.3}$$

In the differential or *non-common mode* operation, the diff-amp does amplify the input signal, and the output signal is *twice the gain times* the input signal.

The diff-amp is normally operated with a *double-ended input* as shown in Fig. 2.21. However, it may also be used with a *single-ended input* as shown in Fig. 2.22. Moreover, the output signal can be taken between the two collectors, as in Fig. 2.21 or there can be two separate outputs, $v_{out1}$ and $v_{out2}$, from each collector to ground, as in Fig. 2.22.

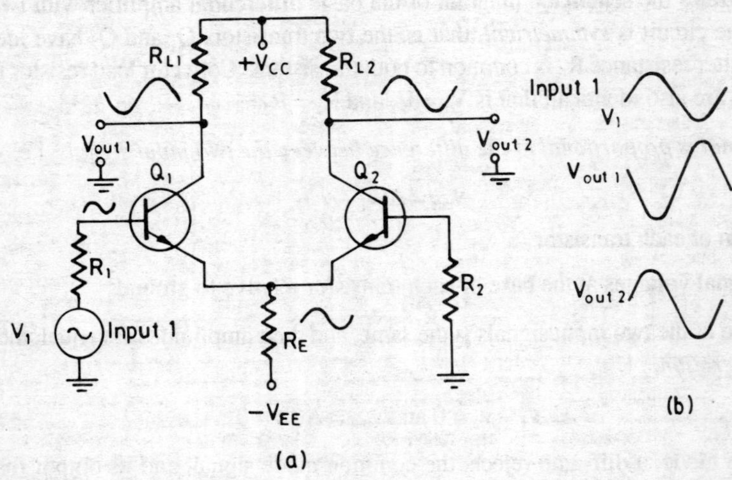


**Fig. 2.22** Differential amplifier with single-ended input

Consider a sinewave $v_1$, with the polarity shown in Fig. 2.22(b), applied to input 1. The amplified signal $v_{out}$ appearing at the collector of Q1 to ground will be 180° out of phase with the input. Moreover, because emitter resistor $R_E$ is unbypassed, a sinewave will develop across $R_E$ which is in phase with the input $v_1$. Since the base of $Q_2$ is returned to ground, there will be no signal voltage from the base of $Q_2$ to ground.

*A diff-amp with a single-ended input and dual output produces two output waveforms, equal in amplitude but 180° out of phase.*

If a single output from the diff-amp in Fig. 2.22 is taken between the collector of $Q_2$ and the collector of $Q_1$, as in Fig. 2.23, the output waveform $v_{out}$ is in phase with the input $v_1$. Input to $Q_1$ is therefore called a *non-inverting input*.

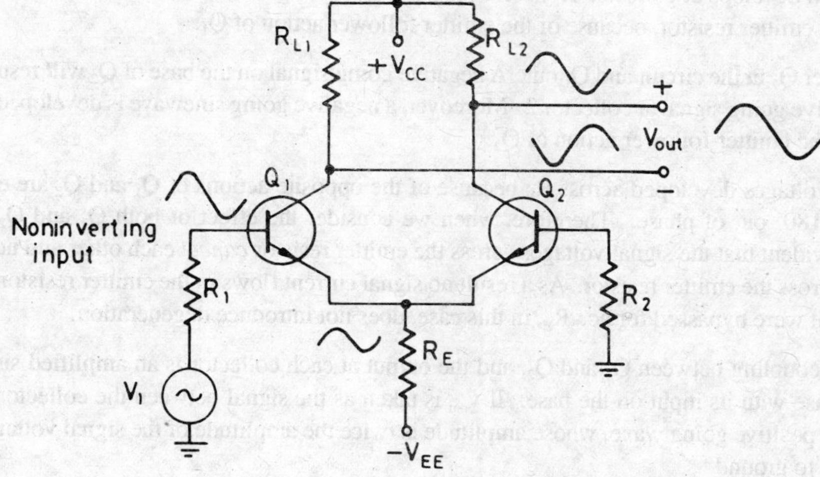


**Fig. 2.23** Single-ended input to $Q_1$ and output taken between collectors of $Q_2$ and $Q_1$

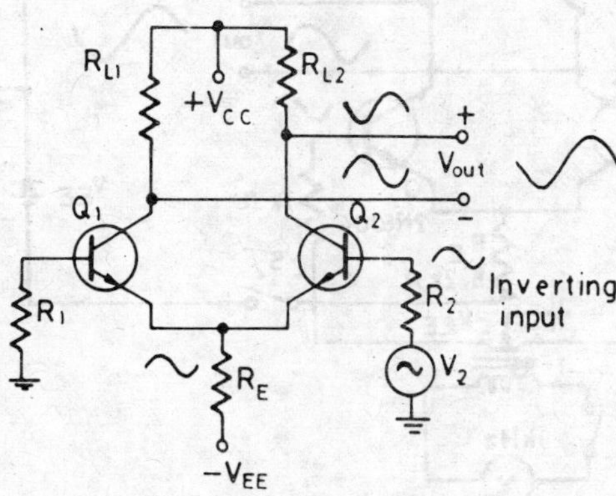


**Fig. 2.24** Single-ended input to $Q_2$ and output taken between collectors of $Q_2$ and $Q_1$

If the base resistor of $Q_1$ is returned to ground and the input signal is applied to the base of $Q_2$, as in Fig. 2.24, and a single output waveform is taken between the collector of $Q_2$ and the collector of $Q_1$, the output waveform $v_{out}$ is 180° out of phase with the input. Input 2 to $Q_2$ is called *an inverting input.*

## 2.10 Differential Mode Input

A diff-amp with differential mode input is shown in Fig. 2.25. Transistors $Q_1$ and $Q_2$, connected as a differential amplifier, receive input sinewaves equal in amplitude but 180° out of phase, the condition for differential mode operation. Assume an instant in time when the sinewave on base 1 is positive going, that on base 2 negative going, as shown in Fig. 2.25. Now consider the action of transistor $Q_1$, as though $Q_2$ were not in the circuit. With a positive going signal in the base of $Q_1$, an amplified negative going waveform develops at collector 1. Moreover a positive going sinewave is developed across $R_E$, the unbypassed emitter resistor, because of the emitter follower action of $Q_1$.

Now consider $Q_2$ in the circuit and $Q_1$ out. A negative going signal on the base of $Q_2$ will result in an amplified positive going signal at collector 2. Moreover, a negative going sinewave is developed across $R_E$ because of the emitter-follower action of $Q_2$.

The signal voltages developed across $R_E$ because of the opposite actions of $Q_1$ and $Q_2$ are equal in amplitude but 180° out of phase. Therefore, when we consider the effect of both $Q_1$ and $Q_2$ acting together, it is evident that the signal voltages across the emitter resistor *cancel* each other and no signal is developed across the emitter resistor. As a result no signal current flows in the emitter resistor, which acts as though it were bypassed for ac. $R_E$, in this case, does not introduce degeneration.

There is no coupling between $Q_1$ and $Q_2$, and the output at each collector is an amplified sinewave 180° out of phase with its input on the base. If $v_{out}$ is taken as the signal between the collectors of $Q_2$ and $Q_1$, $v_{out}$ is a positive going wave, whose amplitude is twice the amplitude of the signal voltage from either collector to ground.

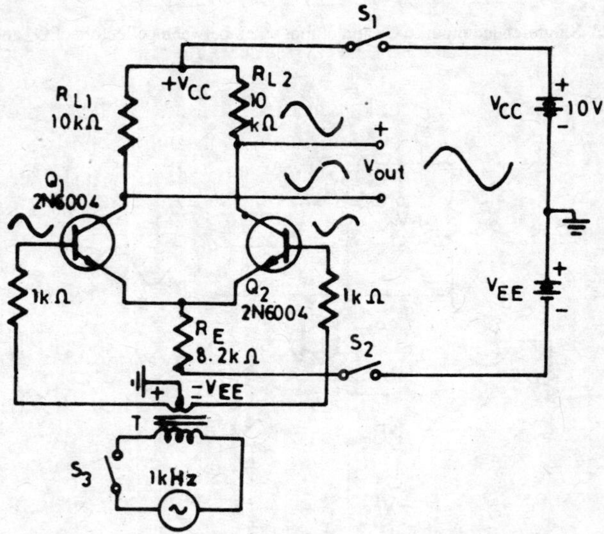


Fig. 2.25 Amplifier with differential mode input

It is possible to take two outputs from the diff-amp; from the collector of $Q_1$ and from the collector of $Q_2$, with respect to ground. These signal voltages are equal in amplitude but 180° out of phase.

## 2.11 Common Mode Input

The circuit of Fig. 2.26 shows the diff-amp connected for common mode input. The secondary of transformer T supplies the same amplitude and the same phase signal voltages $v_1$ and $v_2$ to the bases of $Q_1$ and $Q_2$. The in phase signals at the bases of $Q_1$ and $Q_2$ cause in phase signal voltages to appear across $R_E$ which add together. The phase of the combined signal across $R_E$ is the same as the phase of the signals at the bases of $Q_1$ and $Q_2$.

$R_E$ provides degeneration or *negative feed back* to each amplifier. Since $R_E$ is connected to both $Q_1$ and $Q_2$, *its degenerative effect is twice as great as it would be if each transistor were acting alone*. Therefore, the effective base-to-emitter signal which each amplifier sees is negligibly small, and the output at each collector is negligibly small Ideally, the output would be zero. In practice, this is never realised.

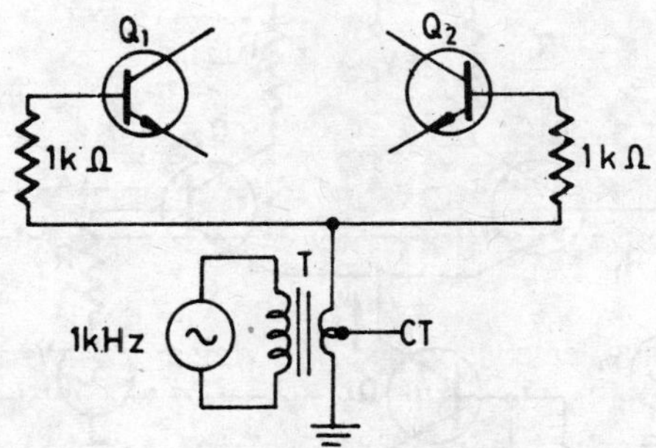


**Fig. 2.26** Amplifier with common mode input

The basic diff-amp circuit is shown in Fig. 2.26. It can operate in both the differential and the common mode. *Out of phase, equal amplitude, signals fed to the bases of both the transistors ensure differential mode operation. In phase, equal amplitude, signals fed to the bases of both the transistors ensure common mode operation*. Both types of signals may appear at the same time on the bases of $Q_1$ and $Q_2$ and the *diff-amp will reject the common mode signal while amplifying the differential mode signal.*

The operation of the diff-amp with differential input is very useful in *amplifying desired signals* (non common mode) and *rejecting undesired signals* (common mode). The common mode signals may result from introduction into the diff-amp inputs of stray negative fields or from power supply ripple or hum.

## 2.12 Common Mode Rejection Ratio

A diff-amp should have *high gain* for differential mode input signals and *very low gain* for common mode input signals. *The ratio of the gain of the amplifier for differential mode signals ($A_{DM}$) to the gain of the amplifier for common mode signals ($A_{CM}$) is called the common mode rejection ratio (CMRR).*

$$\text{CMRR} = A_{DM}/A_{CM} \tag{2.4}$$

*The CMRR value is, therefore, an index of the effectiveness of a diff-amp, the higher this value, the better the diff-amp.*

For common mode signals, the unbypassed resistor $R_E$ provides degeneration or negative feedback. Therefore, the larger the resistance of $R_E$, the higher the negative feedback voltage with common mode, and the more effective the amplifier is in rejecting common mode signals.

In the differential mode, $R_E$ acts as though it were completely bypassed and does not reduce the gain of the amplifier for differential mode signals.

A large value of resistor $R_E$ is, therefore, desirable for a diff-amp because it insures the least common mode gain without affecting the differential mode gain, that is, *the higher the value of $R_E$, the higher the value of CMRR.* There is a practical limit to the value of $R_E$, since $R_E$ determines the value of emitter current $I_E$.

$$I_E = V_{EE}/R_E \tag{2.5}$$

The emitter supplies equal currents to both the collectors in diff-amp. By increasing $R_E$, the emitter and collector currents are reduced.

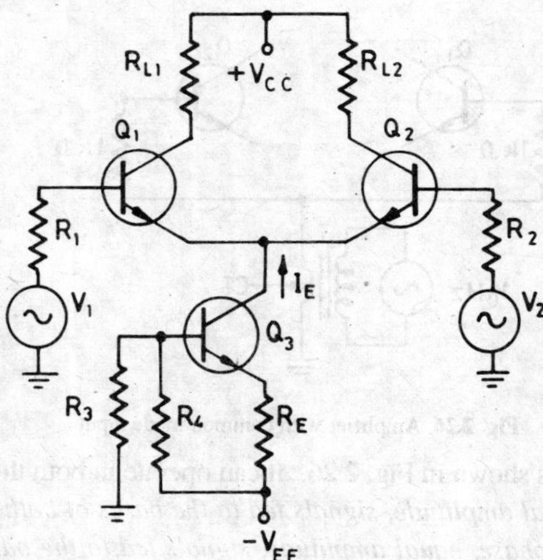


**Fig. 2.27** **Improved differential amplifier using constant current transistor Q3 to replace the emitter resistor RE in Fig. 2.21**

It is possible to replace $R_E$ with a *constant current transistor* $Q_3$, in Fig. 2.27. The collector emitter resistance of $Q_3$ is high without unduly limiting the collector current of $Q_3$, and therefore without limiting the emitter current $I_E$ which supplies collector currents for $Q_1$ and $Q_2$.

The emitter resistor $R_E$ in Fig. 2.27 can be relatively small in value. DC bias for $Q_3$ is determined by the voltage divider $R_3$ and $R_4$. The addition of $Q_3$ to the circuit, therefore, ensures a high effective emitter resistance $R_E$ without the restriction on the collector currents for $Q_1$ and $Q_2$.

Diff-amp circuitry may be more complex than the circuit in Fig. 2.27. It is simple and more convenient to represent the diff-amp as a block (Figs. 2.28 and 2.29) rather than by conventional circuit symbols. This is particularly true of IC differential amplifiers when terminals of the IC represent input

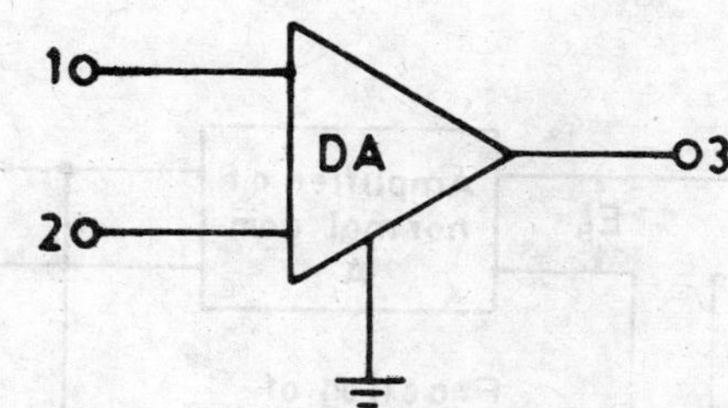


**Fig. 2.28** Differential amplifier circuit symbol; 1 and 2 are inputs, 3 is output

connections to and output connections from the diff-amp. Fig. 2.28 represents a diff-amp with two inputs and a single output. Fig. 2.29 is the block symbol for a two input diff-amp with two separate outputs. The single output in Fig. 2.28 is in fact the output voltage from output 3 to 4 in Fig. 2.29.

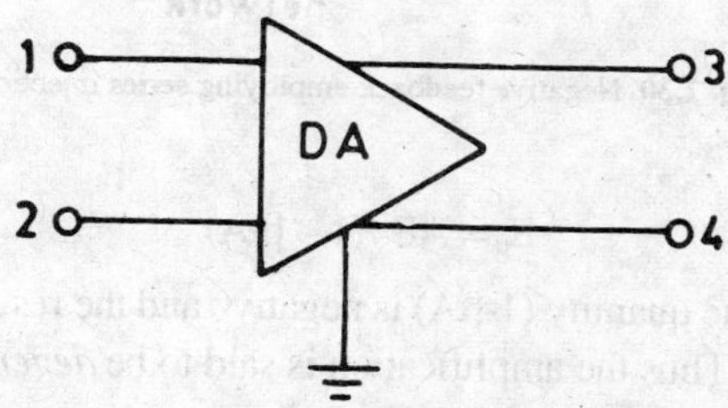


**Fig. 2.29** Differential amplifier circuit symbols; 1 and 2 are inputs, 3 and 4 are dual outputs taken with respect to ground

## 2.13 Negative Feedback

The amplification of an integrated transistor amplifier depends not only on the design of the amplifier itself but on the supply voltage and ambient temperature as well. *Negative feedback can be used to stabilise the amplification to a precise value.*

Assume that the signal voltage $E_S$ is applied to the input terminals, as shown in Fig. 2.30. Let a portion $\beta E_0$ of the output voltage $E_0$ be fed back in series with $E_S$ in such a way that the signal $E_B$ appearing between the base and emitter is of the form

$$E_B = E_S + \beta E_0 \tag{2.6}$$

where $\beta$ is a fractional part of $E_0$.

Since the normal gain A of the amplifier is defined as

$$A = E_0/E_B \tag{2.7}$$

then by transposing we have,

$$E_0 = AE_B \tag{2.8}$$

Substituting the value of $E_B$ from Eq. 2.6 in the above equation, we have

$$E_0 = A(E_S + \beta E_0)$$

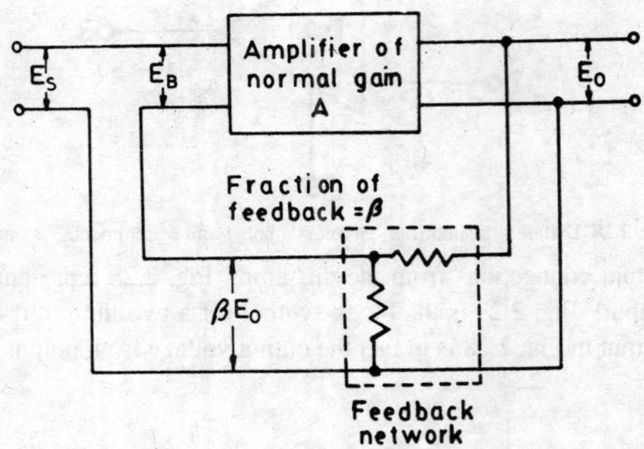


**Fig. 2.30** Negative feedback employing series injection

Solving for $E_0$,

$$E_0 = AE_S/(1 - \beta A) \tag{2.9}$$

If βA is greater than unity, the quantity (1-βA) is negative and the resulting amplification is less than it would be without feedback. Thus the amplification is said to be *degenerative*.

The resulting gain $A_R$ with feedback incorporated is,

$$A_R = E_0/E_S \tag{2.10}$$

Substituting Eq. 2.9 in the above expression, we have

$$A_R = \frac{AE_S}{E_S(1 - \beta A)} = \frac{A}{1 - \beta A} \tag{2.11}$$

The resulting amplifier gain is expressed in terms of the gain without feedback (A) and the fraction of the output (β) fed back to the input. $A_R$, A and β may be complex quantities. Generally the feedback portion (βA) is so much larger than unity that the resulting amplification ($A_R$) for all practical purposes may be expressed as

$$A_R = 1/\beta \tag{2.12}$$

## 2.14 Advantages of Negative Feedback

The following is a summary of the effects of the application of negative feedback on the performance of an amplifier:

1. the *gain* of the amplifier is reduced;
2. the *attenuation distortion*, due to the variation in gain of the amplifier with frequency, is reduced;
3. the *phase shift* through the amplifier is reduced;
4. *distortion*, due to the non linearity of the amplifier, is reduced;

5. *noise* in the output of the amplifier is reduced;
6. change in *overall amplification*, due to change in inherent gain, is reduced;
7. frequency response may be adjusted to any desired value by choice of a suitable network; and
8. both *input and output impedances* of the amplifier are modified the output impedance may be so modified that the voltage gain becomes independent of the output load.

## 2.15 Operational Amplifiers

In its original conception, the operational amplifier (*op-amp*) was a high gain dc amplifier whose operational behavior could be precisely controlled by the proper selection of components used in the negative feedback loop. Such amplifiers normally had but one input and one output, meaning that the output was inverting or out of phase with the input. To day, however, the term operational amplifier is generally used to apply to *any high gain dc amplifier having single or differential inputs, single or differential outputs, and whose gain characteristics are controlled by selection of feedback components.*

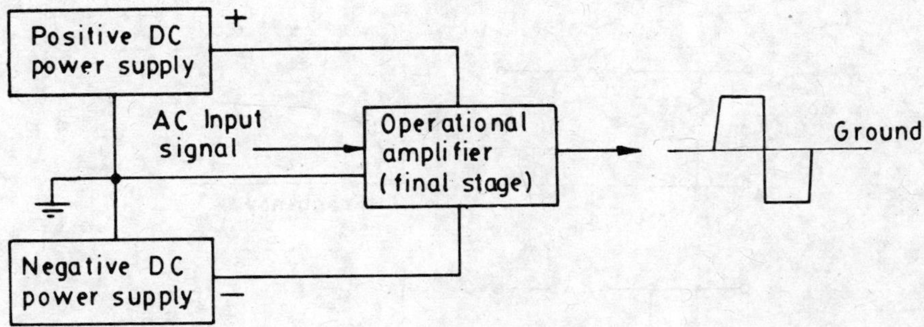


**Fig. 2.31** Basic block diagram of the operational amplifier, showing the method used to obtain positive and negative voltages at the input

An important feature of op-amps is that the input signal is usually referred to ground or zero volts. The input signal is then allowed to swing *either above or below* ground level— either positive or negative with respect to ground. *The gain of the op-amp is determined by the components in the feedback loop.* When simple resistive feedback components are used, the op-amp becomes a precision dc amplifier whose gain is equal to the ratio of the feedback resistance, $R_f$, to the input resistance, $R_{in}$ that is

$$A_R = R_f/R_{in} \tag{2.13}$$

The above gain relationship remains fairly constant provided that the *closed loop gain* ($A_R$) remains relatively small compared to the *open loop gain* (A) of the amplifier without feedback.

The gain of the amplifier can be easily controlled using various combinations of resistances and capacitances. The frequency behavior of the feedback circuits are often exploited to shape the response of the op-amp.

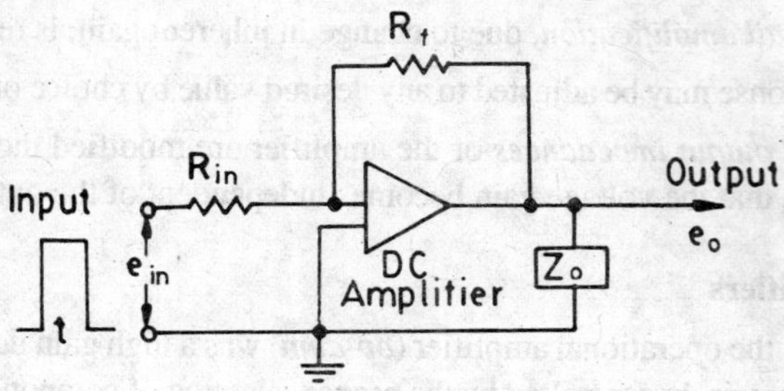


Fig. 2.32 Operational amplifier using resistive feedback circuit

When it is desired to have the amplification vary in some specific manner with respect to frequency, the negative feedback network through which β is obtained may be designed to *attenuate* specific frequencies. For example, if the high frequencies are to be amplified more than low frequencies, only the low frequencies will be fed back to the input. Some of the *response shaping networks* are shown in Fig. 2.33.

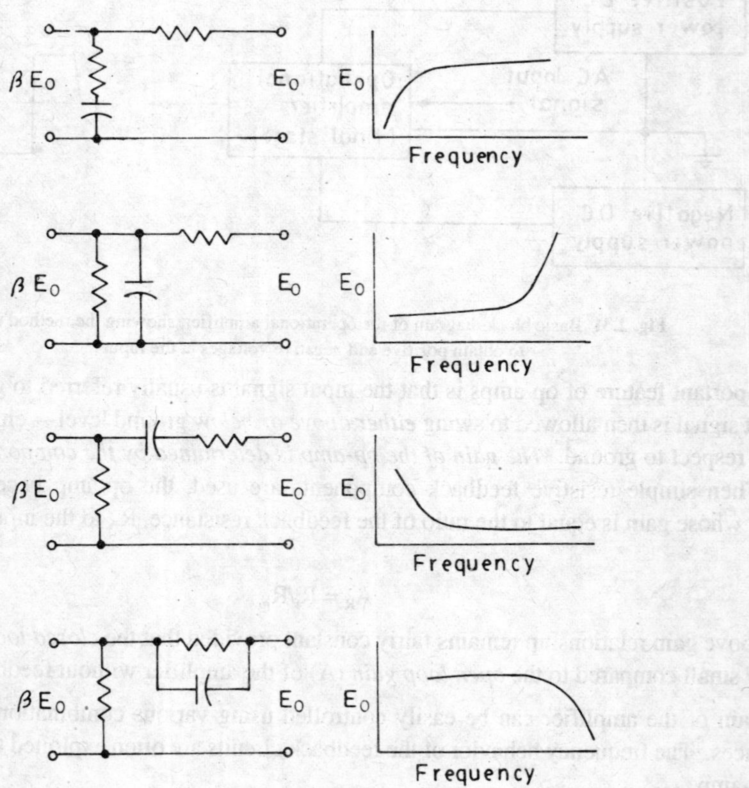


Fig. 2.33 Response shaping by means of negative feedback networks

OP-amplifiers, widely used in computers, can perform *mathematical operations* such as summing, integration and differentiation. OP-amplifiers are also used as video and audio amplifiers, oscillators etc, in communication equipment. Because of their versatility op-amps are widely used in all branches of electronics both in digital and linear circuits.

*OP-amplifiers lend themselves readily to IC manufacturing techniques.* Improved IC manufacturing techniques, the op-amps adaptability and extensive use of op-amps in the design of new equipment have brought the price of IC op-amps from very high to very reasonable levels. This fact ensures a very substantial role for the IC op-amps in electronics.

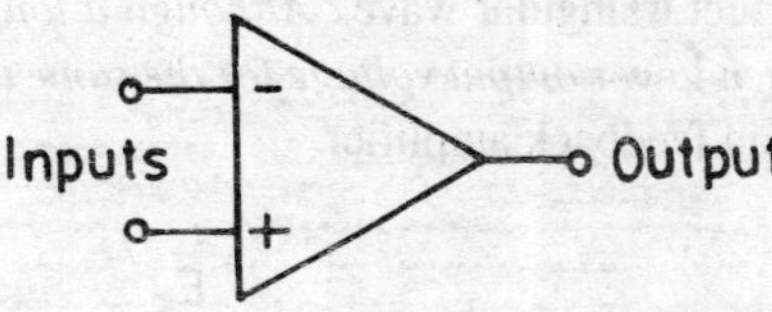


**Fig. 2.34** Symbols for operational amplifier

Fig. 2.34 shows the symbol for an op-amp. The op-amp, like the differential amplifier, has two inputs marked (–) and (+). The minus input is the *inverting input*. *A signal applied to the minus terminal will be shifted in phase 180°* at the output. The plus input is the *non-inverting input*. *A signal applied to the plus terminal will appear in the same phase at the output as at the input*. Because of the complexity of the internal circuitry of an op-amp, the op-amp symbol is used extensively in circuit diagrams.

## 2.16 Operational Amplifier as an Integrator

A simple and commonly used integrator consists of two circuit elements, a resistor and capacitor, (see Fig. 2.35). *The voltage across the capacitor is proportional to the integral of the charging current*. The voltage across the capacitor V = Q/C. For any given capacitor C, the voltage depends directly on charge Q, which is the imbalance of electrons on the two capacitor plates. The amount of this charge depends both on the amount of current and the time for which this flow exists.

The fact that the voltage is proportional to the integral of the charging current allows the RC network to be used to make an integrator. *The capacitor voltage is the integrator output*. Provision is then made to supply a charging current that is proportional to the input signal. The purpose of the resistor is to

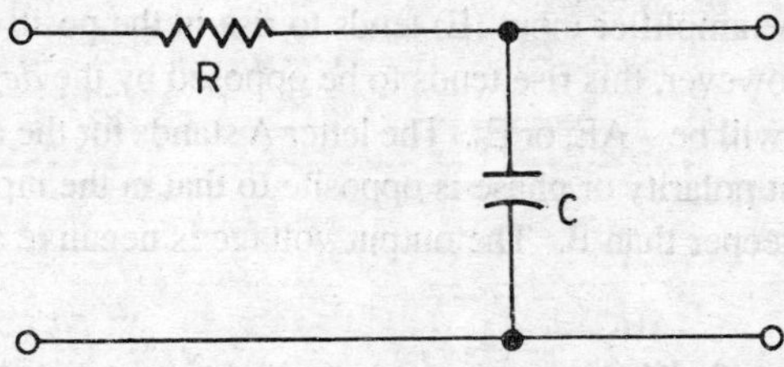


**Fig. 2.35** Simple integrator circuit

produce this proportional current from an input signal voltage $V_T$. At the instant this voltage is applied, the charging current becomes $I = V_T/R$.

Unfortunately, this proportionality does not continue to exist. As the capacitor becomes charged the capacitor voltage opposes the charging current and the charging current becomes less proportional to the input signal. This results in an *error* in the output. The ideal output for a constant input signal is a steadily increasing output. This steady increase is attained only when the signal voltage is first applied and the capacitor has not become appreciably charged.

A remedy to this error in the RC integrator is to use a circuit with a *long time constant*, at least with respect to the signal frequencies of interest. Such a circuit delays the charging of the capacitor. The result is a more accurate integration of an input signal. The ideal output response to the square wave pulse in Fig. 2.36 would be a perfect triangular wave. *Although a long time constant produces better integration, it also produces a much lower output voltage for the same input voltage*. Better integration is possible by the use of a high gain feedback amplifier.

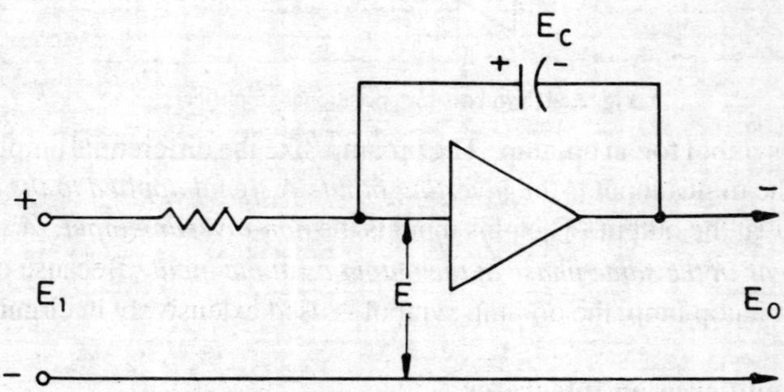


**Fig. 2.36** Using a high gain amplifier, Miller integrator, to improve integration

The circuit arrangement in Fig. 2.36 uses a high gain amplifier and is known as the *Miller integrator*. The amplifier produces an output which is not limited by the input signal, as it is in the simple RC integrator. The amplifier also supplies any energy which is required in the output. The function of the input signal is to control the charging current.

The operation of the circuit can be explained by assuming a constant input, as shown in Fig. 2.36. Assume that to start with the initial condition is zero, that is $E_I = E_0 = 0$. Also assume that the capacitor is discharged. The positive voltage to be integrated is then applied. The capacitor charges with the polarity shown, since electrons are attracted from the left plate. The charging path is shown in Fig. 2.37.

A voltage measured at the amplifier input (E) tends to rise in the positive direction, since this point is directly coupled to $E_I$. However, this rise tends to be opposed by the *degenerative feedback* voltage from the output. The output will be – AE, or $E_0$. The letter A stands for the amplifier gain and the minus sign indicates that the output polarity or phase is opposite to that in the input. The output voltage now changes A times faster or steeper than E. The output voltage is negative and *aids* the charging of the capacitor.

For a certain input voltage, the charging current is limited to a particular value which tends to keep E practically zero. If the current should exceed this value, E would decrease a small amount due to the

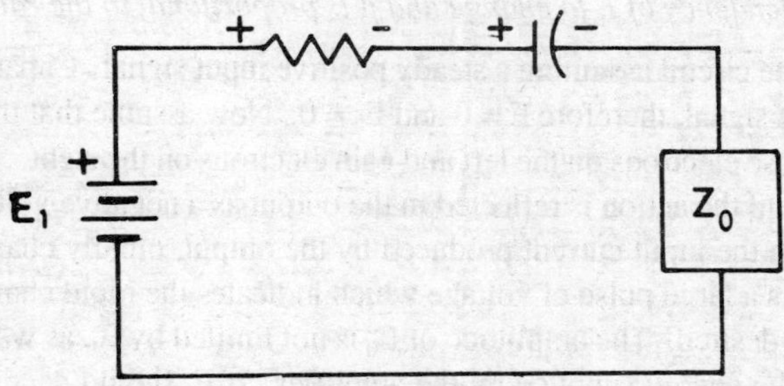


**Fig. 2.37** The charging path of the capacitor in Fig. 2.36

increased voltage drop across R. Then $E_0$ would decrease and the charging current would decrease to the original value. If the initial charging current would decrease, the opposite action would occur. The value of the charging current is, therefore, *stabilised* to a specific value proportional to the input voltage. This eliminates the error caused by E and the charging current not remaining proportional, as in the simple RC integrator.

The *constant charging current* must be produced by $E_0$ despite the fact that the steadily increasing capacitor voltage opposes the charging current. To do this, $E_0$ must also steadily increase, and this steady increase in $E_0$ is exactly the integrator output voltage desired for a constant input signal.

The limits for $E_0$ are determined by the amplifier and not by $E_I$ or the range of $E_C$. The output range would be designed to produce an increasing output for any probable input amplitude and period of application.

## 2.17 Operational Amplifier As A Differentiator

An RC differentiator, Fig. 2.38, is just the opposite of an integrator. *The differentiator circuit has a capacitor input and resistive feedback*. The amplifier can be a modulator type so as to reduce drift and noise.

It is normal to consider such an amplifier to have infinite gain. On this basis, E can be considered to be zero. Although infinite gain is impossible, the amplifier does have a very large gain. If $E_I$ should

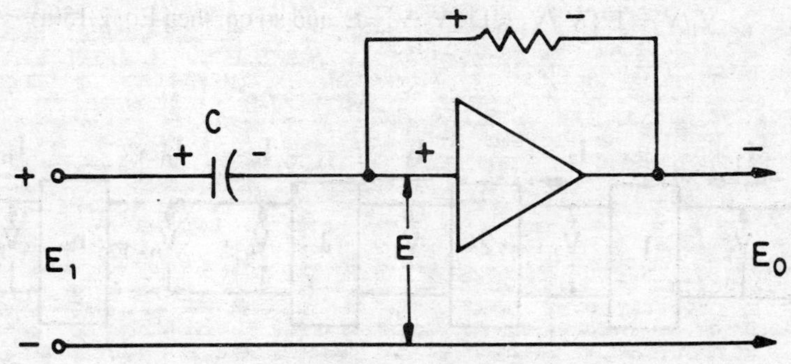


**Fig. 2.38** RC op-amp differentiator

change, E tends to change also but the inverse feedback voltage *counteracts* the change in E. *The output voltage depends on the tendency of E to change and it is proportional to the rate of change of $E_I$.*

Fig. 2.38 illustrates the circuit assuming a steady positive input signal. Capacitor C is charged to the steady value of the input signal, therefore E = 0 and $E_0 = 0$. Now assume that the error signal suddenly increases. C begins to lose electrons on the left and gain electrons on the right. The charging of C tends to make E go positive, and the action is reflected in the output as a negative voltage, which increases at a rate A times faster than the input current produced by the output, rapidly charges C to the new value of $E_I$. Thus, the output is a large pulse of voltage which indicates the rapid change in the input voltage $E_I$, and this is the output desired. The amplitude of $E_0$ is not limited by $E_I$, as was the case in the simple RC circuit, because of the energy supplied by the amplifier. If $E_I$ should decrease, an output pulse of opposite polarity would be produced.

## 2.18 Decibels

In filter circuits and other electric networks, it is frequently convenient to appraise the performance of a circuit in terms of the ratio of input current to output current magnitude. If the ratios of voltage to current at input and output of the network are equal, then the ratios of input to output currents, or input to output voltages may be written as:

$$I_1/I_2 = V_1/V_2 \qquad (2.14)$$

If several networks are used in succession, the overall performance may be expressed as;

$$V_1/V_2 \times V_2/V_3 \times V_3/V_4 \ldots \times V_{n-1}/V_n = V_1/V_n \qquad (2.15(a))$$

It may also be written as:

$$A_1 \times A_2 \times A_3 \times A_4 = A_1A_2A_3A_4 \qquad (2.15(b))$$

both processes employing *multiplication* of magnitudes. In general the process of addition or subtraction may be carried out with greater ease than the process of multiplication or division. It is, therefore, of interest to note that

$$E^a \times E^b \times E^c \times \ldots.\ E^n = E^{a+b+c+\ldots n} \qquad (2.16)$$

is an application in which *addition is substituted for multiplication.*

If the voltage ratios of Eq. 2.15(a) are defined as

$V_1/V_2 = E^a$; $V_2/V_3 = E^b$; $V_3/V_4 = E^c$ and so on, then Eq. 2.15(a)

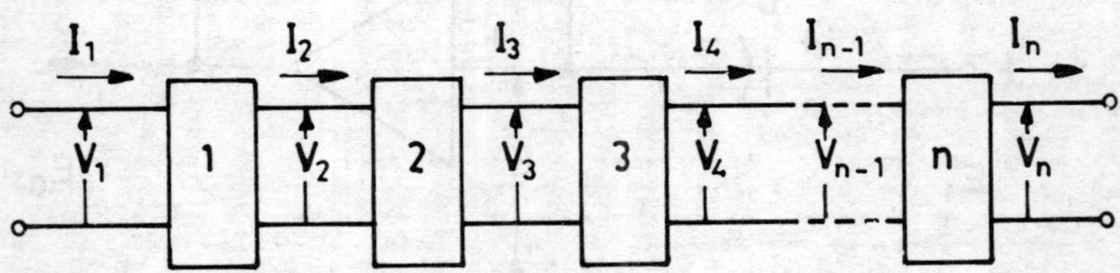


**Fig. 2.39** A succession of n networks in cascade

becomes

$$V_1/V_n = E^{a+b+c+...n} \qquad (2.17)$$

and if the natural logarithm (In) of both sides is taken, then

$$\text{In } V_1/V_2 = a+b+c+...n \qquad (2.18)$$

Consequently, if the ratio of each individual network is given as E to an exponent, the logarithm of the current or voltage ratio for all the networks in series is very easily obtained as the simple sum of the various components. It has become common, for this reason, to define

$$V_1/V_2 = I_1/I_2 = E^N \qquad (2.19)$$

*under conditions of equal input and output impedances.* The unit N has been given the name *neper* and defined as:

$$N \text{ nepers} = \text{In } V_1/V_2 = \text{In } I_1/I_2 \qquad (2.20)$$

*Two voltages or currents differ by one neper when one of them is ε times as large as the other.*

Obviously, ratios of input to output power may also be expressed as

$$P_1/P_2 = E^{2N} \qquad (2.21)$$

*The number of nepers represents a convenient measure of the power loss or gain of a given network.*

The telephone industry proposed a similar unit based on logarithms to the base 10, naming the unit a *bel* after Alexander Graham Bell. *The bel is defined as the logarithm of a power ratio.*

$$\text{Number of bels} = \log P_1/P_2 \qquad (2.22)$$

It has been found that a unit one-tenth as large is more convenient and the smaller unit is called the *decibel* abbreviated 'db' and defined as:

$$\text{db} = 10 \log P_1/P_2 \qquad (2.23)$$

*For equal impedances in input and output circuits.*

$$\text{db} = 20 \log I_1/I_2 = 20 \log V_1/V_2 \qquad (2.24)$$

The relation between neper and decibel can now be established.

$$8.6\,86 \text{ N} = \text{db or 1 neper} = 8.6\,86 \text{ db} \qquad (2.25)$$

Most of the energy transmitted through eleçtric networks, lines and filters, is ultimately converted to acoustic energy and heard by the human ear as sound. An important property of the human ear is that *the ear hears logarithmically.* The logic of the logarithmic units, the neper and the decibel, is then apparent.

A *low pass filter* transmits low frequencies but stops high frequencies. Fig. 2.40 shows one way to build a low pass filter. At very low frequencies the inductive reactances approach 0 and the capacitive reactances approach infinity. This is equivalent to saying that *the inductors appear shorted and the capacitors appear open.* Therefore, the output voltage equals the input voltage at very low frequencies.

As the frequency increases, the inductive reactances increase and the capacitive reactances decrease. At some point, the output voltage starts to decrease. For very high frequencies *the inductors appear open and the capacitors appear shorted.* Therefore, the output voltage approaches 0.

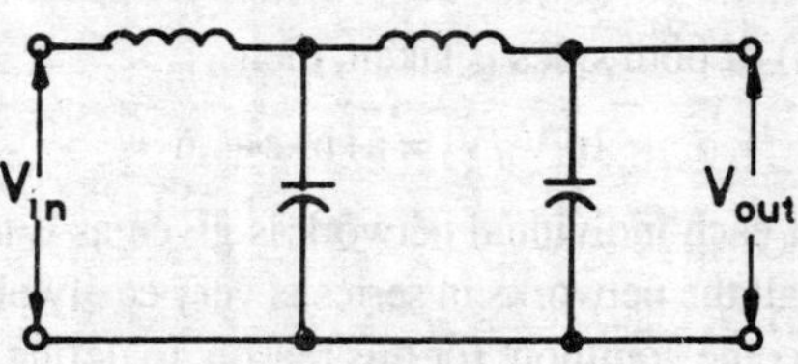


Fig. 2.40 Low pass filter

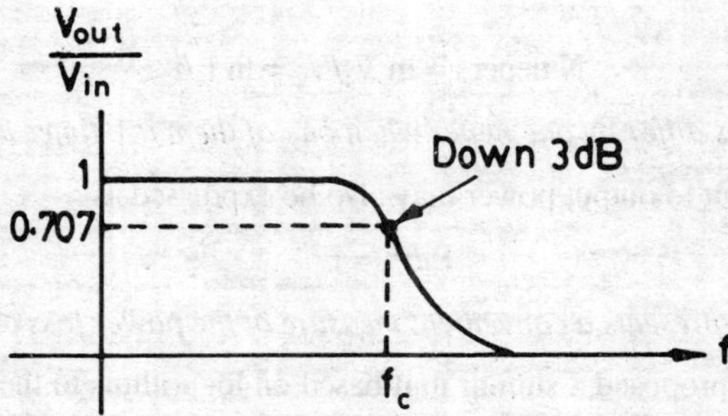


Fig. 2.41 Voltage gain of a low pass filter

*The cut-off frequency is where the voltage gain equals 0.707 (equivalent to the half-power point)*. This is shown in Fig. 2.41.

A *high pass filter* transmits high frequencies but stops low frequencies. Fig. 2.42 shows one way to build a high pass filter. At very low frequencies the inductive reactances approach 0 and the capacitive reactances approach infinity. This is equivalent to saying that *the inductors appear shorted and the capacitors appear open*. Therefore the output voltage approaches 0.

As the frequency increases the inductive reactances increase and the capacitive reactances decrease. At some point, the output voltage starts to increase. For very high frequencies *the inductors appear open and the capacitors appear shorted*. Therefore, the output voltage equals the input voltage at very high frequencies.

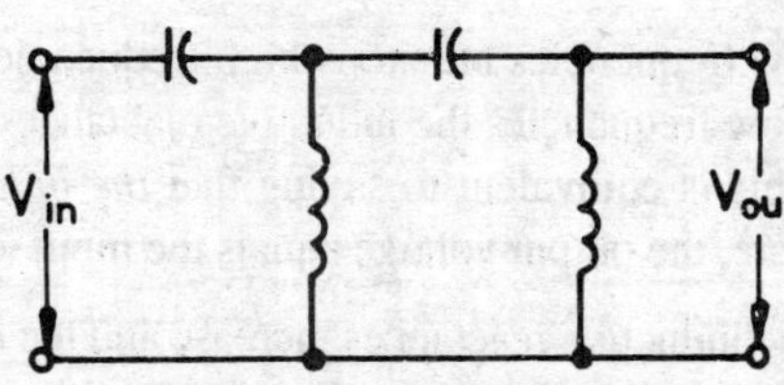


Fig. 2.42 High pass filter

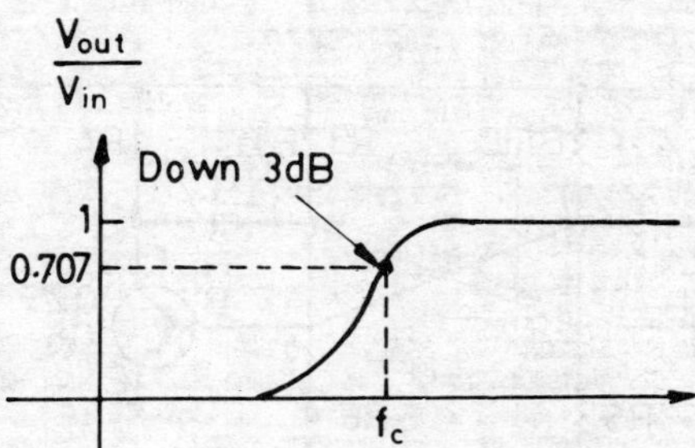


**Fig. 2.43** Voltage gain of a high pass filter

The *cut-off frequency* is where the voltage gain equals 0.707 (equivalent to the half-power point). This is shown in Fig. 2.43. Table 2.1 lists some voltage gains and their decibel equivalents.

**Table 2.1**

Voltage gains and their decibel equivalents

| A | $A_{db}$, db | A | $A_{db}$, db |
|---|---|---|---|
| 1,000 | 60 | 0.707 | –3 |
| 100 | 40 | 0.5 | –6 |
| 10 | 20 | 0.25 | –12 |
| 8 | 18 | 0.125 | –18 |
| 4 | 12 | 0.1 | –20 |
| 2 | 6 | 0.01 | –40 |
| 1 | 0 | 0.001 | –60 |

## 2.19 Multivibrators

The multivibrator is a *two stage amplifier with positive feedback* and may be biased for free running or triggered operation. To obtain feedback, the multivibrator uses RC coupling.

The *astable or free running multivibrator* is used for generating square waves, timing frequencies and frequency division. The astable multivibrator has *no stable state.*

The *monostable multivibrator* is suitable for standardising pulses of random width, or for generating time delayed pulses. The monostable multivibrator has *one stable state*.

The *bistable multivibrator* is actually a variation of the astable multivibrator. It has *two stable states.*

There are many variations of the three multivibrator types depending upon the purpose for which they are used. One of the main applications of the multivibrator is in computers.

## 2.20 Astable Multivibrator

Fig. 2.44 illustrates a practical astable multivibrator circuit with the transistors dc biased so that both can conduct simultaneously except for *cross-coupling capacitors* which prevent such action and insure alternate conduction of the transistors. A synchronising pulse would *lock* the multivibrator to an external oscillator fundamental or subharmonic as desired.

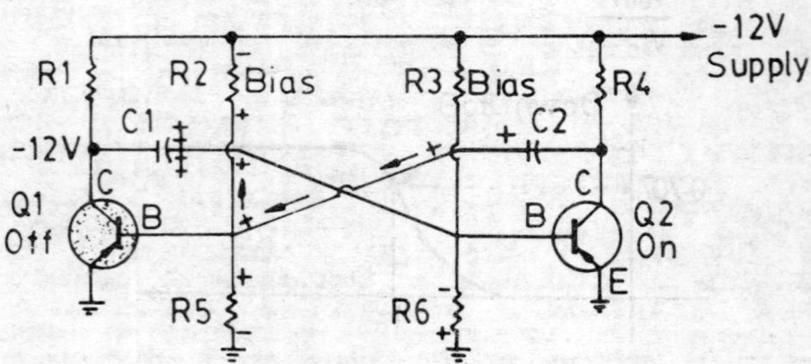


Fig. 2.44 Astable multivibrator

*Once it is turned on, one transistor will conduct more than the other.* Let's start with transistor Q1 in the non-conducting state-the voltage on the collector of transistor $Q_1$ will approach the supply voltage and this minus signal will be coupled to the base of transistor $Q_2$. Capacitor C1 will be charged through the base-to-emitter junction of transistor $Q_2$. This condition is shown in Fig. 2.44.

Capacitor $C_2$ is discharging through resistors $R_2$ and $R_5$. When the capacitor is discharged, the voltage across $R_5$ will be controlled by the bias current in $R_2$ and $R_5$. Transistor $Q_1$ will then turn on due to the negative voltage at the base, the collector voltage will drop toward zero and $C_1$ will start to discharge through resistors $R_3$ and $R_6$. Transistor $Q_2$ will be cut off and transistor $Q_1$ will conduct. *The conducting and cut off states will continue to switch so long as power is applied to the circuit.*

The circuit in Fig. 2.45 uses a unijunction transistor. When power is applied, the transistor is reverse biased and $C_1$ charges through $D_1$ and $R_1$. $C_1$ can charge only to the maximum voltage drop across $R_2$, and $C_1$ voltage must be capable of *firing* the transistor. As this takes place, $D_1$ is cut off and the UJT remains fired for the time constant provided by $(R_2 \times C_1)$. At firing time, the emitter voltage drops near ground potential, permitting reverse bias across the diode (cathode more positive than anode) which produces an *open switch* and isolation. During the diode switch open period, the transistor is still fired and current flows from negative through base 1, the p-n junction, and $R_1$ to positive. Now $C_1$ discharges

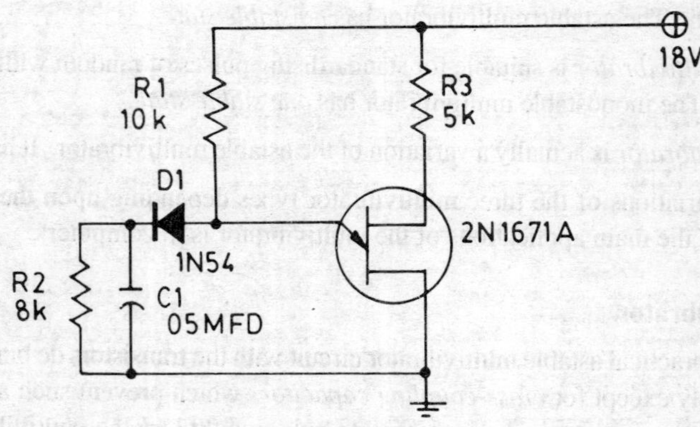


Fig. 2.45 Astable multivibrator using a unijunction transistor

through $R_2$ and $R_2$ voltage drop decreases slowly. As the cathode of $D_1$ becomes more negative than the anode, the diode conducts and *the switch is closed.* The UJT is now reverse biased and $C_1$ recharges once more. The period when the diode switch is open determines the *pulse width* of the generator and is governed by $R_2$-$C_1$ time constant.

### *Stability of the Astable Multivibrator*

As a result of the frequency of the multivibrator being dependent on the resistance-capacitance time constant, it is but natural that any change in these values as a result of temperature variation, will cause the frequency of the circuit to change. In order to operate under variable temperature conditions, it is necessary *that the components used are capable of maintaining precise values under adverse temperature conditions.* Although the overall change as a result of these conditions may be reduced to comparative insignificance, *the $I_{CBO}$ increase with temperature cannot be avoided.*

The timing capacitor discharges through the base biasing resistor and controls the frequency, but in doing so furnishes the leakage current $I_{CBO}$, which varies considerably with temperature. As the discharge time of the capacitor is directly affected, frequency changes, as a result, may be held to a minimum by the use of a large capacitor and a low value base biasing resistor. By selecting a base biasing resistor of much lower value than the base-collector reverse bias resistor, the leakage current will amount to a mere fraction of the discharge current of the capacitor and frequency variations will be slight

In the event of extremely accurate frequency requirements, a multivibrator may be *locked in step by a synchronising pulse*, originating in a crystal controlled oscillator. By applying sync pulses to the off transistor base (as the sum of sync and bias voltages reaches a negative value), the transistor switches on. During the time the transistor is off, the base is slightly positive and the addition of the negative going sync pulses causes the total voltage to swing to the negative polarity, allowing the transistor to conduct and switch on. Any definite arrangement of the sync pulse will be sufficient to maintain reasonable stability in the astable multivibrator. In other words, it is not necessary to pulse each cycle of the square wave or even every other cycle; every fifth, tenth, or other suitable *regular triggering interval* should be enough.

## 2.21 Monostable Multivibrator

By simple triggering action the monostable multivibrator switches to the unstable state and remains there for a prescribed time before returning to the original stable condition. The circuit parallels the flip-flop or stable multivibrator except that *one cross-coupling network permits ac only to be coupled.*

The flip-flop can remain in its unstable state only until the circuit reactive components discharge. Fig. 2.46 illustrates a monostable circuit with capacitor timing. Fig. 2.47 shows the use of inductor timing. *The inductor gives much better pulse width stability at high temperatures.*

The circuit of Fig. 2.48 is a monostable multivibrator in which transistor $Q_1$ is conducting and transistor $Q_2$ is turned off. This type of multivibrator requires a trigger pulse to initiate a change in conduction status after which the circuit will return to the condition shown in Fig. 2.48.

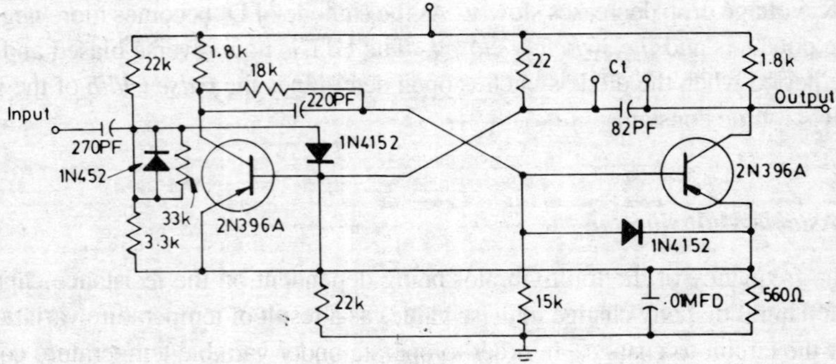


Fig. 2.46 Monostable multivibrator with capacitor timing

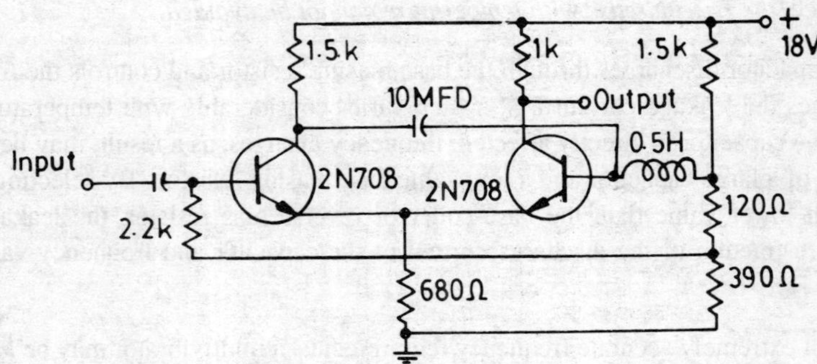


Fig. 2.47 Monostable multivibrator with inductor timing

The bias for the base of transistor $Q_2$ is obtained from the collector of the on transistor. The voltage at the collector of transistor $Q_1$ is nearly zero and not enough current is available to turn $Q_2$ on unless $Q_1$ is turned off.

Rectangular pulses for a predetermined duration, which begin with a triggering pulse, offer many useful applications when a pulse generator with good synchronisation is required. The monostable

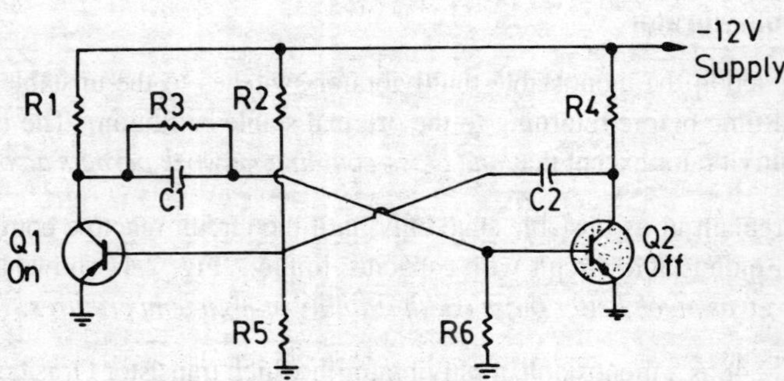


Fig. 2.48 Monostable multivibrator

multivibrator is frequently used to *unblank* the crt of a scope during sweep time, as the sync pulse that starts the sweep also triggers the one shot. This causes an output pulse to be applied to the grid of crt, reducing the cut-off bias and allowing the electron beam to reappear on the screen during horizontal sweep. As the pulse of the monostable is accurately timed to coincide with the sweep time, the beam is cut off at the end of the horizontal sweep during retrace and as the next triggering pulse starts the sweep again it also triggers the multivibrator to again *blank* the crt and thus permits the trace to appear on the screen.

The monostable finds its most important use in *time delay functions* and, compared to delay lines offers the features of small size, lower cost and easier pulse duration adjustment.

*Gating circuit* applications in the computer field offer additional functions for the monostable device when the passing of a pulse within a specific period of time following a timing pulse is a fundamental requirement.

## 2.22 Bistable Multivibrator

In the bistable multivibrator shown in Fig. 2.49, the base biasing resistors are returned to a reverse bias voltage and the capacitors from collector to base are not necessary for switching. However, *the capacitors do speed up switching in the bistable multivibrator and are normally included for that reason.* The resistors and bias values of the bistable are chosen so that the actual application of dc power causes cut-off of one transistor and saturation of the other.

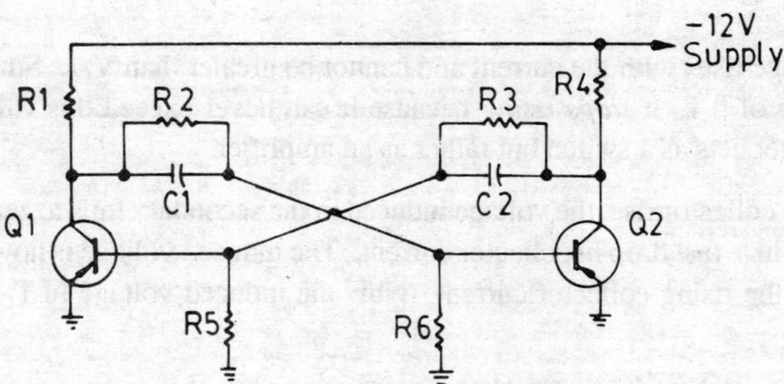


**Fig. 2.49** Bistable multivibrator

In the bistable multivibrator in Fig. 2.49, bias for the base of each transistor is derived from the collector of the *opposite* transistor. Either transistor can remain in the on condition and the other will remain off. This effect is caused by having the bias for both transistor bases supplied from the opposite transistor.

## 2.23 The Blocking Oscillator

The blocking oscillator, although similar to the multivibrator, requires only a *single transistor stage.* The blocking oscillator may be biased for *free running or triggered operation,* but to obtain feedback the blocking oscillator utilises inductive coupling.

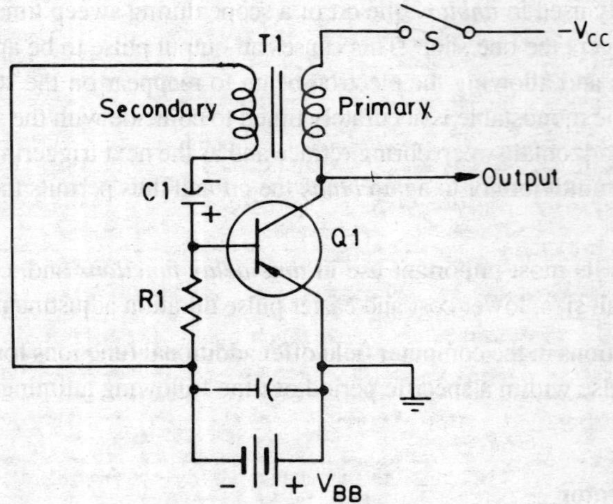


**Fig. 2.50** Astable blocking oscillator

In the astable blocking oscillator in Fig. 2.50, $Q_1$ is forward biased. By closing the switch the transistor is forward biased through $-V_{CC}$ and the primary of $T_1$. Although the collector current is zero momentarily, it must vary quickly when the circuit to $V_{CC}$ is completed. This causes a heavy current flow in the secondary which in turn increases $I_b$ and drives $Q_1$ wide open, simulating a closed switch. As $Q_1$ is such a low resistance, the voltage at the collector is about zero and that across the primary equals $V_{CC}$.

The induced voltage rises with the current and cannot be greater than $V_{CC}$. Since the collector does reach a current value of $\beta I_b$, it *stops rising* because it can never exceed this value. $Q_1$ ceases to be saturated and no longer acts as a switch but rather as an amplifier.

When $\beta I_b$ halts the collector rise, the voltage induced in the secondary falls to zero and drops the base current which results in a fast drop in collector current. The induced voltage is now opposite in polarity to that induced with the rising collector current. Now the induced voltage in $T_1$ secondary collapses completely.

Capacitor C1 becomes charged during the time of base current flow; when it discharges, a voltage drop appears across $R_1$, reverse biasing the base until $C_1$ discharges enough for $V_{BB}$ to again take control and forward bias $Q_1$ to begin another cycle.

As is the case with the monostable multivibrator, the *monostable blocking oscillator* represents a single output pulse for each triggering pulse. Fig. 2.51 shows a one-shot blocking oscillator circuit identical to the astable type except that $Q_1$ is off due to reverse bias.

Applying a triggering pulse to the base of $Q_1$ through capacitor $C_C$ causes the transistor to start conducting. Current flowing in the primary of $T_1$ induces a voltage in the secondary of that transformer, and positive feedback through $C_1$ to the base of $Q_1$ drives the transistor into saturation. The rest of the cycle parallels the astable blocking oscillator but for the fact that $C_1$ discharges only to the positive voltage level established by $V_{BB}$. $Q_1$ remains at cut-off until the next trigger.

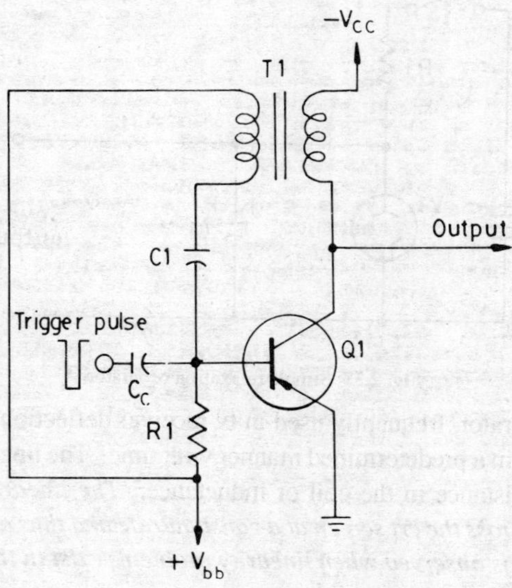


**Fig. 2.51** Monostable blocking oscillator

## 2.24 Time Base Generators

The *linear time base generator* is of major importance since it offers a constant change of voltage or current with respect to time. Fig. 2.52 illustrates the linear time waveshape. The wave shown could represent either a current or voltage change, depending upon the source. Although voltage time base generators are more common, a *current time base generator* finds considerable use in tv magnetic deflection sweep circuits.

The relaxation sawtooth oscillator is one of the simplest generators for producing a sawtooth by merely charging and discharging a capacitor. Fig. 2.53 illustrates a gas tube relaxation oscillator circuit.

As soon as the switch $S_1$ is closed, capacitor $C_1$ starts to charge through resistor $R_1$. Since the gas in $V_1$ is not ionised the tube acts like an open circuit. When the voltage across $V_1$ reaches the *firing level*, the gas *ionises* and presents a very low resistance, causing $C_1$ to discharge quickly through this path. However, $C_1$ discharges only until the *extinguishing voltage* is reached, at which point the gas *deionises* and once again appears to be an open circuit. Now C1 starts charging again and oscillation continues producing fairly linear sawtooth output.

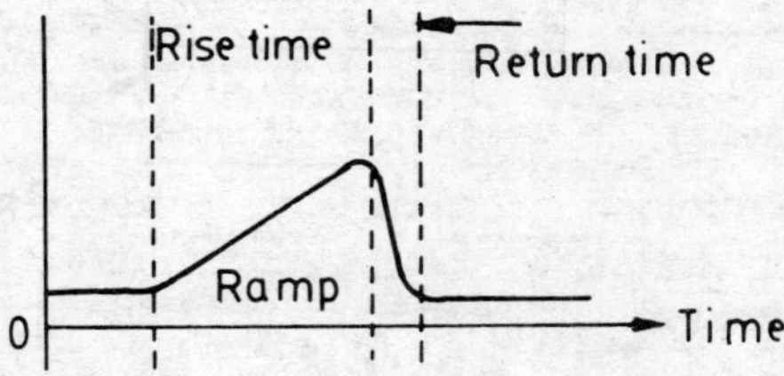


**Fig. 2.52** Linear time base

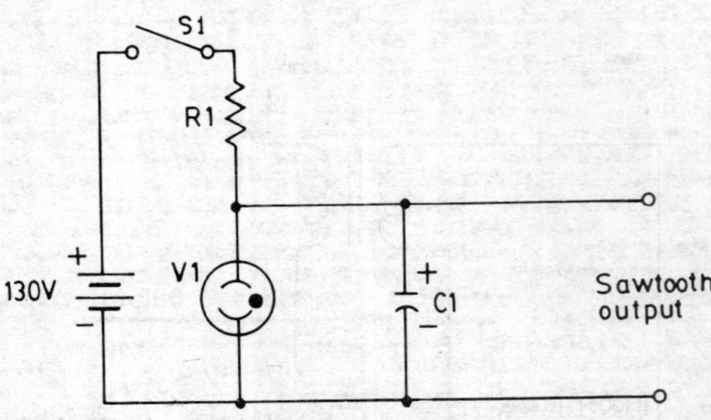


**Fig. 2.53** Simple relaxation oscillator

A current time base generator, frequently used in tv receiver deflection circuits, generates a current through the yoke that varies in a predetermined manner with time. The linearly increasing current flows as long as there is little resistance in the coil or inductance. *The linearly increasing magnetic field sweeps the electron beam across the crt screen at a constant rate and thus avoids stretching or shrinking of portions of the picture as is observed when linearity problems exist in these circuits.*

The Darlington circuit shown in Fig. 2.54 eliminates difficulties encountered in less sophisticated circuitry. In order to avoid problems involved in simple switching, special circuits are used to force the current to flow through the yoke at a predetermined rate. By virtue of its high impedance, which does not distort the input ramp, the Darlington circuit is merely *two cascaded emitter followers* having an extremely high input impedance equivalent to the product of betas of $Q_1$ and $Q_2$ times $R_e$. *The input pulse is trapezoidal, consisting of a step and a ramp.* The *step* does reduce the effect of the coil resistance and the current through $Q_2$ is controlled as desired by the input wave. The current through $Q_2$

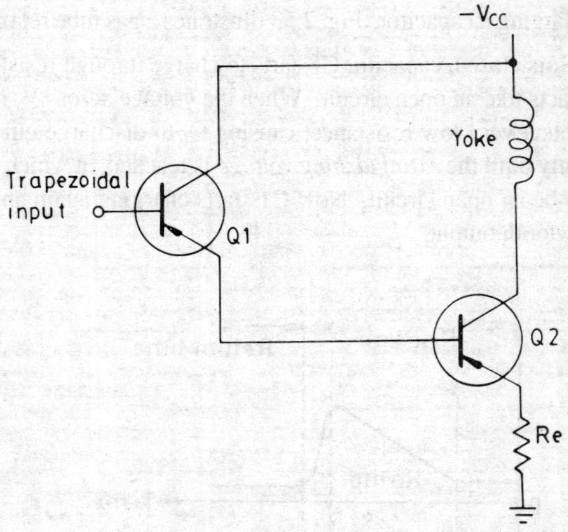


**Fig. 2.54** Basic Darlington circuit

controls the current through the yoke, and the rate of change of the input voltage *ramp* directly regulates the rate of change of current through the deflection coil.

## 2.25 Phase Shift Oscillators

The phase shift oscillator, Fig. 2.55, is simple to construct, stable in operation and produces a good sine wave output.

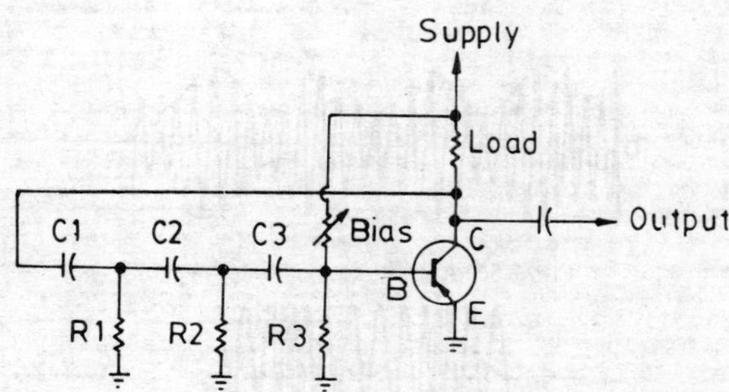


**Fig. 2.55** Basic phase shift network

The phase shift oscillator consists of an amplifier and a *phase shift network*. The amplifier must be capable of overcoming the losses in the network. This requires that the gain of the transistor be approximately 30 times or better depending on the network used.

The network composed of $C_1$, $C_2$, $C_3$ and $R_1$, $R_2$ and $R_3$ provides a phase shift of 180° in order to produce positive feedback. In this case each of the three capacitor/resistor combinations supplies 60° phase shift for a full 180° total.

The *requirements of an oscillator* include, amplifier with feedback from the output to the input circuit in proper phase to overcome the circuit losses and sustain oscillations.

The manner in which feedback is accomplished is unimportant, as long as it is in *proper phase* and of *sufficient amplitude* to overcome the energy losses in the circuit. *The feedback path simply distinguishes one type of oscillator from another.*

The basic phase shift oscillator circuit shown in Fig. 2.55 is probably one of the simplest circuits for producing a sine wave oscillation.

## 2.26 Modulation

A radio frequency wave may be *modulated* with intelligence in accordance with the intelligence to be transmitted by varying either the *amplitude* or *frequency* of RF oscillations.

Modern radio transmitters radiate waves which may be of two general types. One is the *continuous wave* (cw) or unmodulated radio wave, Fig. 2.56 - *the amplitudes of all the cycles are equal and the frequency is constant*. There is nothing to distinguish one cycle from the next.

The other type of radio wave is the modulated wave. In one type of modulated wave Fig. 2.57, the amplitudes of the peaks may vary from one cycle to the next. This is called the *amplitude modulated*

**Fig. 2.56** Continuous (unmodulated) wave

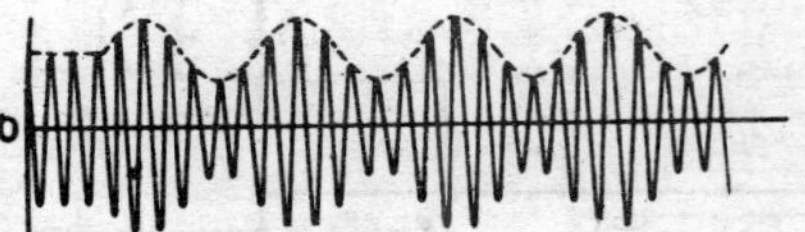


**Fig. 2.57** Amplitude modulated wave

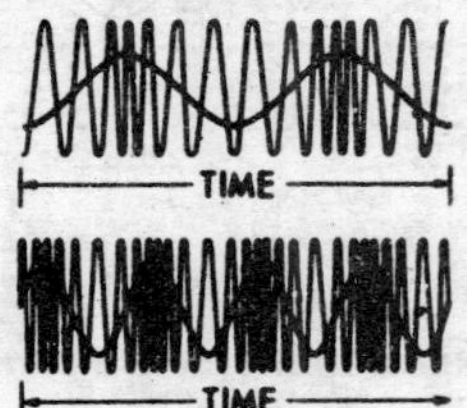


**Fig. 2.58** Frequency modulated wave

*wave*. In another type of modulated wave, Fig. 2.58 - the amplitudes of the cycles remain constant but the frequency is constantly changing. This is called the *frequency modulated wave*.

Yet another type of modulation, Fig. 2.59, is called the *pulse position modulation*. It is the system where encoding is carried out by positioning the pulses. In the PPM method of signal encoding used in the television remote control, the transmitted signal takes the form of a binary coded word formed from succession of short bursts of ultrasound. Ultrasound frequencies are not included in the code.

The code is determined by time interval, or space, between successive ultrasound bursts. The principle is illustrated with the help of Fig. 2.59. Here the frequency of ultrasound is 40 kHz.

In Fig. 2.59, a regular train of six ultrasound pulses, each of 3 ms duration, is repeated at 27 ms interval. The separation rate is, therefore, 1000/27 or 37 Hz and the separation time from the trailing

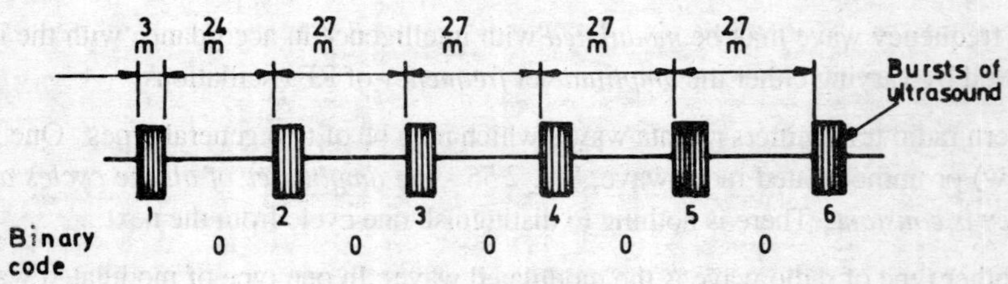


**Fig. 2.59** Pulse position modulation encoding

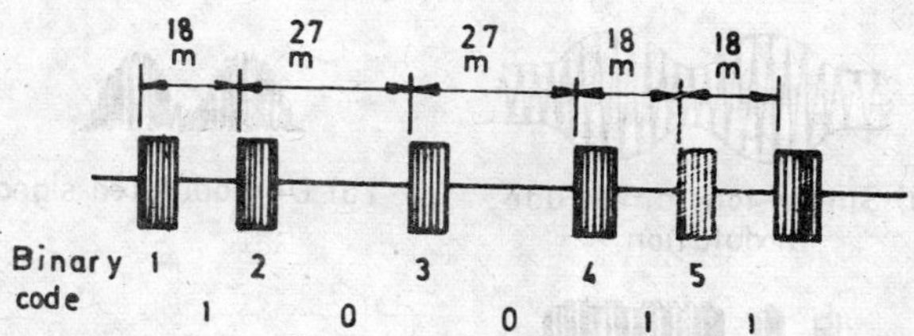


**Fig. 2.60** Five bit binary coded word identified by notation 10011

edge to the leading edge is 24 ms. Suppose the pulse separation interval of less than 20 ms is recognised as *binary 1* and of more than 20 ms as *binary 0*. Six equally spaced ultra sound bursts, with in between spacing more than 20 ms, represent five bit binary code 0 and, therefore, identified by notation 00000.

In Fig. 2.60, five bit binary coded unit is identified by notation 10011, since the first, fourth and fifth pulse spacings are *less than 20 ms*, and the second and third spacings are *more than 20 ms*.

## 2.27 Amplitude Modulation

In the process of modulation, the initial frequencies must be impressed on a *nonlinear element*, otherwise they will not combine to form the desired composite waveform which, in this case, is the modulation pattern. *Once the modulation pattern is obtained, a non linear-element is again necessary to recover all of the original signals.*

The distortion of a sine wave by a nonlinear element and the modulation of a sine wave with the aid of a nonlinear element are mathematically different. The combining of harmonics to form a non sinusoid is an *additive* process. The modulation of a sine wave by another is, in essence, the *multiplication* product of the two.

Modulation patterns are *composite waveforms* in that they are composed of two or more frequencies. However, not all composite waveforms can be said to be modulation patterns. For example, pulses and non sinusoidal waves composed of harmonically related sine waves do not qualify as modulation patterns.

Fig. 2.61 illustrates some examples of modulation patterns. In Fig. 2.62 some composite waveforms which are not modulation patterns are depicted.

For example, suppose that a 1 MHz carrier is amplitude modulated by a 10 kHz sine wave. The modulation pattern will appear as shown in Fig. 2.63. The modulation spectrum will appear as shown in Fig. 2.64, where the *upper side band frequency* will be 1 MHz plus 10 KHz, or 1010,000 Hz. The *lower side band frequency* will be 1 MHz minus 10 kHz or 990,000 Hz.

Two facts are noteworthy; first, *the modulation pattern does not reveal the presence of the side band frequencies;* second, *the modulation frequency which is seen in the modulation pattern is not actually present in the modulation spectrum.*

The intelligence to be transmitted usually consists of an *audio modulating signal*, with frequencies in the range of 20 to 20,000 Hz, or a *video modulating signal*, with frequencies from dc to several megahertz. The modulating signal is superimposed on the radio frequency *carrier* by applying both to

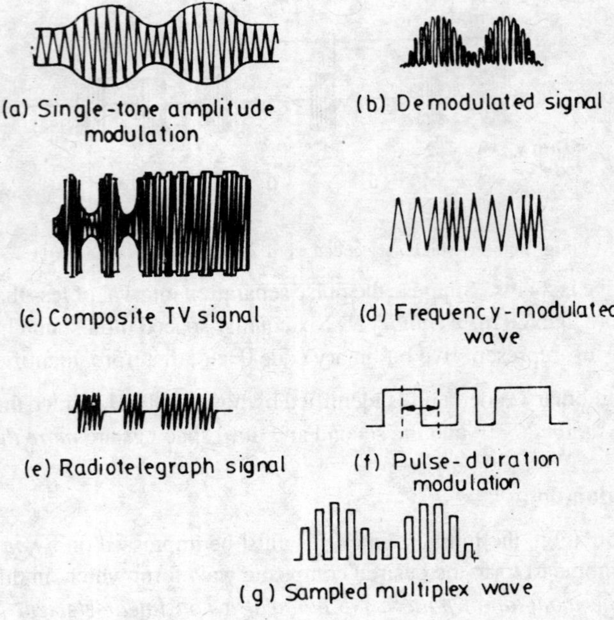


**Fig. 2.61** Typical modulation patterns

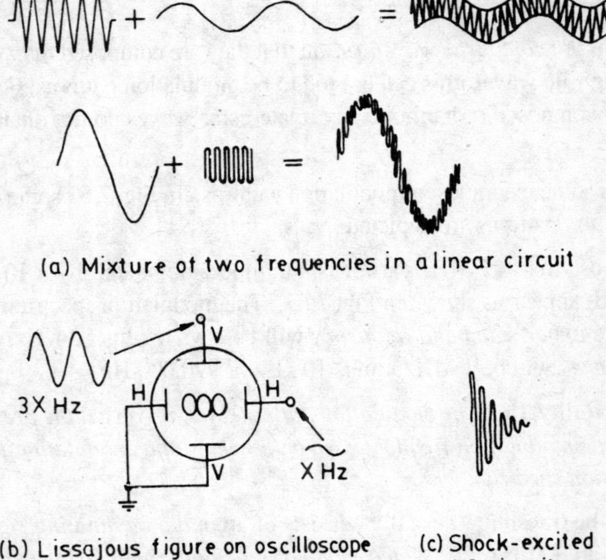


**Fig. 2.62** Composite waveforms which are not modulation patterns

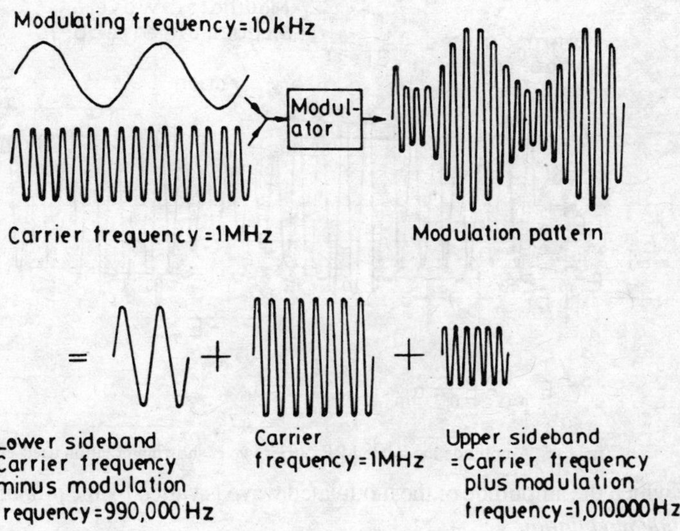


**Fig. 2.63** The constituent frequencies of the modulation pattern

a *nonlinear impedance* (e.g. the modulator). When this is done, the modulating and carrier frequencies beat together or *heterodyne* to produce additional frequencies equal to the *sum* and the *difference* of the applied frequencies. The resulting modulated wave, therefore, consists of the original *carrier frequency*, various *sum frequencies* which are collectively known as the *upper side band*, and various *difference frequencies* which are collectively known as the *lower side band*. The total *band width* of the modulated wave is twice the value of the highest modulating frequency. This band width must be passed by all transmitting and receiving circuits.

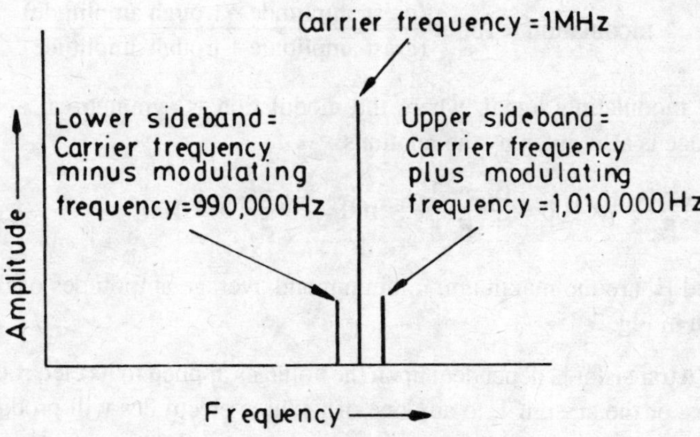


**Fig. 2.64** Modulation spectrum due to modulation of a 1 MHz carrier by a single 10 kHz signal

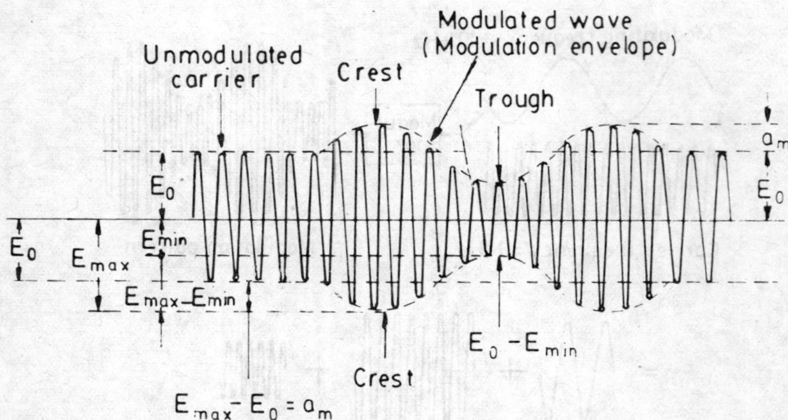


**Fig. 2.65** Amplitude modulated RF carrier wave, showing notation used

The extent by which the amplitude of the modulated wave is varied by the modulating signal is called *degree of amplitude modulation.*

$$m_a = \frac{a_m}{E_0} \; (0 \le m_a \le 1) \tag{2.26}$$

where $a_m$ is the maximum amplitude of the modulating signal, Fig. 2.65, and $E_0$ is the amplitude of the unmodulated carrier, equal to the average amplitude of the modulated wave or modulated envelope.

The degree of modulation is expressed in percent, as follows:

$$\%\ \text{modulation} = 100\ m_a = 100\ \frac{a_m}{E_0} \tag{2.27}$$

For any kind of modulating signal, sinusoidal or other

$$\%\ \text{modulation} = 100 \times \frac{(\text{crest amplitude} - \text{trough amplitude})}{(\text{crest amplitude} + \text{trough amplitude})} \tag{2.28}$$

For a sinusoidal modulating signal, where the modulation is symmetrical about the carrier, the modulation percentage is also given by the relations:

$$\%\ \text{modulation} = \frac{E_{max} - E_{min}}{2E_0} \times 100 = \frac{E_{max} - E_0}{E_0} \times 100 = \frac{E_0 - E_{min}}{E_0} \times 100 \tag{2.29}$$

where $E_{max}$, $E_{min}$ and $E_0$ are the maximum, minimum and average amplitudes of the modulated wave (envelope) as shown in Fig. 2.65.

Since the gain of a transistor is dependent upon the voltages applied to its electrodes, it is evident that changing the dc bias, or the ac signal, to any one of its three electrodes will produce a corresponding change in gain. Therefore, it is possible to produce amplitude modulation of a transistor by any one of the three basic methods; base injection, emitter injection or collector injection.

## 2.28 Balanced Amplitude Modulator

Two thirds of the average radiated power of a fully modulated AM transmission is contained in the RF carrier. Since all of the intelligence contained in the modulating signal is transposed to the upper and lower side bands, which are identical to each other, *the carrier, and one of the side bands, may be eliminated from the output without changing the remaining side bands.* This allows the available transmitter power to be utilised to a much greater advantage in a single side band transmission. This is generally known as *suppressed carrier single side band transmission. In SSB transmitters, the carrier is usually suppressed, or eliminated, by the use of a balanced modulator.*

The basic principle of a balanced modulator is to introduce the RF carrier to the modulator in such a way that it does not appear in the output. *There is only an output signal when both the audio modulation and the RF carrier are present simultaneously at the modulator input.* This output signal consists of only the upper and lower side band frequencies generated in the balanced modulator by the mixing of these two input signals across the nonlinear resistance of the transistor. The original modulation and the RF carrier inputs are suppressed because of the operational characteristics of the circuit.

The basic balanced modulator produces amplitude modulated upper and lower side bands and suppresses the RF carrier. This is achieved by coupling the RF carrier, in phase, to the bases of the two transistors whose output is connected in push-pull. The RF carrier is kept 8 to 10 times as large as the modulating voltage to keep distortion to a minimum. In push-pull amplifiers, an input signal must be out of phase to produce an output since any in phase inputs cancel in the output. The modulating signal is applied to the base of each transistor in push-pull through a centre tapped transformer; see, Fig. 2.66.

When both RF carrier and audio modulating signals are applied simultaneously to the bases of the balanced modulators, sum and difference frequencies are produced by the modulating frequencies *beating* against the carrier, since any AM process is essentially the same as *heterodyning*. As in a frequency converter, any modulation which exists on one of the mixing frequencies is *linearly transposed* to the resultant sum and difference frequencies.

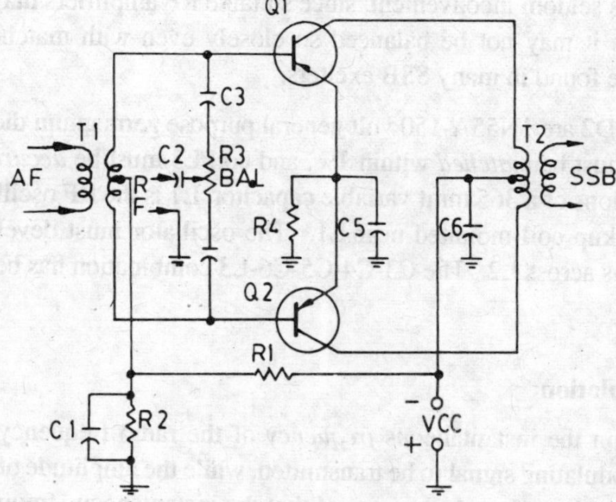


**Fig. 2.66** Balanced modulator

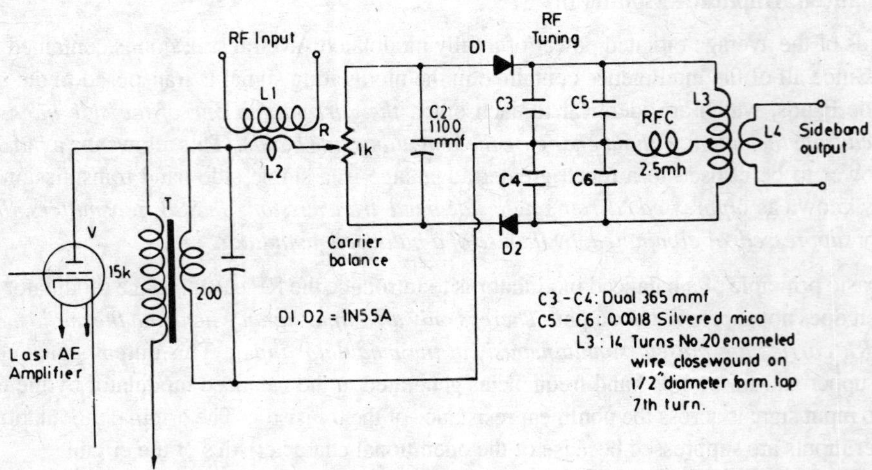


**Fig. 2.67** Two-diode balanced modulator

The collector circuits contain the upper and lower side bands which are the sum and difference frequencies, the RF carrier, and the modulation. *The carrier is cancelled out by push-pull action in the output transformer* and the output transformer also presents a low impedance to the audio modulating signal. Therefore, the original modulating signal also is not developed in the output. The generated side bands are out of phase with each other at the collectors of the transistors since the modulating signal is out of phase at the bases. These out of phase signals add in the output transformer rather than cancel as in phase signals do, and they are inductively coupled to the following stages through the output transformer. The circuit diagram of a typical balanced modulator is shown in Fig. 2.66.

Balanced modulators may employ transistors as in Fig. 2.66 or diodes, as in Fig. 2.67. The diode circuit is simple, compact and *requires no power supply*. It has the disadvantage that it must operate at *low power level*. This is seldom inconvenient, since suitable RF amplifiers may be provided following the modulator although it may not be balanced so closely even with matched diodes. Diode type balanced modulators are found in many SSB exciters.

In Fig. 2.67, D1 and D2 are 1N55A-150 volt general purpose germanium diodes. For good balance, capacitors C5 and C6 must be *matched* within 1%, and coil L3 must be *accurately centre tapped*. C3 and C4 are the two sections of a 365 mmf variable capacitor, L1 is the RF oscillator tank coil, and L2 is a two or three turn pickup coil mounted near L1. The oscillator must develop an RF signal of approximately 2.5 volt rms across L2. The C3-C4-C5-C6-L3 combination has been chosen for operation in the 75 meterband.

## 2.29 Frequency Modulation

In frequency modulation the instantaneous *frequency* of the radio frequency oscillation is varied in accordance with the modulating signal to be transmitted, while the amplitude of the oscillations is maintained constant. The number of times per second that the instantaneous frequency is varied about the average or carrier frequency is called the modulating frequency, while the amount of frequency varia-

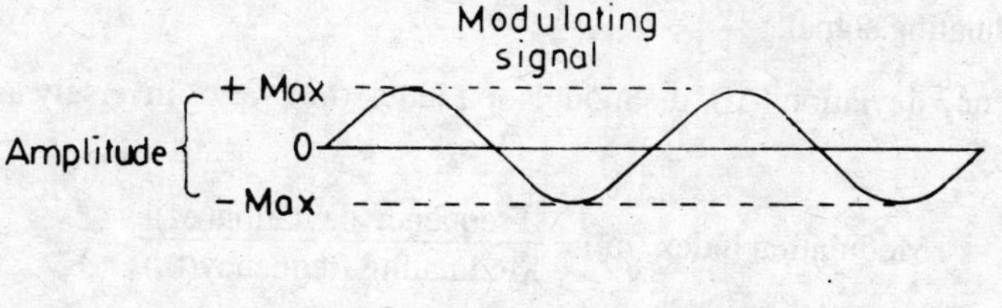


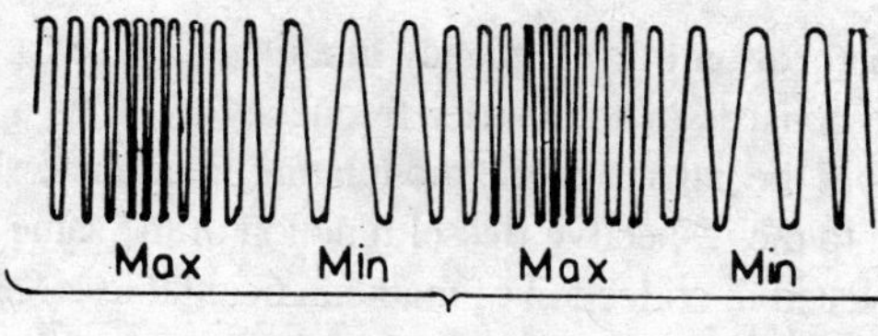


**Fig. 2.68** Frequency modulation

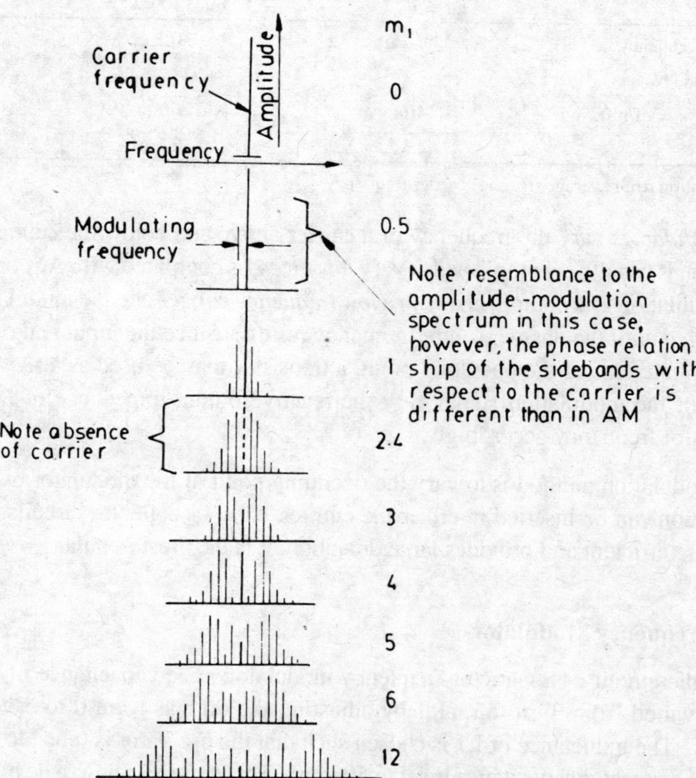


**Fig. 2.69** Examples of modulation spectra produced by frequency or phase modulation

tion from the average is called the *frequency deviation*. The frequency deviation is proportional to the amplitude of the modulating signal.

For a given frequency deviation ($\Delta f$), the modulation index ($m_f$) varies inversely as the modulating frequency ($f_m$).

$$\text{Modulation index } (m_f) = \frac{\text{Frequency deviation}(\Delta f)}{\text{Modulating frequency}(f_m)} \tag{2.30}$$

By expanding the equation of a frequency modulated wave in terms of Bessel functions it can be shown that the wave consists of the carrier frequency ($f_c$) plus a series of *side bands of decreasing amplitude* spaced about the carrier by the modulating frequency. The amplitude of each of these side bands is proportional to the respective Bessel function of the same order, and to the modulating index. Bessel functions are listed in collection of mathematical tables. *Side bands in the outer regions whose amplitudes are less than 2% of the peak amplitude may be neglected.*

**Table 2.2**

FM side bands and required bandwidth

| | Modulation index, $m_f = \Delta f/f_m$ | | |
|---|---|---|---|
| | mf = 5 | mf = 10 | mf = 20 |
| Modulation (signal) frequency $f_m$ | 0.2 $\Delta f$ | 0.1 $\Delta f$ | 0.05 $\Delta f$ |
| Number of pairs of side bands | 7 | 13 | 23 |
| Bandwidth required ($f_m = \Delta f/m_f$) | 14fm | 26fm | 46fm |
| | 2.3 $\Delta f$ | 2.6 $\Delta f$ | 2.3 $\Delta f$ |

$\Delta f$ = Frequency deviation from average (resting) carrier frequency

*Frequency modulators* vary the frequency of a carrier above and below the centre frequency. Frequency modulation is usually accomplished at very *low levels* as compared with AM operation. In many instances, the modulation is accomplished with a *low frequency carrier* and the modulated carrier is then *frequency multiplied* until the desired centre frequency results. Since the input and output impedances of a transistor are functions of the operating point, a transistor may be used as a *reactance modulator*. Voice variations of the modulation signal vary the reactive output impedance of the modulator and change the oscillator frequency accordingly.

Yet another modulation method is to vary the operating point of the modulator by varying the bias. Then the modulation can be inserted in either the emitter, base, or collector circuits. Since *reactance modulation* is more efficient and provides large deviation, it is the most popular.

## 2.30 Varactor Frequency Modulator

Fig. 2.70 shows the circuit of a varactor frequency modulator. The capacitance of varactor C3, V33 varicap, may be varied from 39 to 85 mmf by adjusting the dc bias form 0 to 3 volts, by means of potentiometer R2. The inductance of L1 is chosen such that the oscillator is tuned to the desired centre frequency when the varactor capacitance is set to 50 mmf. This corresponds to a dc bias of 1.5 volts and places the operating point along the *steepest portion* of the varactor voltage capacitance curve. The tank circuit is isolated from the transistor collector by capacitor C1, which blocks the dc voltage coming from

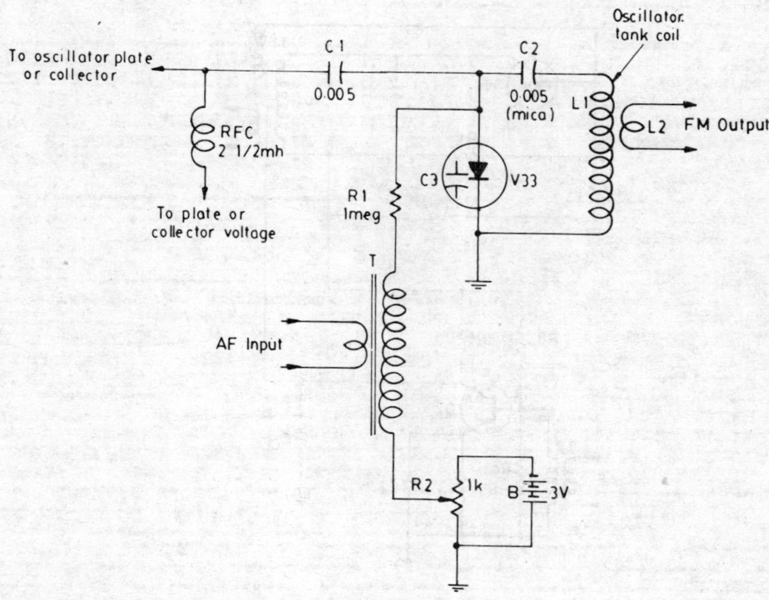


**Fig. 2.70** Varactor frequency modulator

the transistor. Similarly, a second blocking capacitor C2 prevents L1 from short circuiting the varactor dc bias. L2 is a one or two turn link-coupled output coil.

Audio voltage is applied to the circuit through transformer $T_1$, which has a 2:1 or 3:1 step-up turns ratio. The maximum voltage developed across the secondary of T should be between 2 and 3 volts rms. This voltage, being in series with the dc bias, swings the latter above and below the 1.5 volt level at which the tank is tuned to the desired centre frequency. This *detunes* the oscillator above and below the centre frequency at the AF rate, thus producing FM output. The FM swing is proportional to the AF amplitude and may be set at a desired width by means of the audio-gain control. AF may be obtained from a transistorised speech amplifier or directly from a carbon microphone and battery. *Because of the extremely high resistance of the varactor, virtually no AF power is required.*

## 2.31 Reactance Modulator

In the reactance modulator shown in Fig. 2.71, transistor Q1 is the FM oscillator. Resistors R1 and R2 are base bias voltage dividers. The modulator's load resistor is R6. Primary L1 of transformer T1 together with C1 forms the tuned tank adjusted to the desired oscillator output frequency. Secondary L2 of T1 is the collector-to-base feedback winding of the oscillator. Tertiary winding L3 is an inductively coupled output winding which couples the FM output signal to the next stage.

Transformer T2 is an audio transformer which inductively couples the modulating signal to the base-emitter junction of Q2, the modulator stage. Resistors R3 and R4 are base bias voltage dividers, while R5 is the emitter decoupling resistor bypassed by C2. The output capacitance of Q2 (shown dotted as $C_{CE}$) shunts the lower portion of RF oscillator coil L1. As the modulator operates, the output capacitance

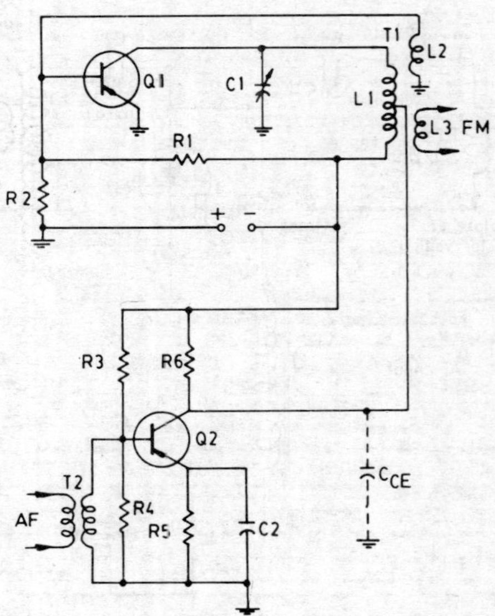


**Fig. 2.71** Reactance modulator

of Q2 is varied by the AF modulation, causing the frequency of the oscillator to be shifted in accordance with the modulation, just as if C1 were varied.

When the modulation is applied to the primary of T2, it is coupled into the Q2 base circuit, causing the emitter-base bias to change constantly at the modulating rate. Since the bias is increasing and decreasing at modulating rate, the collector voltage of Q2 also increases and decreases accordingly. When the collector voltage decreases, $C_C$ increases. An increase in voltage has the effect of *spreading the capacitor plates (depletion region) further apart* by increasing the width of the pn barrier. Conversely, the reduction of collector voltage reduces the width of the pn junction and has the same effect as *pushing the capacitor plates together* to provide more capacitance.

This ultimiately results in the oscillator frequency increasing and decreasing at the modulation rate and the output is a frequency modulated carrier.

Since the audio modulation causes the collector voltage to increase and decrease, there is also an AM component induced in the output. This produces both an FM and AM output. By introducing a *limiter* stage after the reactance modulator, the amplitude variations are removed and only the frequency modulation remains, with a constant amplitude output. Frequency multipliers are then used to increase the oscillator frequency to the desired output frequency.

For high power, *nonlinear* classC RF amplifiers may be used to increase the signal strength. This contrasts with the AM requirement for *linear* RF amplification once modulation is applied. With the initial modulation occurring at *low level*, FM represents a *saving in power* as compared with conventional AM, with the FM *noise-reducing properties* providing a better signal-to-noise ratio than is possible with AM.

## 2.32 Frequency Changers

A frequency changer is any device, or circuit, which changes frequency. Such a definition is quite simple, however, it is deceptive. Defined in this way, frequency changers encompass practically the whole spectrum of electronic techniques.

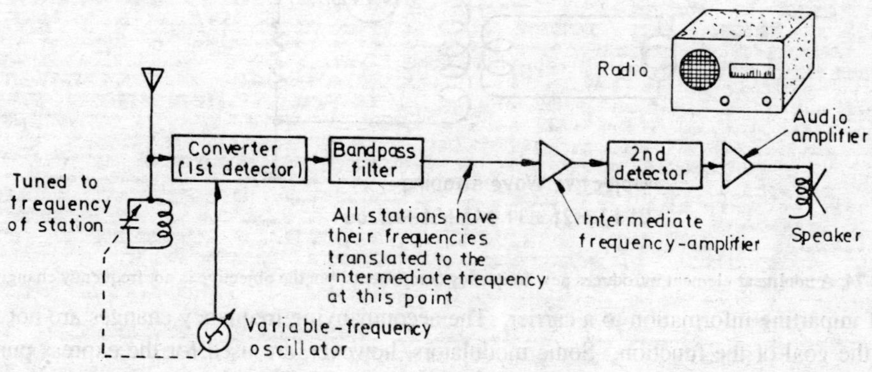


**Fig. 2.72** Frequency changing is an important process in the superheterodyne radio receiver

A few examples will illustrate this situation. First consider the phenomenon of oscillation. Are oscillators frequency changers? Apparently, for oscillators change dc (zero frequency) to some other frequency. Rectifiers perform the converse function. They, likewise, seem to be frequency changers as denoted by the simple definition.

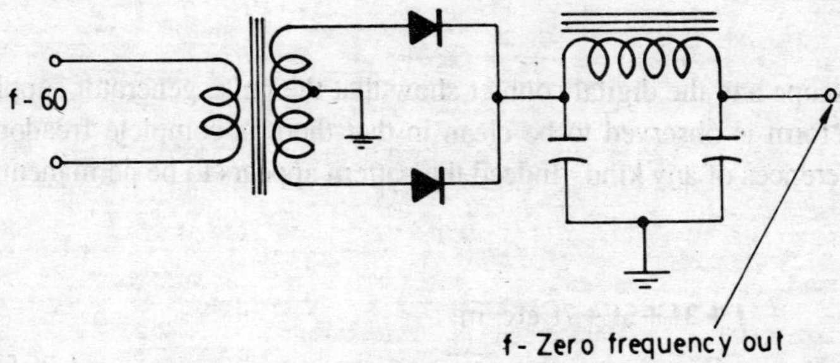


**Fig. 2.73** In rectification, the objective is not frequency changing

*The process of modulation always introduces new frequency components.* The same is the case with nonlinear elements. It appears that the easiest answer to this question is provided by yet another question, what is not a frequency changer?

Most electronic techniques involve, or are accompanied by, frequency changes, but the prior purpose is something other than frequency change. For example, modulation is generally employed for the

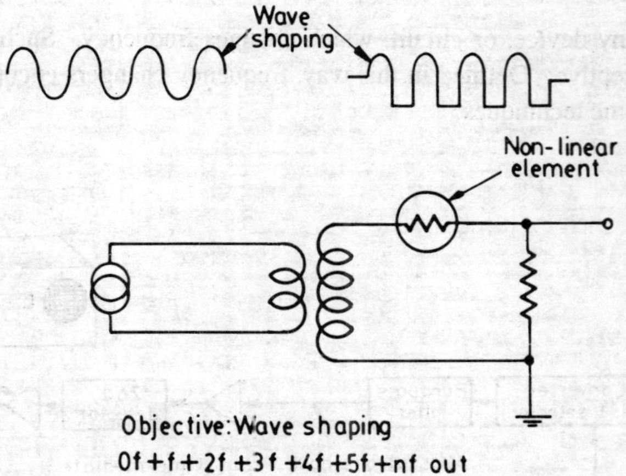


**Fig. 2.74** A nonlinear element introduces new frequency components, but the objective is not frequency changing

purpose of imparting information to a carrier. The accompanying frequency changes are not initially sought as the goal of the function. Some modulators, however, are used for the express purpose of bringing about a change in frequency. This duality furnishes the clue; *those functions, no matter how accomplished, which are intended to change frequency as an end result in itself clearly operate as frequency changers.*

A frequency changer is a device, or circuit, which serves the primary purpose of changing frequency. Typical frequency changing techniques are shown in Figs. 2.76 and 2.77.

In electrical parlance, a periodic occurrence of voltage, or current, is not necessarily described completely by saying that it has a certain frequency. To illustrate this, suppose an apparatus is connected as shown in Fig. 2.78.

Both the oscilloscope and the digital counter show that the pulse generator supplies a frequency of 100 Hz. The waveform is observed to be clean in that there is complete freedom from overshoot, transient, and interferences of any kind. Indeed the pattern appears to be permanently engraved on the

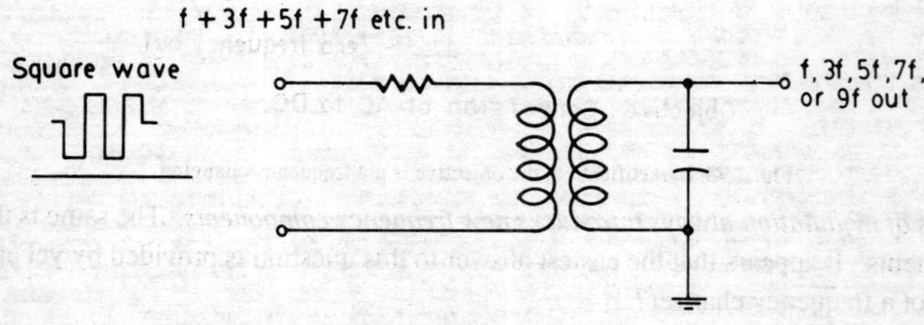


**Fig. 2.75** In frequency selective circuits the objective is not frequency changing

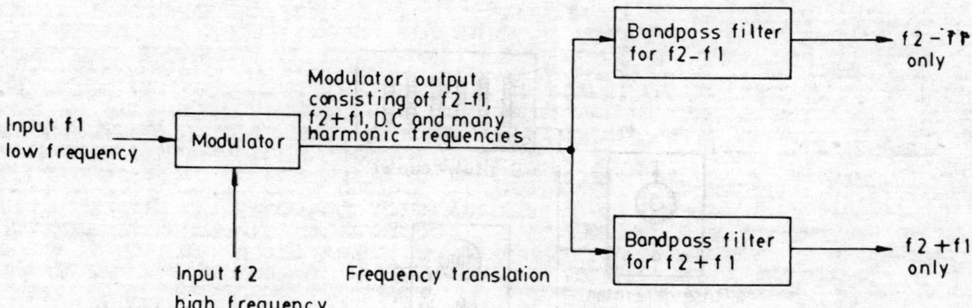


**Fig. 2.76** Heterodyne frequency changer

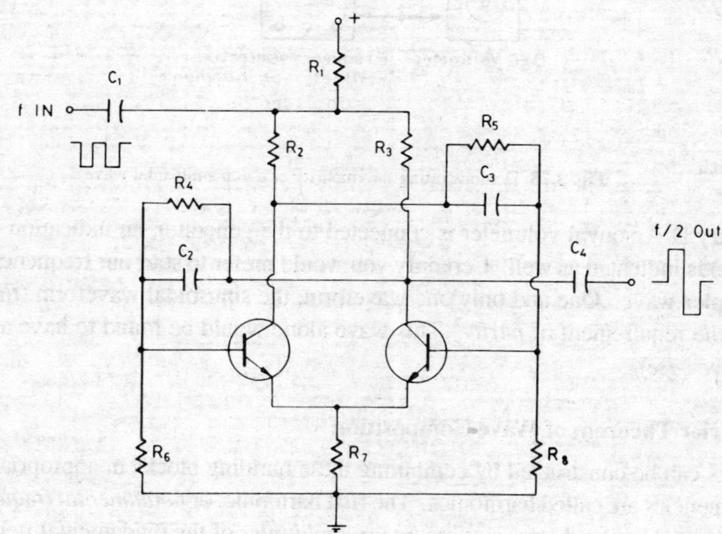


**Fig. 2.77** Binary frequency divider

scope screen. Yet this waveform, despite its *symmetry and stability* is far from a phenomenon which can truly be described by its frequency alone.

To demonstrate the fact, connect a frequency selective voltmeter, or *harmonic analyser*, to the generator. You will find that neither the digital counter nor the oscilloscope tells the whole truth. Again an indication of 100 Hz frequency is indicated, but in addition to this measurement, frequency components at *integral multiples* of 100 Hz (that is 200 Hz, 300 Hz, 400 Hz, etc) are shown in diminishing strength upto the highest multiple to which this instrument will tune. These frequencies are completely cancelled in the pattern of the square wave. Identifying this wave by its 100 Hz frequency merely provides a first name acquaintance. There is much more to know about it.

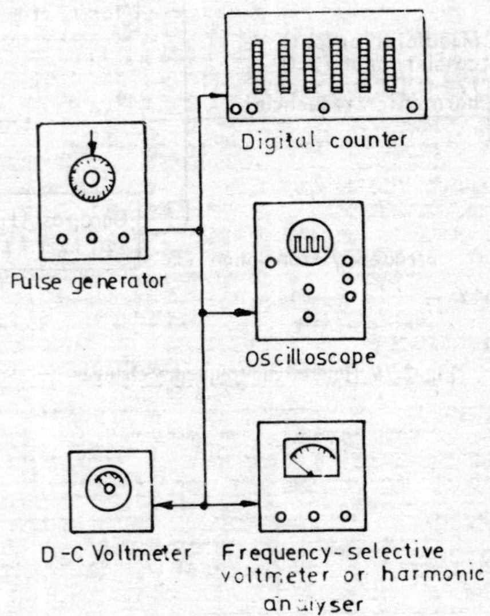


**Fig. 2.78** Demonstrating the impurity of a non sinusoidal wave

If an ordinary D'Arsonval voltmeter is connected to the generator, an indication of zero-frequency component (dc) is indicated as well. Certainly you would prefer to start our frequency changing operations on a simpler wave. One and only one waveform, the sinusoidal waveform (*the building block*), would satisfy the requirement of *purity*. This wave alone would be found to have a frequency of 100 Hz (*in the above case*).

## 2.33 The Fourier Theorem of Wave Composition

All waveshapes can be constructed by combining these building blocks in appropriate ways. All such sine wave frequencies are called harmonics. The first harmonic, or *fundamental frequency*, is the lowest of the frequencies. All other harmonics are *integral multiples* of the fundamental frequency. In electrical parlance we say that they are *harmonically related*. The harmonic frequencies can be further classified as *odd harmonics* and *even harmonics*. This definition establishes the fundamental frequency as the first harmonic. Zero frequency is the mathematical representation of dc. A dc component may, or may not, be present in a waveform. It is not, however, spoken of as a harmonic.

Jean Baptist Joseph Fourier, a French physicist, established the mathematical law which shows exactly how various harmonically related sine waves can be combined to form any waveform whatsoever.

The significant aspect of this law is that *waveforms differ by virtue of the amplitude distinction of their harmonics and the phase relationship these harmonics bear to one another*. Fig. 2.80 and 2.81 illustrate some aspects of the composition of a sawtooth wave and a square wave.

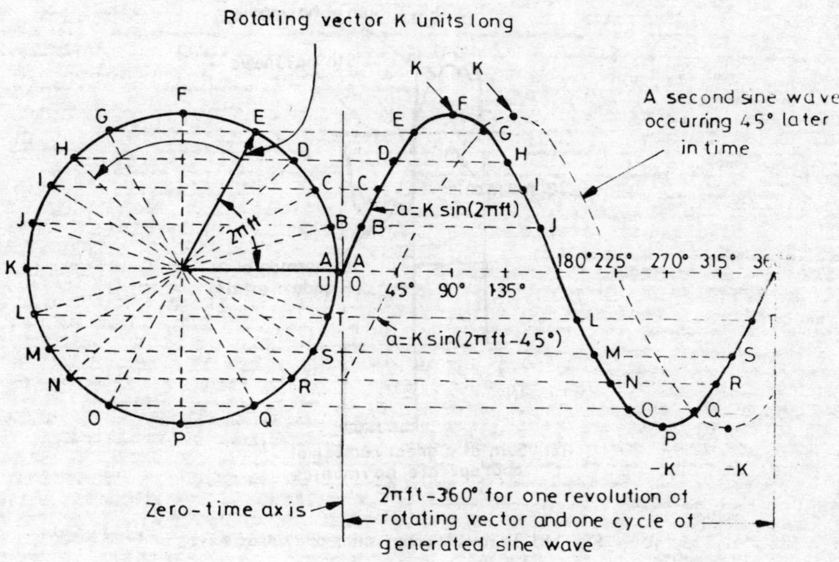


**Fig. 2.79** Evolution of a sine wave, the building block of all composite waves

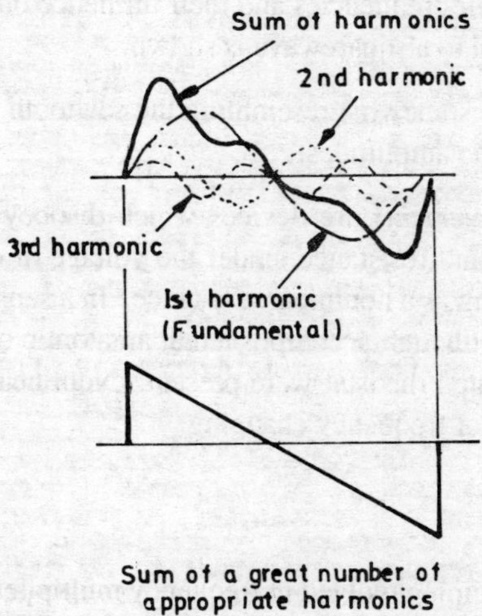


**Fig. 2.80** Harmonic composition of a sawtooth wave

Fourier theorem is generally expressed as the sum of both sine and cosine waves. If a picture of a wave having a shape dictated by the sinusoidal function is observed, it would be correct to call it either

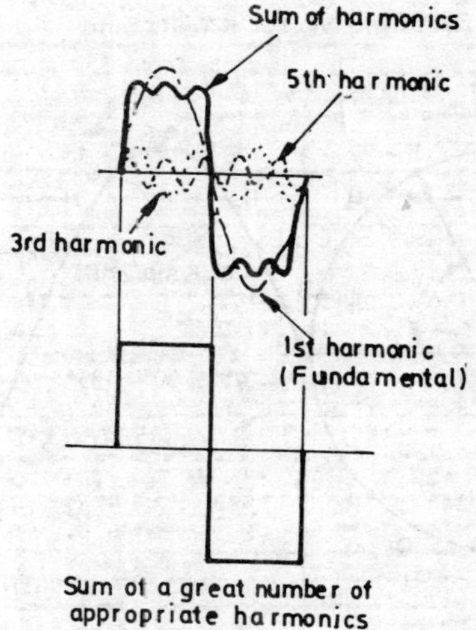


**Fig. 2.81** Harmonic composition of a square wave

a sine wave or a cosine wave. However, it follows that *the generally accepted practice of simply calling such time functions sine waves is valid.*

The other factors governing waveshape are the *number and the amplitude of the various harmonics*. Waves can have any combination of harmonics and can be devoid of certain harmonics.

The relative phasing of the harmonic frequencies and their influence on the resultant pulse shape is shown in Fig. 2.84. An approximation to a squarewave is shown.

Fig. 2.85 illustrates a peaked wave somewhat resembling the sawtooth wave. Both of these waves contain the same harmonics at the same amplitudes.

In technical literature, *nonlinear elements* are devices which disobey Ohm's law. Any resistive device which does not maintain constant resistance, under the voltage or current excursions it is subjected to, must be said to be functioning as a nonlinear resistance. In a sense, such a device displays an active rather than a passive nature. Although accomplished at an atomic or molecular level, the result is much the same as if a motor-activated rheostat were present. Nonlinear resistances are playing an ever increasing role in the techniques of frequency changing.

## 2.34 Frequency Multipliers

Harmonic generation is the basic principle involved in frequency multiplies. For this reason frequency multipliers provide new frequencies which are *integral multipliers* of an impressed frequency. Means, other than frequency multiplication, must be employed in order to achieve multiplying factors which are *not integral multiples*. For example, if it is desired to produce a frequency one and a half times that of a given signal, this would be accomplished by first multiplying by three (producing and extracting the third harmonic), then dividing the frequency by two.

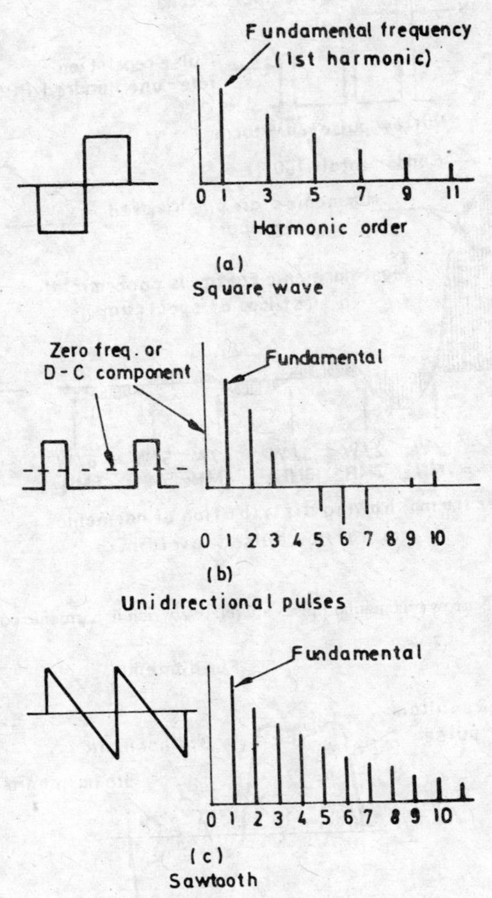


**Fig. 2.82** Typical waveforms and harmonic spectra

*The association of a nonlinear element with frequency selective circuits provides the basis for an extensive variety of simple frequency multipliers.* The nonlinear element can be a semiconductor diode, zener diode, tunnel diode, neon bulb, or thyrite, or thermistor element.

Any device which, over a cyclic excursion of the applied sine wave, changes its resistance or provides a switching action qualifies. Non-linear reactances can also be used. Included in this category are inductors and transformers driven into the saturation region of the core, certain barium titanate and other ceramic dielectric capacitors and the junction capacitance of reverse biased semiconductor diodes. The main index, in any event, is *to so distort the impressed sine wave that the desired harmonic frequency will be available.*

Because of its simplicity the parallel resonant *tank circuit*, Fig. 2.86, finds widespread application as a frequency selective circuit. Tapping can often be used to provide a better impedance matching

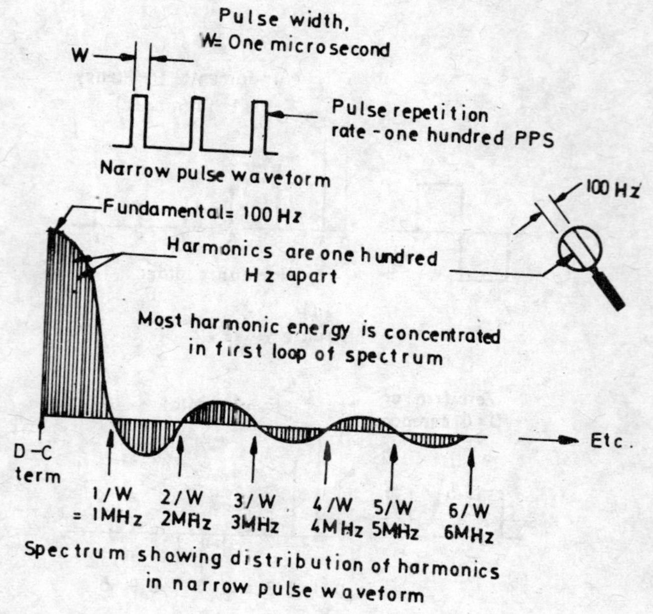


**Fig. 2.83** Narrow rectangular pulses are relatively rich in harmonic contents.

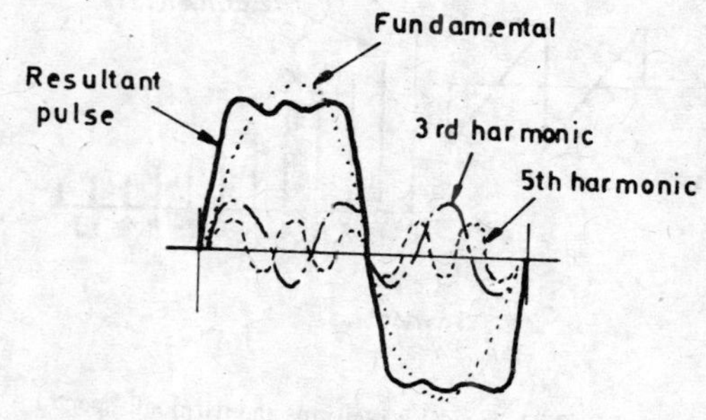


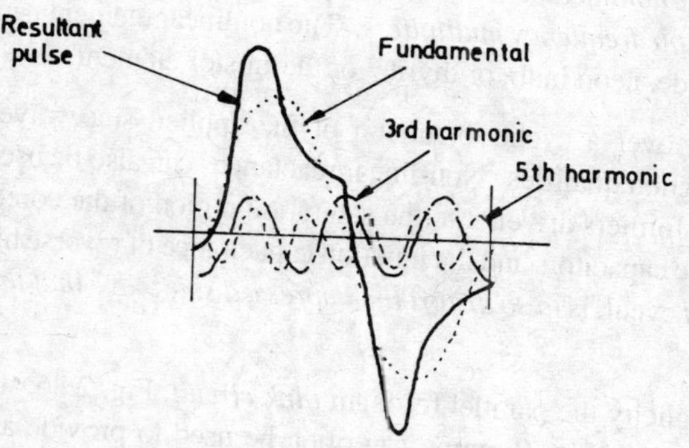


**Fig. 2.84** Approximation to a square wave

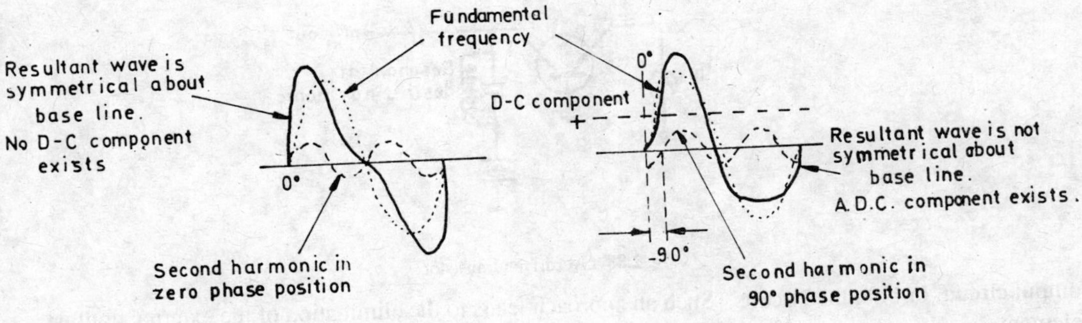


**Fig. 2.85** A peaked wave somewhat resembling the sawtooth

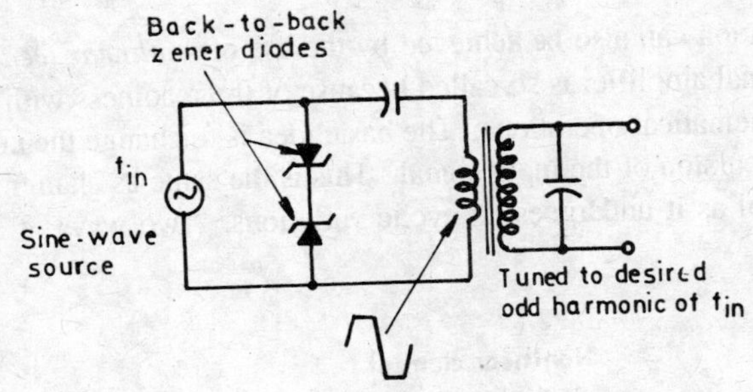


**Fig. 2.86** The parallel resonant tank circuit finds widespread applications as a frequency selective circuit.

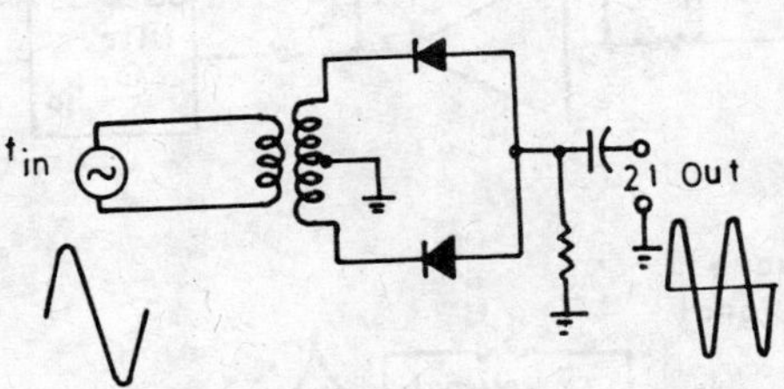


**Fig. 2.87** Balanced rectifier circuit

situation for the nonlinear element, load, or both. Purity of the extracted harmonic can only be achieved by the use of multiple tuning, or electric wave filters. Trap circuits can be used to further attenuate undesired harmonics.

Some *departure from class A operation* of an amplifier also improves harmonic generation. In such instances the amplifier also functions as an additional nonlinear element either in the input circuit

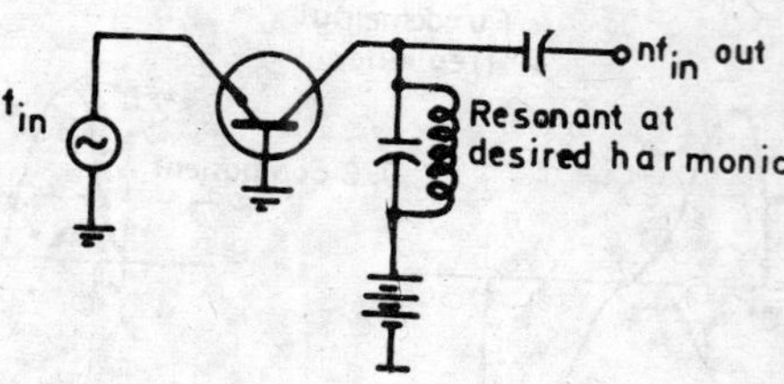


Fig. 2.88 Overdriven transistor

output circuit, or both the circuits. Such an approach leads to the elimination of the external nonlinear element.

Pronounced advantage can be derived by *using a transistor in switching mode.* This has led to the evolution of class C frequency multipliers. *Class C operation is capable of providing harmonic power at high levels,* for this reason it is widely used in radio transmitters.

Frequency multiplication can also be achieved by the use of *nonlinear feedback in an operational amplifier.* The operational amplifier is so called because of the readiness with which it lends itself to the performance of mathematical operations. The basic idea is to change the negative feedback factor during the amplitude excursion of the input signal. This is the same as changing the voltage gain imparted to the input signal as it undergoes its cyclic variations. Two ways of accomplishing this are shown in Fig. 2.89.

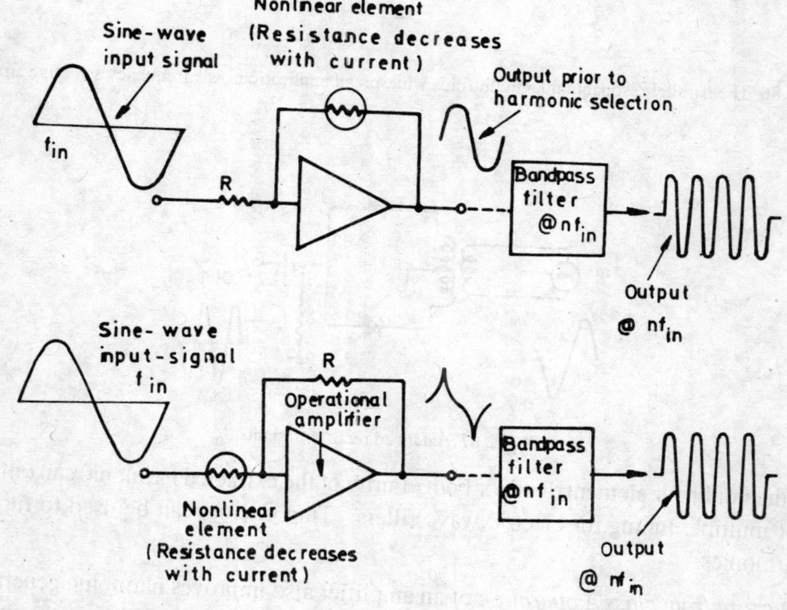


Fig. 2.89 Multiplication caused by nonlinear feedback in an operational amplifier

## 2.35 Frequency Translators

*In frequency translators, the change in frequency can have any relationship to the original signal or signals.* This is in contrast to both frequency multipliers and frequency dividers in which a whole number factor relates the input frequency to the output frequency.

Frequency translation is extremely useful. Not only can single tone signals be shifted throughout the practically usable portion of the electronic spectrum but *modulation, or information bearing features are, in essence, transferred from one frequency to another.*

Frequency translation is at the heart of superheterodyne radio receivers and tv systems. An example of relatively recent application is found in stereo systems.

Despite great evolution in the technology of radio receivers over a span of several decades, *the superheterodyne circuit endures.* Not even the advent of television, or solid-state elements, has changed this basic approach.

The superheterodyne translates all incoming station frequencies to a fixed frequency, the so called *intermediate frequency* (IF). This makes it much easier to impart high amplification and to achieve desired response characteristics than does the more direct approach of tuned RF amplification.

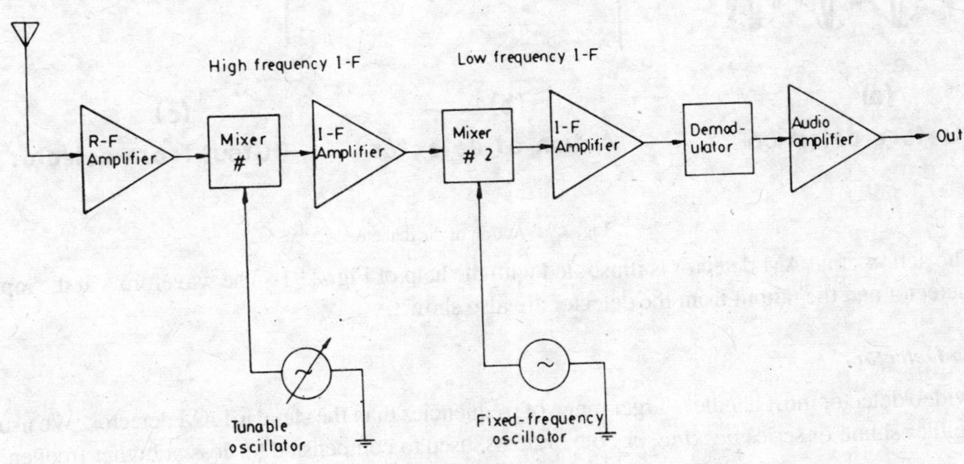


**Fig. 2.90** Conceptual approach of a double superheterodyne circuit

The frequency translation is due to the *interaction* between the incoming station frequency and the frequency generated by a *local oscillator* (LO). The local oscillator frequency can be either above, or below, the carrier frequency of the incoming signal. In either case, the local oscillator must be tuned along with the resonant circuits that select the station. The tuning must be done in such a way that a *constant difference frequency* is maintained between the resonant circuits and the local oscillator frequency. The accomplishment of this is called *tracking*.

The modulator is known in communication practice as the *mixer*. It provides the required non-linearity to generate modulation products which include the desired intermediate frequency. Often, a single transistor is made to provide the *dual functions* of oscillator and mixer. The single element is then termed a *converter*.

## 2.36 Detectors

*Detectors are used to remove the modulation from a received RF signal and convert it back to its original signal form.* Detectors can be classified as AM detectors and FM detectors. There are several variations of the circuitry in each type. Both diode and transistor type detectors are in common use in communication equipment.

### AM Detectors

AM detectors are used in one of two types of circuits, the voltage output circuit called the *series diode detector* and the current output circuit called the *shunt diode detector*. Because of the *lack of gain* in diode circuits, transistors are frequently used in the all solid-state receivers.

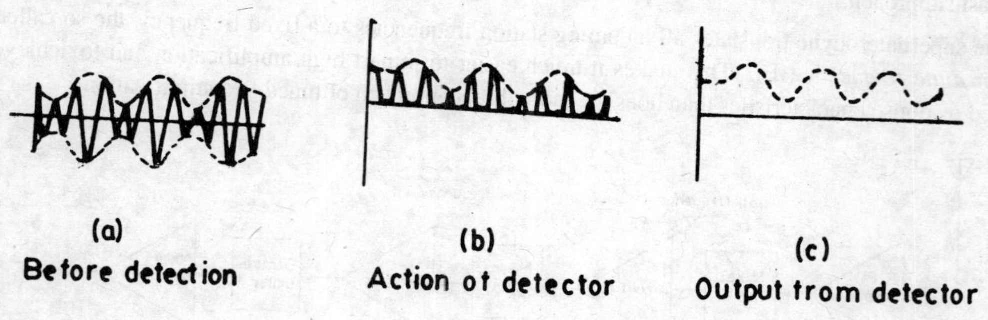


**Fig. 2.91** Action of the detector

The action of an AM detector is illustrated with the help of Fig. 2.91. The waveforms at the input to the detector and the output from the detector are also shown.

### Video Detectors

The video detector must handle a larger range of frequencies than the standard AM detector. We usually find either shunt, or series *peaking*, or both systems, used to compensate for loss at higher frequencies.

The simplest circuit, of course, is that of the diode video detector; however, this provides little or no gain since inherently the diode provides no amplification. On the other hand, by using a transistor type of detector, the emitter-base junction can provide the detection while amplification is obtained from the collector-base junction. Thus both *detection and amplification are obtained from a single stage* and fewer driver circuits are needed to boost the amplitude.

If a diode is used, it is necessary to keep the input impedance high to maintain rectification efficiency of the diode. On the other hand a transistor can serve efficiently as a video detector when looking into a relatively low value of impedance. The high base impedance provides the necessary high impedance input while the output at medium or moderately low impedance matches the following video amplifier stage.

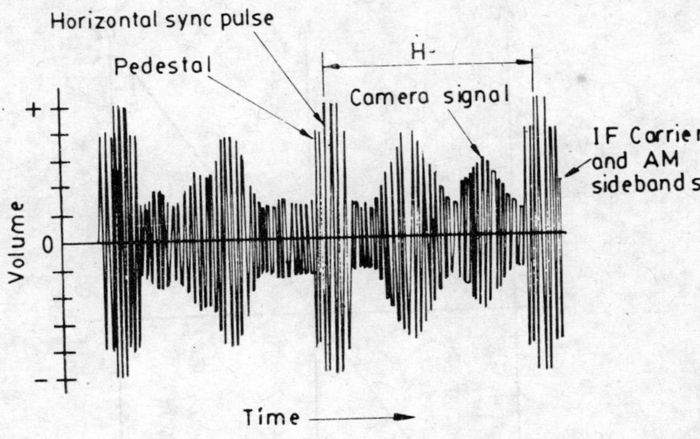


**Fig. 2.92** Amplitude modulated video

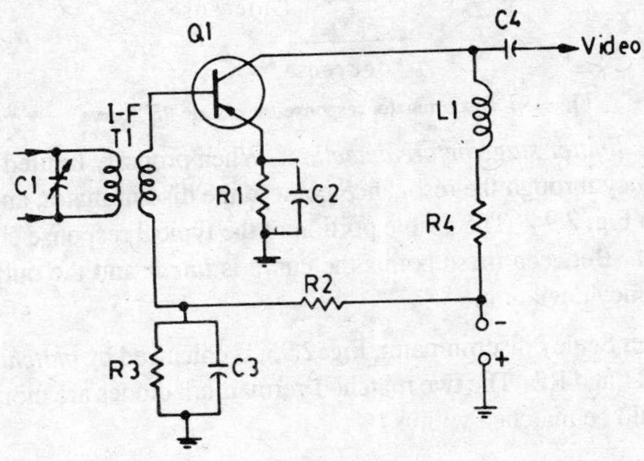


**Fig. 2.93** Basic video detector

### *FM Detectors*

The FM detector is used to demodulate frequency modulated signals in communication equipment. FM detectors are of two types: the *discriminator* and the *ratio-detector*.

### *The Foster Seeley Discriminator*

The Foster Seeley, or *phase shift discriminator* uses a double tuned RF transformer to convert the frequency variations of the received FM signal into amplitude variations. The amplitude variations are then rectified and filtered to provide a dc voltage which varies in amplitude and polarity as the input signal varies in frequency. The output voltage is *zero* when the input frequency is equal to the centre frequency (the unmodulated carrier frequency), the output *increases in one direction* when the input frequency rises above the centre frequency; the output *decreases in the opposite direction* when the input frequency *falls below* the centre frequency.

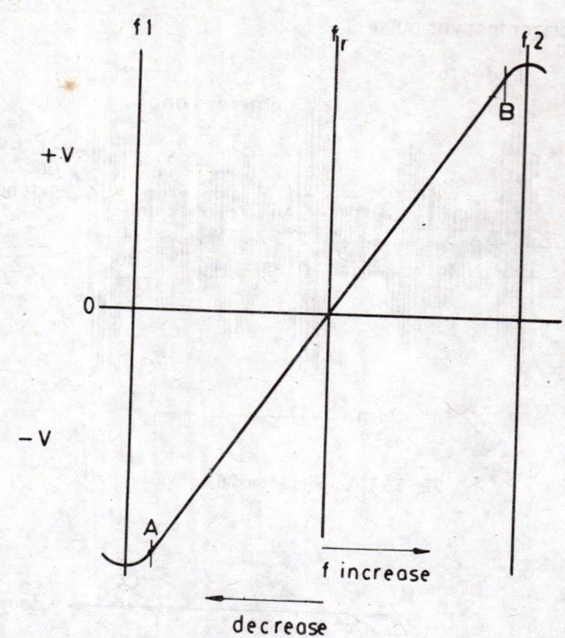


**Fig. 2.94** Discriminator response-the classic "S" curve

*It is necessary to use a limiter stage before detection.* When properly limited the input frequency is varied from a low frequency through the resonance point of the discriminator, and is then raised higher in frequency, as shown in Fig. 2.94. The usable portion of the typical response curve is from point A to point B in the illustration. Between these points the curve is *linear* and the output voltage is directly proportional to the frequency deviation.

The quality of the Foster Seeley discriminator, Fig. 2.95, is enhanced by *matching* germanium diodes D1 and D2 and resistors R1 and R2. The two matched germanium diodes are mounted on a single base. Resistors R1 and R2 should be matched within 1%.

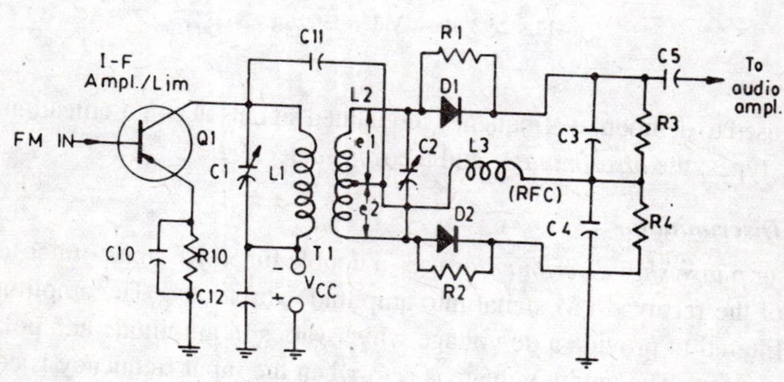


**Fig. 2.95** Foster Seeley discriminator

***The Ratio Detector***

The semiconductor ratio detector uses a double tuned transformer connected so that the frequency variations of the FM input signal are converted into amplitude variations. These amplitude variations are rectified by the diodes to provide a dc output voltage which varies in amplitude and polarity as the input signal varies in frequency.

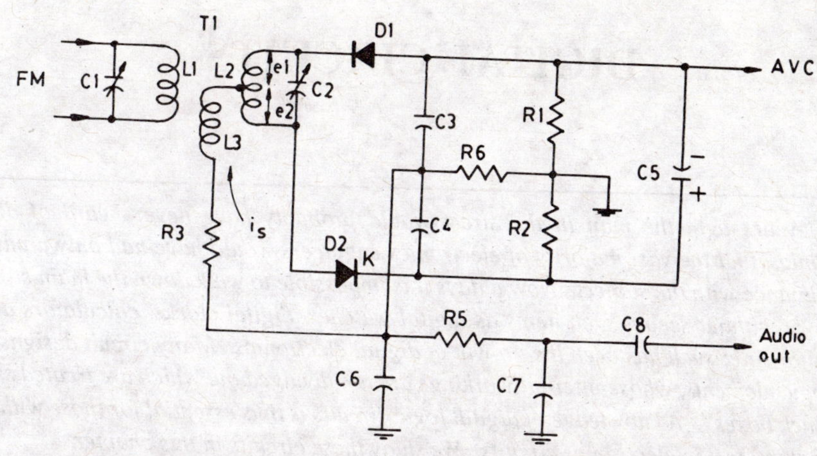


The specific polarity of the output voltages obtained for an increase or decrease in input frequency is determined by the design of the circuit. The significant difference between this type of circuit and the discriminator is that *the ratio detector "kills" any AM*, an advantage when receiver layout must be simple and economical.

## Chapter 3

# DIGITAL CIRCUITS

*Twenty years ago, the man in the street would probably have never heard of digital electronics, and the vast majority of electronics engineers would have had only a passing acquaintance with the subject. Now a days it is impossible to walk down the main street of any town without seeing such items as digital watches, digital clocks, calculators and tv games for sale; such has been the growth of digital electronics. Many circuit designs contain logic elements, whose internal workings are not divulged but which are treated simply as "black boxes". A knowledge of digital logic circuits is thus essential for those with even a casual interest in electronics circuits. We study these circuits in this chapter.*

### 3.1 Analog and Digital Circuits

What are digital circuits? This is probably best explained by contrasting digital circuits with analog circuits. In an analog circuit, a *continuously variable input signal* (voltage, current etc.) gives rise to a *continuously variable output*, which is related to the input by some mathematical expression. For example, in an amplifier this might be, $V_o = AV_i$ where $V_i$ and $V_O$ are the input and output voltages and A is the gain of the amplifier.

In a digital circuit, the input and output signals are assigned *only two values;* a so called logic zero or *low state*, which is frequently zero volts in practical circuits, and a so called logic one or *high state*, which is often some defined positive voltage. The output of a digital circuit element can, in general, assume only one of these two levels in response to the input, or inputs, also having one of these two levels. The performance of a digital circuit element for voltages between these two levels is generally not defined.

### 3.2 Analog And Digital Computers

One basic need of business, industry and government, is to *process data* (information) rapidly with little or no error. Computers meet this need. Computers are considered the main core of *automatic data processing* (ADP). There are two basic types of computers — analog and digital. *Analog computers represent data with a continuous range of signal levels.* The signal level in an analog computer might range from –15 to +15 volts, with an infinite number of actual signal levels. *Digital computers represent data with only two signal levels.* These two signal levels might be 0 and +5 volts in a digital computer.

Analog computers have some basic disadvantages for most computing applications. One problem relates to *drift*. Any slight drift in an analog computer may cause one or more of the voltage levels to change. This upsets the *accuracy* of the system. Digital computers, with only two voltage levels, are not bothered by drift. Another problem is that analog computers are *difficult to program*.

Generally, all computers contain the following five basic elements: input, output, storage, arithmetic (logic) and control. This is illustrated in Fig. 3.1.

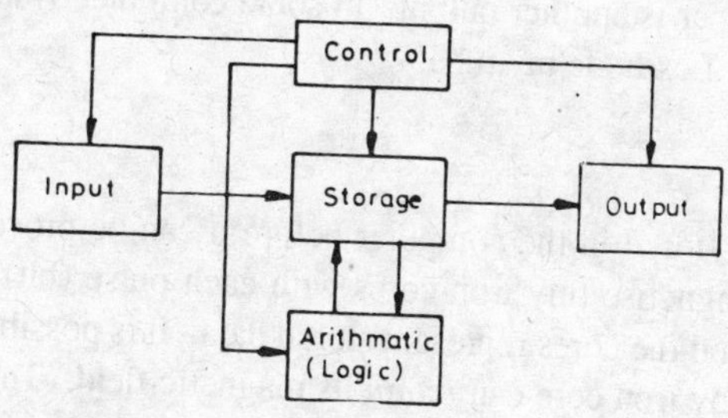


**Fig. 3.1:** An automatic data processing system.

***Computer Input***

The purpose of the input device is to *feed data* into the system. The primary input devices are paper tape, punched cards, and magnetic tapes, see Fig. 3.2. The punched holes in the paper tape and card, and the magnetic patterns imprinted on the tape cause pulses of electricity to appear in the system.

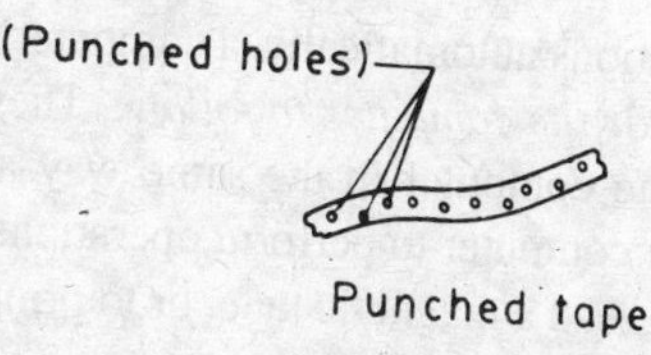


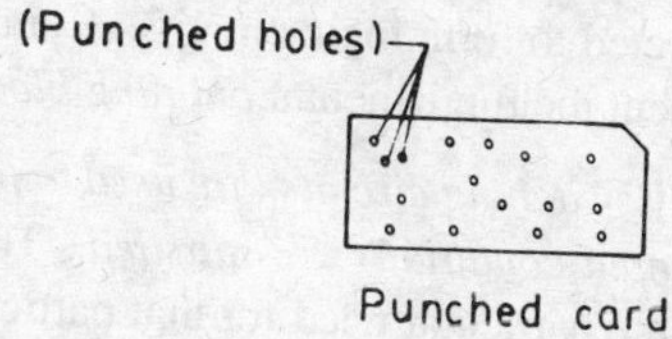


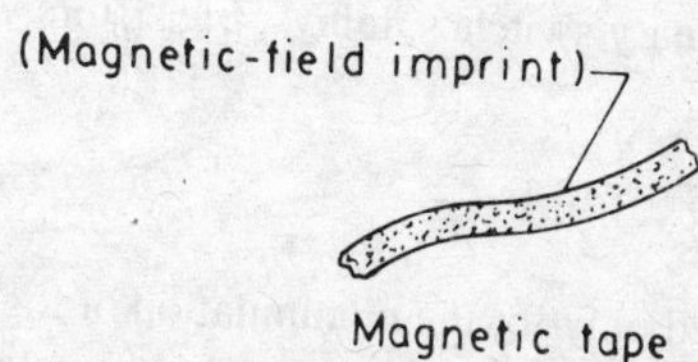


**Fig. 3.2:** Primary input devices.

These pulses are *encoded* to represent alphanumeric characters. The computer *senses* these pulses of electricity and *decodes* them. *In computer work, the pulses of electricity are called bits.* Several bits make up the alphanumeric character. This coding system is similar to that used in tele-typewriters.

### Computer Output

The output of a computer is often *in the same form as its input* — paper tape, punched cards or magnetic tapes. Printing directly on paper is another output. In some computer systems the output is displayed graphically on a special type of cathode ray tube.

### Storage

Data or information must be stored in the computer before it can be processed. One of the common ways of storing data is to magnetise tiny iron cores with each pulse (bit) of electricity. The various magnetic fields which surround the cores represent stored data. It is possible to sense the presence of a magnetic field because the tiny iron core can *retain* its magnetic field. This gives to it the property of *memory*.

### Arithmetic (Logic)

The arithmetic element of the computer has the ability to do comparisons, make decisions, add and subtract. The computer performs these operations by rapidly counting and adding electrical pulses with specially designed *logic circuits*.

### Control

In order to perform its various operations automatically, a computer must be programmed. *A computer program is a series of instructions for the computer to follow.* These instructions are called operation codes. The program is stored in the computer in the same way as the information it controls The program, for example, can cause the computer to perform operations on the stored information, such as to add, shift bits to left, to divide, to read a magnetic tape, or to print on paper.

A computer consists of a large array of transistors, diodes, resistors, capacitors, wires, connectors, and magnetic cores which are connected to form logic circuits. Common logic circuits are AND, OR, NOT and FLIP-FLOP. These represent the fundamental *building blocks* of a computer.

*Logic circuits, which are in reality switching circuits, are used to provide a prescribed output condition when one or more prescribed input conditions are satisfied.* The selection of input conditions is established by the designer to meet the particular need for that particular logic element. The output of a logic circuit is one of two conditions or states, ON or OFF. The switching function to achieve these conditions can be performed by ordinary switches, relays, transistors, diodes, and by a combination of these devices.

## 3.3 Number Systems

Logic circuits are, of course, not used only for the manipulation of logic functions. They are also used for the manipulation of numbers in systems ranging from digital voltmeters to computers. To understand the application of logic circuits to number manipulation, an understanding of number systems and in particular the *binary system* is necessary. Fig. 3.3 shows some of the binary elements used in digital computers.

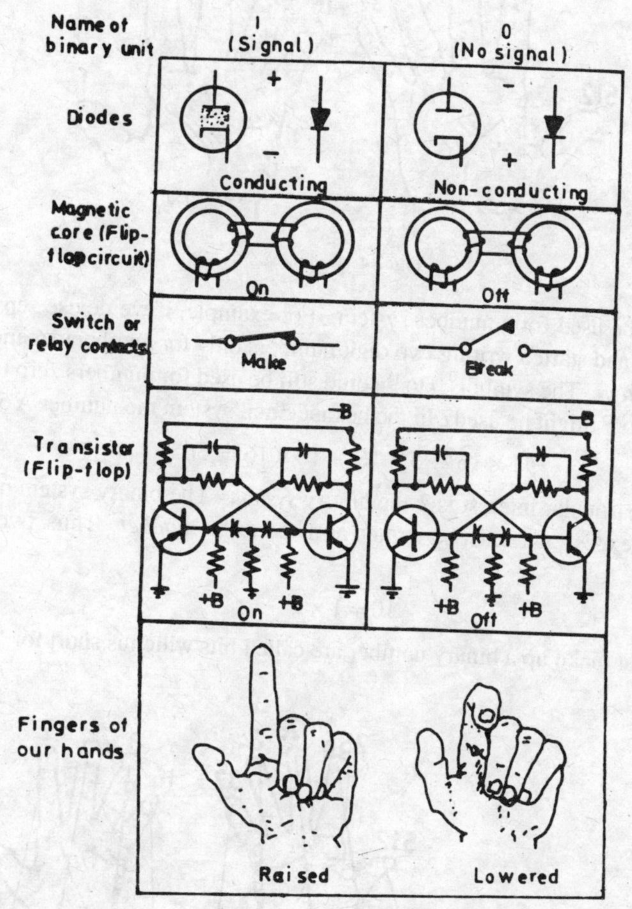


**Fig. 3.3:** Binary elements

The number system with which we are all familiar is the *decimal system*. In the decimal system there is a separate symbol for the ten numbers 0 to 9. To express a number greater than 9 we use two or more of these symbols in a row, using the convention that the extreme right position represents 'units', the next position 'tens', the next position 'hundreds' and so on. A decimal number such as 195 is made up of $5 \times 1 + 9 \times 10 + 1 \times 100$ or to write it in another way $5 \times 10^0 + 9 \times 10^1 + 1 \times 10^2$. Ten is known as the *base* of the decimal system, and any symbol in a decimal number has a value equal to the value of the symbol multiplied by a power of ten, which depends on its *position* in the number.

By using only the fingers of our two hands, we can count any number between 0 and 1023. We can do this by assigning values to fingers of each hand. Illustrated in Fig. 3.4, we call this number assignment process *coding*.

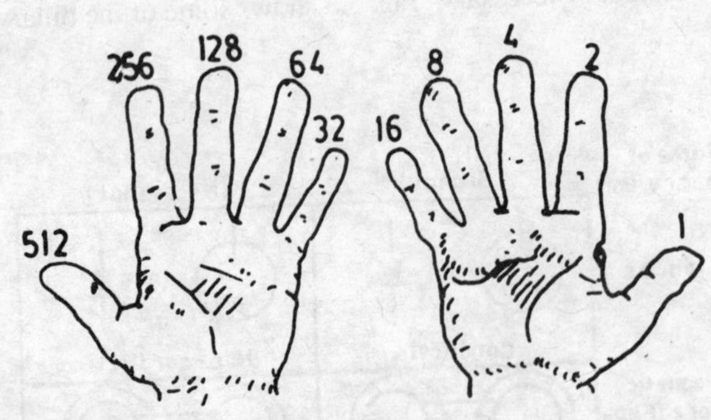


**Fig. 3.4:** Coding

Any other base can be used for a number system. For example, if we devise separate symbols for every number up to 15, and started writing two digit numbers only for numbers 16 and over, we should have a *hexadecimal system*. The symbols 0 to 9 could still be used for numbers zero to nine, but for ten to fifteen the letters A to F might be used. In the hexadecimal system the number A 5 B would be:

$$11 \times 16^0 + 5 \times 16^1 + 10 \times 16^2 = 3651$$

The base that is of particular interest is 2-the *binary system*. The binary system has but the digits 1 and 0. Any number in excess of 1 must be written as a *multidigit number*. Thus, two would be written as

$$10 = 1 \times 2^1$$

The '1s' and '2s' that make up a binary number are called bits which is short for 'binary digits'.

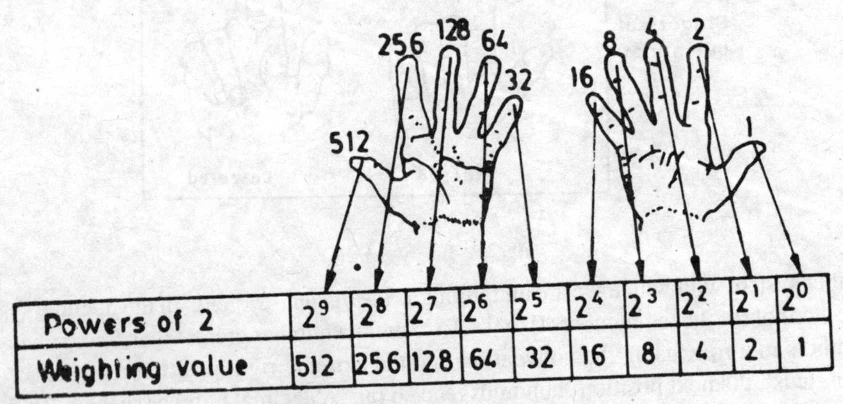


**Fig. 3.5:** The digit position weighting value

The extreme left hand bit of a binary number is the *most significant digit* (MSD) and the extreme right hand bit the *least significant digit* (LSD). The origin of these names is obvious. The LSD makes a difference of 1 in the total number whereas the MSD in say, a four bit number makes a difference of 8 in the total.

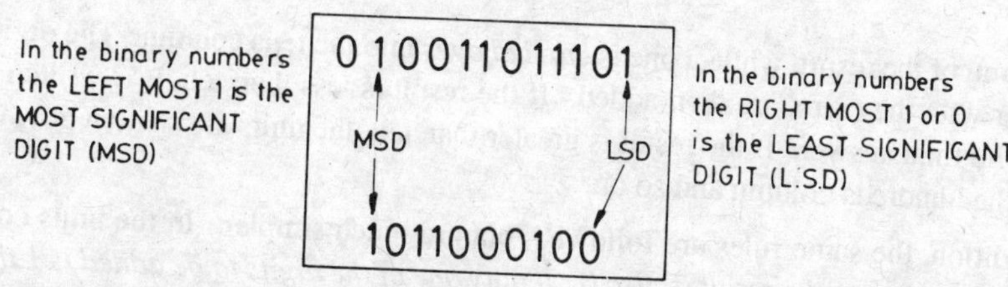


**Fig. 3.6:** The most significant digit (MSD) and the least significant digit (LSD)

## 3.4 Binary to Decimal and Decimal to Binary Conversion

Binary to decimal conversion is extremely simple. *Where a 1 occurs in a binary number then that number contains the relevant power of two.* The powers of two are simply wirtten down in their decimal form and added up to form the decimal equivalent of the binary number as shown in the example below,

$$1011011 = 2^0 + 2^1 + 2^3 + 2^4 + 2^6 = 1 + 2 + 8 + 16 + 64 = 91$$

Converting decimal to binary is a bit more complicated. Basically *it involves dividing two into the number and writing down the remainder after each division. If the remainder at the (n-1)th division is 1 then the number contains nth power of 2. If it is 0 then the number does not contain that power.* The sequence of remainders obtained by dividing the number right down to 0 represents the binary equivalent of the number. This is illustrated in the example below:

*Example:3.1*

| | | | |
|---|---|---|---|
| $150 \div 2$ | = | 75 remainder | 0 |
| $75 \div 2$ | = | 37 remainder | 1 |
| $37 \div 2$ | = | 18 remainder | 1 |
| $18 \div 2$ | = | 9 remainder | 0 |
| $9 \div 2$ | = | 4 remainder | 1 |
| $4 \div 2$ | = | 2 remainder | 0 |
| $2 \div 2$ | = | 1 remainder | 0 |
| $1 \div 2$ | = | 0 remainder | 1 0 0 1 0 1 1 0 |

## 3.5 Manipulating Binary Numbers

Binary numbers can be added, subtracted, multiplied and divided just like any other kind of number. However, since the base of the binary number is only two, this does lead to certain simplifications and short cuts.

### *Binary Addition*

In calculators, computers, and other electronic data manipulation systems, binary numbers are generally added *two at a time*, multiple addition being perfomed by adding the result of the first addition to the third number, the result of the second addition to the fourth number and so on. For this reason, the discussion of binary addition and subtraction will be confined to problems involving two numbers only.

When adding two decimal numbers, the units column is first added. If the result is less than ten it is written down as the units column of the result. If the result is greater than ten, the units are written down as the units column of the result, while a one is *carried over* into the tens column. The digits in the tens column, together with the carry, are then added. If the result is less than ten, it is written down as the tens column of the final answer. If the result is greater than ten, the units are written down and a one is carried over to the hundreds column and so on.

In binary addition, the same rules are followed, but are much simpler. In the units column *if both digits to be added are 0 then the result is also 0. If only one of the digits to be added is 1 then the result is 1; if both the digits to be added are 1 then the result is 0, but the carry is 1.* A few examples may serve to clarify.

*Example 3.2*

```
              1            1 1
             1010          1010
  1010       0010          1110
  0100      -----         -----
 -----       1100         11000
  1110
```

**Binary Subtraction**

Binary subtraction may also be performed in the same way as decimal subtraction i.e. as the reverse of addition. In both cases each digit (or bit) of the subtrahend is subtracted from the corresponding digit in the minuend. Where a digit in the minuend is smaller than the digit to be subtracted then a 1 is *'borrowed'* from the next column, *borrowing being the reverse of carrying.*

However, this method of subtraction would lead to a fairly complicated electronic circuit, especially as the concept of a 'logic -1' does not exist as such, so some other way must be found of representing a negative binary number. Fortunately, there is a method of binary subtraction known as the *complement and add* method, which allows subtraction to be performed by addition. *The complement of a binary number is obtained by inverting each bit of the number.* Thus the complement of 1001 is 0110. To performa a binary subtraction the complement of the subtrahend is added to the minuend and the result from the carry of the most significant bit is noted. *If it is 1 then it indicates two things.* Firstly, the result of the subtraction is a *positive number*, and secondly to obtain the final answer, a procedure known as *end-around carry* must be performed, which involves adding the carry bit to the least significant bit of the result to obtain the correct answer.

*If the carry bit is 0, then this also indicates two things.* Firstly, the result of the subtraction is a *negative number* and secondly the result of the addition is the *complement* of the final answer and it must therefore itself be complemented to obtain the correct result. A few examples will make this procedure clear.

*Example 3.3:* 1000(8) - 0101(5)

```
1000 (8)
1010 (complement of 5)
----
10010 (sum)
↓
+0011 (end around carry)
```

**Answer: +3**

*Example 3.4:* 1001(9) - 0101(5)

```
1001 (9)
1010 (complement of 5)
----
10011 (sum)
    1
----
0100
```

**Answer: +4**

*Example 3.5:* 1001(9) - 1110(14)

1001(9)

0001 (complement of 14)

1010 (sum)

0101 (complement)

**Answer: –5**

There are two important points to note when using this procedure. Firstly, *both numbers must be written down as having the same number of bits*. Secondly, *it has to be noted how many bits are actually being operated upon, so that it can be decided which bit is the carry bit that alternately decides the sign of the answer.*

## 3.6 Switches and Logic

In fact, the basic digital circuit is a single pole double through (SPDT) switch. By itself, the switch forms a single bit of *memory* circuit. If the switch is in one position, current is through one pole, and in the opposite position through the other. The live pole indicates the position of the switch, or to use digital jargon *stores* the position.

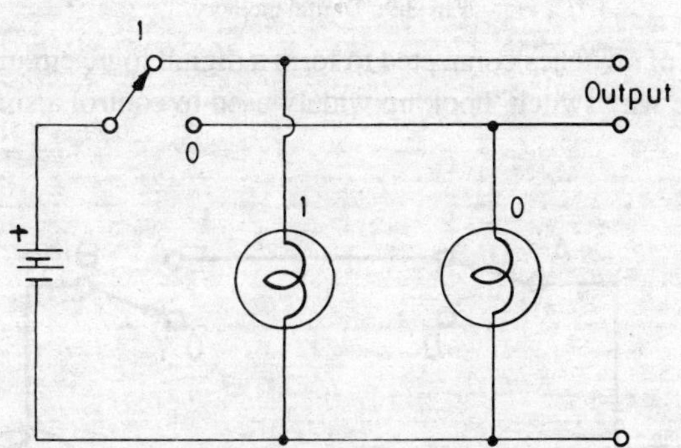


**Fig. 3.7:** Basic memory circuit.

By adding a battery to supply power and two pilot bulbs to indicate which pole has the current, we arrive at the circuit shown in Fig. 3.7. It would also be possible to use two single pole single throw (SPST) switches if they were ganged, Fig. 3.8, so that one was open when the other was closed.

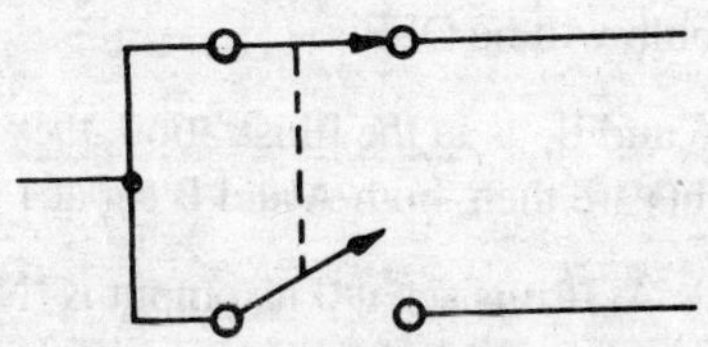

**Fig. 3.8:** Ganged SPST switches.

The pilot bulbs are not essential, they merely indicate the switch position. When the switch is in the position labelled, 1, the 1 bulb glows, when the switch is in position labelled 0, the 0 bulb glows. The switch thus operates as *digital memory* because when the switch is set at one position, it remains in that position until it is operated again. That is to say, it remembers its setting. Toggle, or rocker, switches are widely used in digital equipment to *remember* instructions from the operator to the equipment.

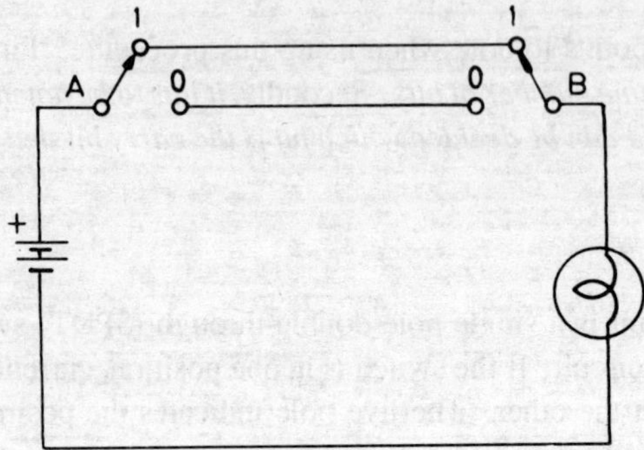


**Fig. 3.9:** Digital memory.

Fig. 3.10 shows a pair of switches connected to form a digital logic circuit. The circuit may be more recognisable as the 'three way switch' hook up widely used to control a single electric lamp from two locations.

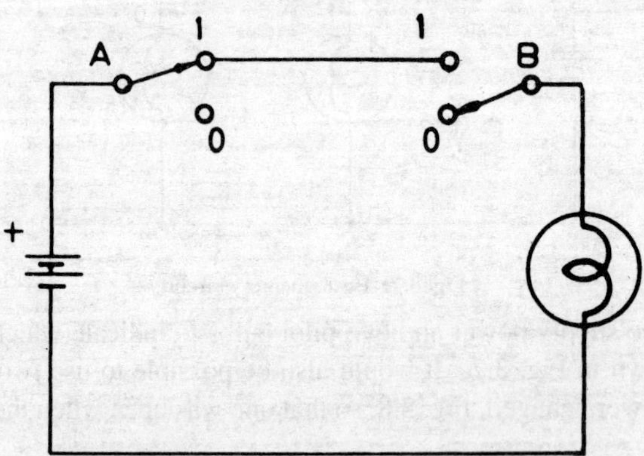


**Fig. 3.10:** Series connected 1 terminals (AB).

If both switches are in the same position, either both at 1 or both at 0, the bulb will be ON; if the positions of the switches differ the bulb will be OFF.

When the switches are labelled A and B, as in the illustration, they can be referred to by name. The conditions under which the bulb lights are then, both A and B set at 1 *OR* both A and B set at 0.

If switch A is set at 1, its output is A. If it is set at 0 its output is 'NOT' A, or A 'NOT'. This verbal shorthand lets us specify with far fewer words the conditions under which the bulb in Fig. 3.10 lights. The bulb lights with (A and B) OR ('NOT' A AND 'NOT' B).

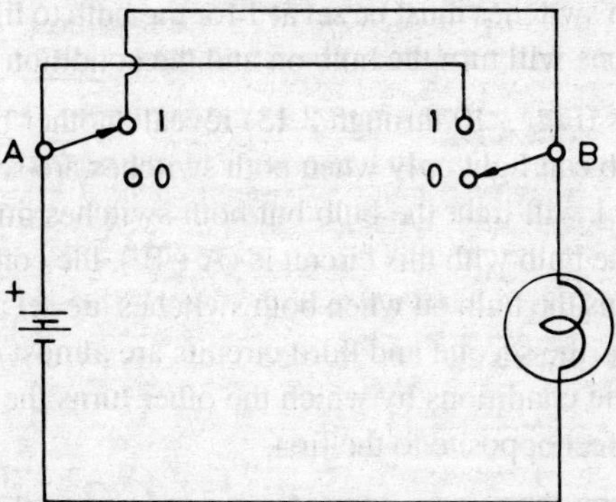


**Fig. 3.11**: Parallel connected 1 terminals (A + B).

The words AND and OR are also shortened as A.B or AB and A+B. When we speak about circuits, the words AND and OR replace these signs. The bulb lighting conditions in Fig. 3.10 become AB + NOT A.NOT B; another way to write it is AB + $\overline{AB}$ since the NOT condition can also be written as a bar over the name.

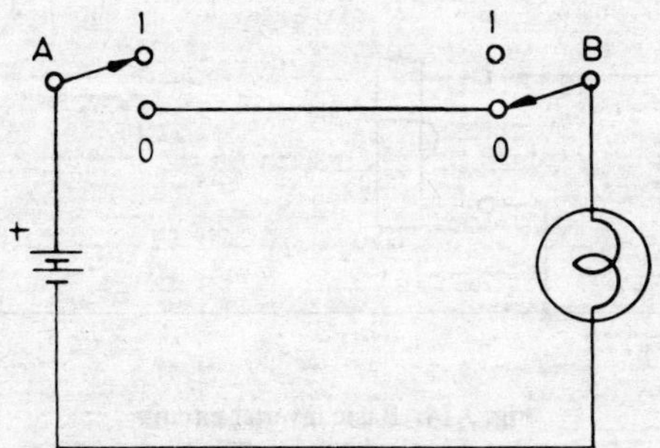


**Fig. 3.12:** Series connected 0 terminals ($\overline{A}\,\overline{B}$ ).

Other possible ways of connecting two switches to control a single light are illustrated in Figs. 3.12 and 3.13. If only the 1 output terminals of the switches are used they can be connected either in series

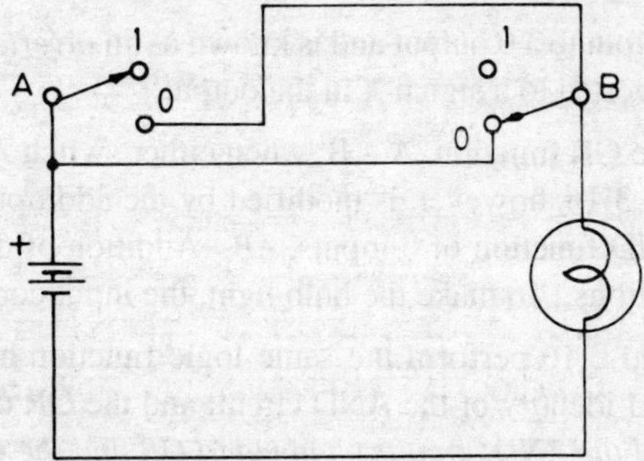


**Fig. 3.13:** Parallel connected 0 terminals ($\overline{A} + \overline{B}$).

or in parallel. If in series, both switches must be set at 1 for the bulb to light, the condition becomes AB. If in parallel, either switch alone will turn the bulb on and the condition is A + B.

The four sets of conditions (Fig. 3.10 through 3.13) reveal another basic fact about digital circuits. In the first case (AB), the bulb can light only when both switches are set at 1. In the second case (A + B), either switch being set at 1 will light the bulb but both switches must be at 0 to turn the bulb off. While the condition to light the bulb with this circuit is (A + B ), the condition to make the bulb dark is $\overline{AB}$. The third circuit ($\overline{AB}$) turns the bulb on when both switches are set at 0; either switch at 1, (A + B), will make the bulb dark. Thus the second and third circuits are almost exact opposites. One turns the bulb on under exactly the same conditions by which the other turns the bulb off. Similarly, the fourth circuit ($\overline{A} + \overline{B}$) produces an effect opposite to the first.

To make the bulb light when there is no current, we need a circuit that will produce current at its output whenever it receives none at the input and vice verse.

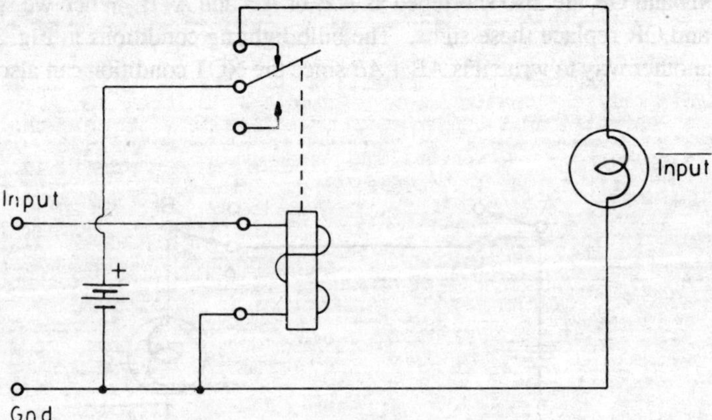


**Fig. 3.14:** Basic inverter circuit.

A normally closed (NC) relay performs this operation if its armature is connected to a power source and the output is taken from the normally closed terminal, as shown in Fig. 3.14. When the current is applied to the coil and the relay is energised, the circuit from power source to output is broken and the output thus becomes zero. If the current is removed from the coil (0 input), the circuit closes and the output receives current, becoming a 1.

Such a circuit, changes a 1 input to a 0 output and is known as an *inverter* since they *negate* the input i.e. a signal X in the input is inverted to a signal $\overline{X}$ in the output.

The circuit in Fig. 3.15 is the OR function, A + B; when either switch A, or switch B, is set at 1, the bulb lights. The circuit of Fig. 3.16, however, is modified by the addition of an inverter to its output. The original circuit was an AND function of 0 inputs, $\overline{AB}$. Addition of the inverter makes the output false under the $\overline{A}\,\overline{B}$ input conditions. To make the bulb light, the input condition must be NOT $\overline{A}\,\overline{B}$.

The circuits in Fig. 3.15 and 3.16 perform the same logic function and are logically equaivalent circuits and bring out the virtual identity of the AND circuit and the OR circuit. *If all the 1 inputs are changed to 0 and vice versa, all the AND circuits changed to OR and the OR circuits changed to AND, and all the outputs are then inverted, the total circuit will remain unchanged.*

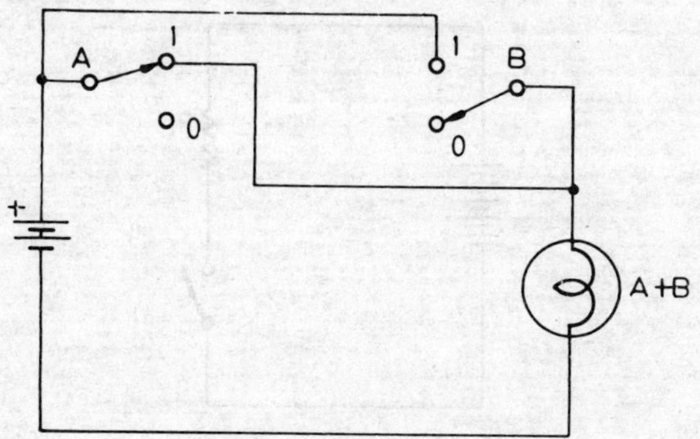


**Fig. 3.15:** Simple OR circuit.

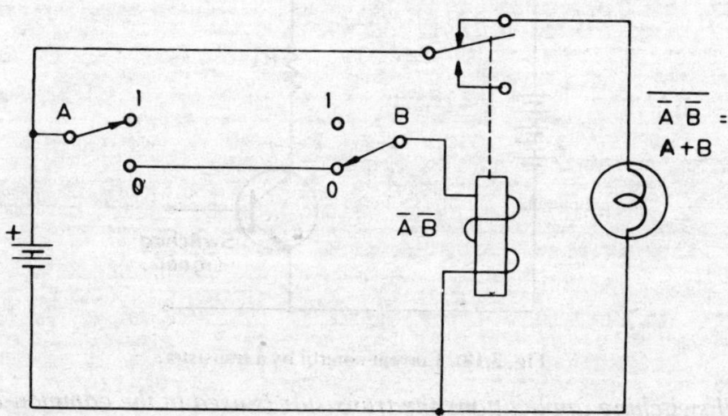


**Fig. 3.16:** Inverted AND circuit.

This fact is of extreme importance with the rise in the use of integrated circuits since one of the simplest digital circuits to produce on an integrated circuit chip consists of an OR circuit combined with an inverter. By suitable interconections, this single circuit known as a *NOR* (NOT OR) gate, can perform all the required operations, functioning as either an OR, an AND, or an inverter. An entire computer can be built with the NOR gate and it operates on exactly the same principle as the circuit of Fig. 3.16.

## 3.7 The Transistor as a Swtich

Fig. 3.17 illustrates a circuit incorporating a conventional switch with mechanically operated contacts. Fig. 3.18 shows a corresponding arrangement in which the current through the load $R_L$ is controlled by a transistor.

With a reverse voltage applied to the base of the transistor, only the very small collector to base leakage current can flow through the transistor, and so there will be almost zero load current, the full battery voltage appearing across the transistor. If the base voltage is made positive, the collector to emitter impedance will fall and the load current can now flow through the transistor and load resistance; the voltage across the transistor falling to almost zero.

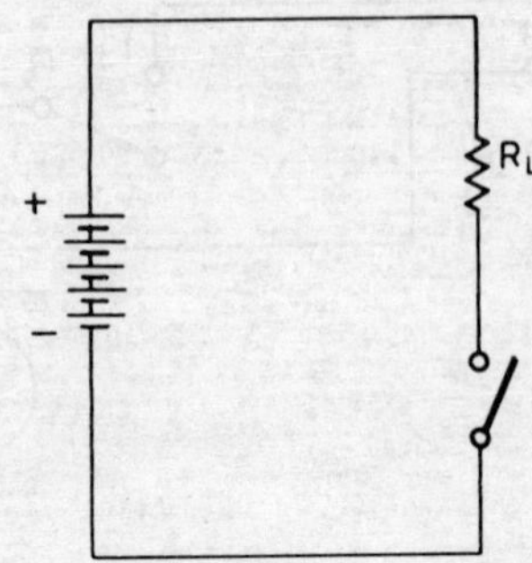


Fig. 3.17: Mechanically operated switch.

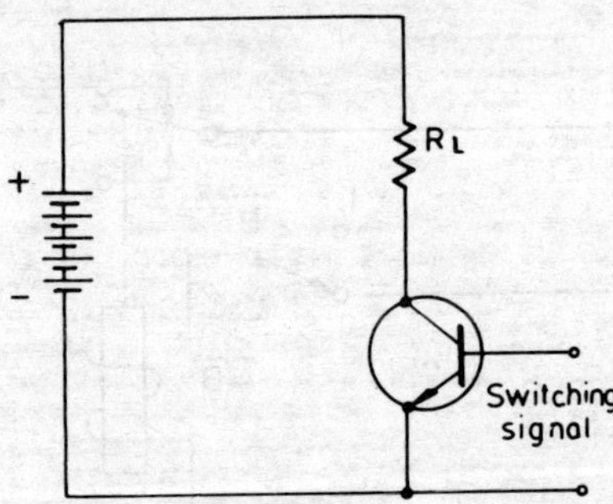


Fig. 3.18: Current control by a transistor.

*In practically all switching applications the transistor is used in the common-emitter configuration since the high current gain allows large collector current to be switched by means of a relatively small base current.*

In addition to the OFF state and the ON state, the *transition* through which the transistor must pass, in going from one to the other, is important. Consider the circuit in Fig. 3.19(a) and the collector characteristics of Fig. 3.19(b). The load line corresponding to the load resistance $R_L$ is shown on the characteristics.

With the base current at zero, collector current will be the collector to emitter leakage current $I_{CEO}$. If a reverse voltage is applied to the base, the collector current is reduced to the collector to base leakage current $I_{CBO}$. This current is very small-with silicon planar transistors of the order of 1 nano ampere-giving an OFF impedance above 1,000 M Ω. *When the base-emitter junction of the transistor is reverse biased the transistor is said to be cut off* and is then in the shaded region of Fig. 3.19(b). With the load resistance $R_L$, the operating point will be at A.

As the base current is increased, the collector current increases, and the operating point moves along the loadline and eventually reaches point B. Then further increase of base current will not result in any more increase of collector current and *the transistor is said to be saturated or bottomed.* The collector to emmiter voltage will have fallen to a very low value, known as the collector saturation voltage $V_{CE(sat)}$. In this condition the resistance between the collector and emitter of the transistor is very low-only a few

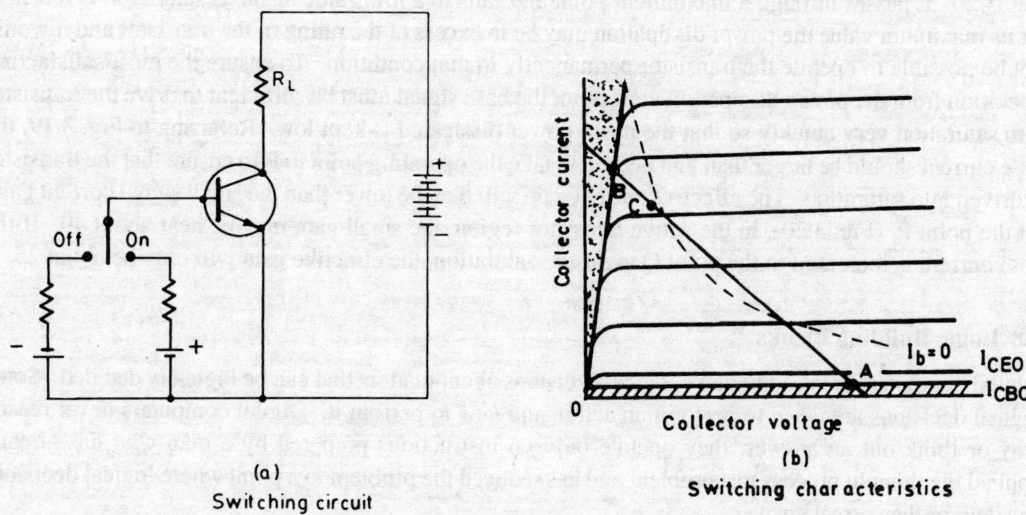


**Fig. 3.19:** Transistor switching characteristics.

ohms. As the collector voltage falls to the low $V_{CE(sat)}$, the collector junction becomes forward biased. Thus *in the ON condition both junctions of the transistor are forward biased* and the current flowing through the load resistance will be $(V_{CC}-V_{CE(sat)})/R_L$ or approximately $V_{CC}/R_L$ which is independent of the base current. The saturation region is shown dotted in Fig. 3.19(b).

Fig. 3.20 illustrates what happens during the transition from the OFF state to the ON state. As the base current is increased, the collector current increases, and the collector voltage falls. In the OFF state the collector dissipation is very low. With a germanium transistor at an ambient temperature of 55°C, the leakage current may be about 30 μA, and with a $V_{CE}$ of 12 volts the dissipation will be 0.36 mW, which is negligible.

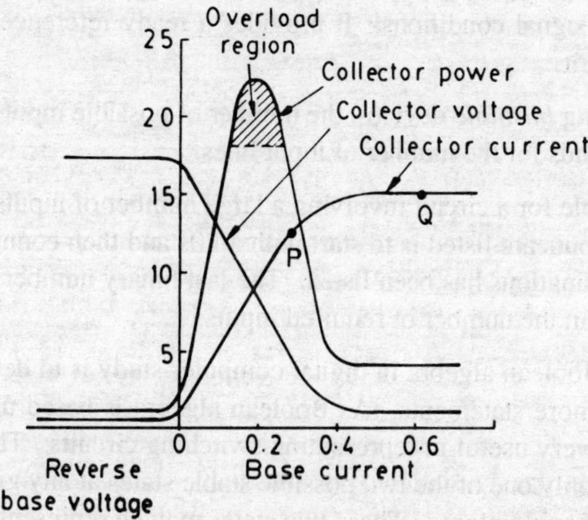


**Fig. 3.20:** Transistor charcteristics.

As the collector current increases from the OFF state the transistor dissipation increases as shown in Fig. 3.20. It passes through a maxiumum point and falls to a low value again as saturation is reached. At its maximum value the power dissipation may be in excess of the rating of the transistor and it would not be possible to operate the transistor permanently in that condition. To ensure the most satisfactory operation from the power dissipation view point the base signal must be sufficient to drive the transistor into saturation very quickly so that the mean power dissipated is kept low. Referring to Fig. 3.19, the base current should be larger than that needed to take the operating point to B to ensure that the transistor is driven into saturation. The effective current gain will then be lower than the small signal current gain. At the point $P_1$ (Fig. 3.20), in the active transistor region, the small gain may be near about 40. If the base current is increased to the point Q to ensure saturation, the effective gain will only be about 25.

## 3.8 Logic Building Blocks

Digital computers are used to make logical decisions about matters that can be logically decided. Some logical decisions are *when* to perform an action and *how* to perfom it. Digital computers never reason why or think out an answer; they operate only on instructions prepared by a man who has already applied the thought process to a problem and has reduced the problem to a point where logical decisions can deliver the correct answer.

We use many logical truths in everyday life without realising it. Most of the simple logical problems are distinguished by words such as *and, or, not, if, else and then.* Once a verbal reasoning process has been completed and results put into statements, the basic laws of logic can be used to evaluate the process. Although simple logic operations can be performed by manipulating verbal statements, the structure of more complex relationships can be usefully represented by the use of symbols. These operations are known as *symbolic logic*.

The symbolic logic operations utilised in digital computers are based on the investigations of George Boole and the resulting algebraic system called *Boolean algebra*. It is similar in some respects to the algebra we are familiar with, however it follows different laws and rules.

A *truth table* is a chart used in connection with logic circuits to illustrate the 'states of the inputs and outputs' under all possible signal conditions. It provides a ready reference for use in analysing the operating theory of the circuit.

In logic circuits employing *blistable* devices, the number of possible input combinations in the truth table will be $2^n$, where n stands for the number of input lines.

When plotting a truth table for a circuit involving a large number of inputs, a convenient method of insuring that all possible inputs are listed is to start with all 0s and then count in *binary sequence* until the correct number of combinations has been listed. The last binary number listed in the input side of the table will be one less than the number of required inputs.

The objective of using Boolean algebra in digital computer study is to determine the truth value of the combnation of two or more statements. As Boolean algebra is based upon elements having two possible states, it becomes very useful in representing switching circuits. The reason for this is that a switching circuit can be in only one of the two possible stable states at any given time, that is, the state of being *open* or the state of being *closed*. These two states may be represented by 0 and 1; we can use these symbols with the Boolean algebra.

In mathematics, there are four basic operations; addition, subtraction, multiplication and division. *In Boolean algebra there are three basic operations-AND, OR and NOT.* These words do not sound mathematical. Logic began with words and not until much later was it transalted into mathematical terms. The basic operations are represented in logic equations by the symbols in Table 3.1.

**Table 3.1**

Logic Symbols

| Operation | Meaning |
|---|---|
| A.B | A AND B |
| A+B | A OR B |
| $\bar{A}$ | A NOT OR NOT A |

Combinations of Logic Symbols

| Operations | Meaning |
|---|---|
| (A+B)(C) | A OR B, AND C |
| AB+C | A AND B, OR C |
| $\bar{A}$.B | A NOT, AND B |
| A+$\bar{B}$ | A OR, NOT B |

The OR operation is indicated by the addition symbol while the AND operation is indicated by the multiplication symbol. The negative function may also be indicated by *a* instead of $\bar{A}$. Examples of combinations of these symbols are given in Table 3.1.

The three basic logic operations (AND, OR and NOT) and four of the simpler combinations of the three (NOR, NAND, INHIBIT and EXCLUSIVE OR) are shown in Fig. 3.21. For each operation a representative switching circuit, a truth table, and a block diagram are given. In all cases, *a 1 at the input indicates the presence of a signal* (corresponding to switch closed) and *0 represents the absence of a signal* (switch open). In all outputs, *a 1 represents a signal across the load, a 0 indicates no signal.*

Logic circuits will always be at one of two specific voltage levels. In either case the two outputs may be considered as being either negative or positive with respect to each other. Normally these output levels are used to represent binary 1s or binary 0s. As far as circuitry is concerned, it is possible to use either voltage state (or level) to represent each binary digit; therefore, it is essential to the understanding of circuitry that some distinction be made regarding the system of logic in any specific application.

The term *positive logic* is used to denote circuitry in which the voltage level utilised to represent a binary 1 is positive with respect to the level which represents a binary 0. The term *negative logic* is applied to the circuit in which the voltage level representing a binary 1 is negative with respect to the voltage level which represents binary 0.

## 3.9 Logic Gates

The circuits which perform the logic operations *sense* the input conditions and provide an output *only if certain input conditions exist.* They can be classified generally as *gate circuits,* and they may employ transistors, semiconductor diodes, or magnetic cores.

### *AND Gate*

An AND circuit, Fig. 3.23, requires that all inputs meet a set of prescribed conditions in order to produce an output. Two versions of AND circuit are given in Figs. 3.23 and 3.24.

In Fig. 3.23, two pnp transistors are connnected together at the collectors and emitters to form an emitter follower circuit. There are two separate input terminals for the base signals. Initially both transistors are biased to saturation condition and nearly all of the collector voltage is dropped across $R_L$.

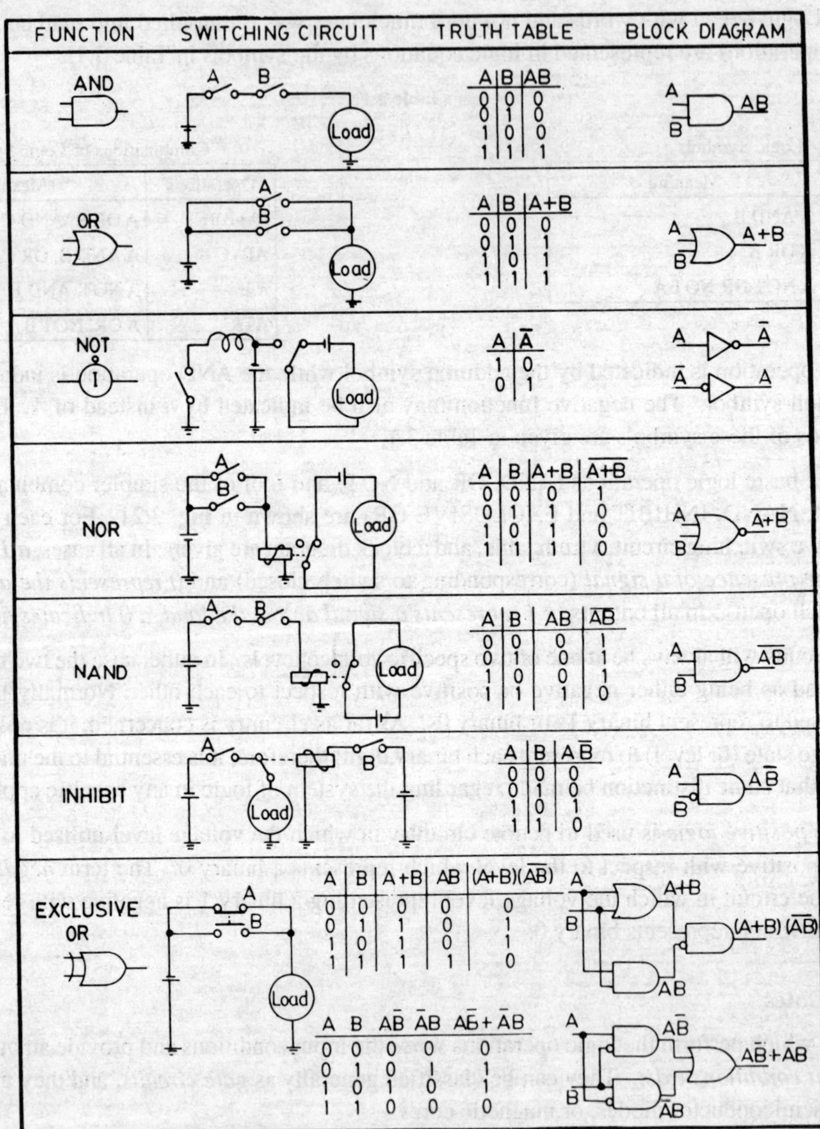


**Fig. 3.21:** Logic operations-comparison chart.

Positive signals fed to the base of each transistor will cut off the respective transistor, but both transistors must be cut off in order to produce an output. When both the transistors are cut off simultaneously, the output terminal voltage will swing in a positive direction.

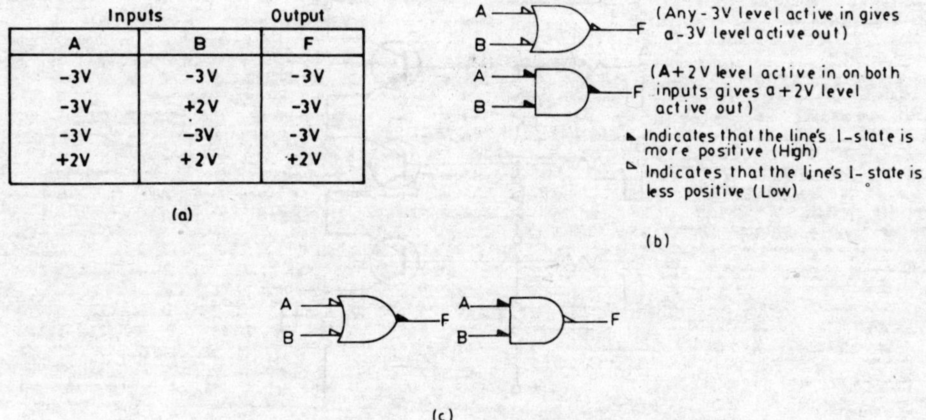

| Inputs | | Output |
|---|---|---|
| A | B | F |
| -3V | -3V | -3V |
| -3V | +2V | -3V |
| -3V | -3V | -3V |
| +2V | +2V | +2V |



**Fig. 3.22:** Logic functions with high-and low-level inputs.

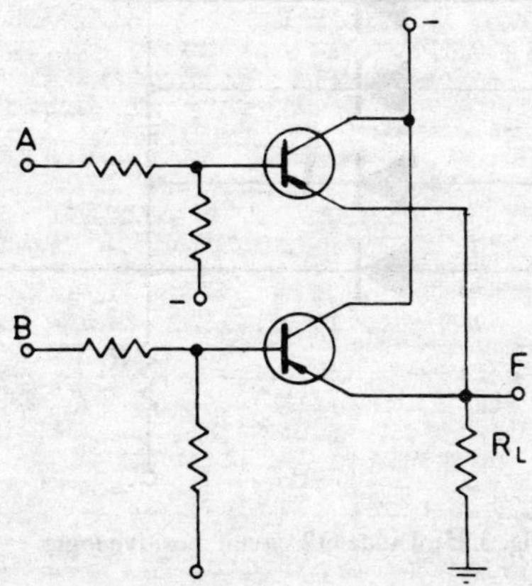



**Fig. 3.23:** Transistorised AND gate, positive logic.

The circuit shown in Fig. 3.24 is a negative logic AND circuit. With no signal applied, each transistor is cut off. During operation all transistors must be utilised together. If any one or more of the transistors remains cut off there is no current path through the series connected emitters and collectors. If all the transistors are driven into conduction, by negative signals applied to the bases, the output at the emitter terminal will swing in a negative direction.

***OR Gate***

An OR circuit is a logic circuit having two or more inputs and producing an output if anyone, or more, or all, of the inputs are placed in a required signal state. Fig. 3.25 illustrates a positive logic diode OR circuit. Any practical number of inputs may be used.

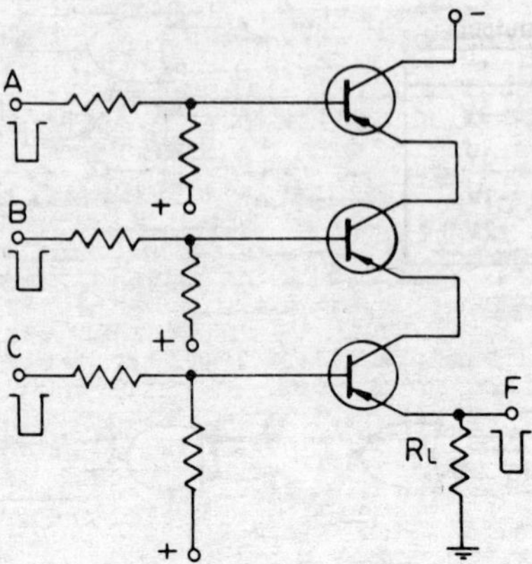


Fig. 3.24: Transistor AND gate, negative logic.

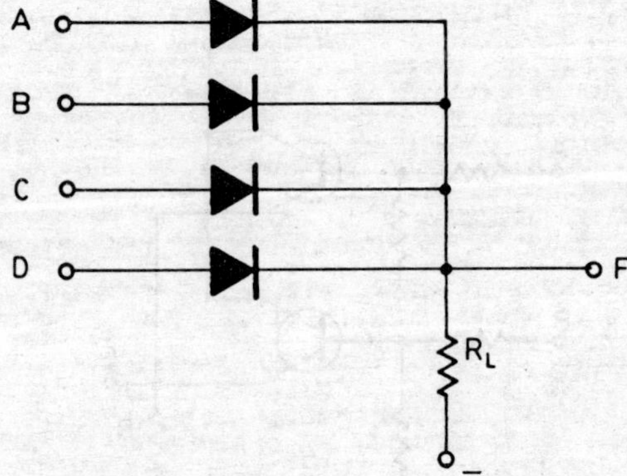


Fig. 3.25: Diode OR circuit, positive logic.

When anyone or more of the inputs receive a positive signal, the diode receiving that input conducts, a signal is impressed across $R_L$, resulting in a positive going waveform of the output.

***NOT Gate***

The NOT circuit, also known as an *inverter*, is nothing more than a normal amplifier stage that produces an inversion (*negation*) of the input logic signal. Fig. 3.26 illustrates a NOT circuit. The waveforms and logic diagram symbols are also given.

In the pnp transistor circuit shown, the transistor is biased to cut off (with no signal applied) by the voltage divider R2 and R3 making the base positive with respect to the emitter. The input signal voltage is divided between R3 and R2 to make the circuit's output swing between the logic high and low voltage levels. The input signal must be negative to cause conduction, therefore, this circuit represents a negative logic NOT circuit. The logic symbol indicates, by the use of a small circle on the triangle, the *low level or negative state* that is, the inversion from a negative going to a positve going signal.

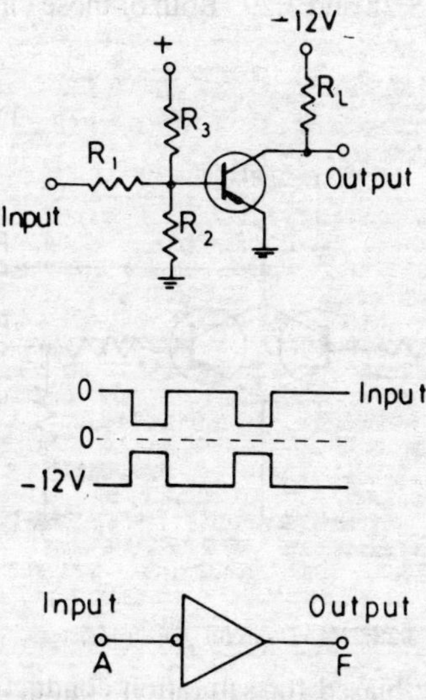


**Fig. 3.26:** NOT circuit negative logic.

### *NOR Gate*

Fig. 3.27 illustrates a transistorised OR circuit which also includes the negative portion of the NOT circuit. A circuit that performs both of these functions is *called* a NOR (NOT OR) circuit.

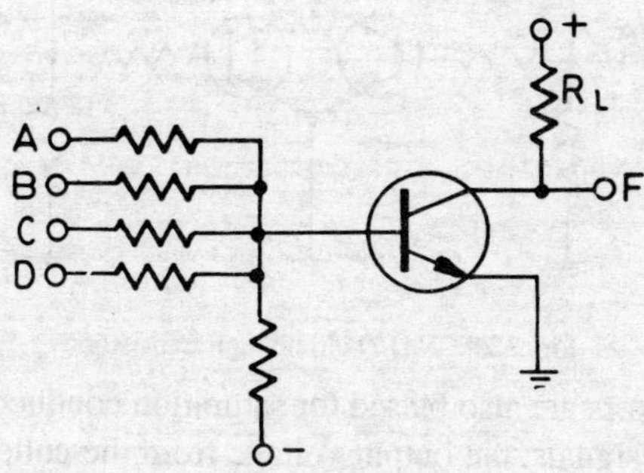


**Fig. 3.27:** Transistor NOR circuit.

With no input signal applied, the npn transistor is biased to cut off by the reverse bias between the emitter and base. A positive input pulse causes conduction of the transistor, producing a negative going output pulse at the collector.

### *NAND Gate*

Two *NAND* circuits are shown in Figs. 3.28 and 3.29. Both of these circuits combine the AND function and the NOT function.

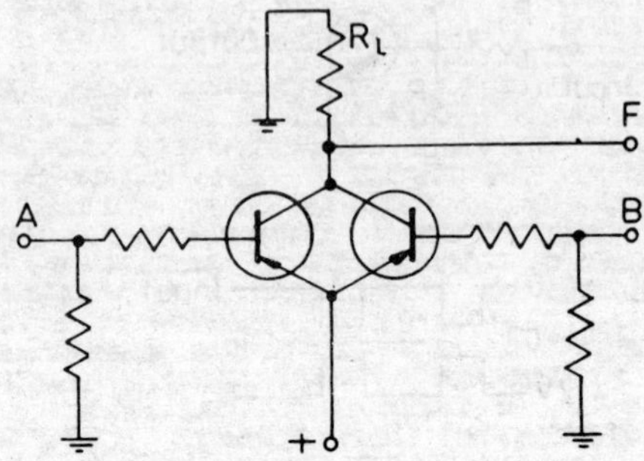


**Fig. 3.28:** *NAND* circuit, pnp transistors.

In Fig. 3.28, the pnp transistors are biased for saturation conduction in the absence of signals. If positive signals are applied simultaneously to both bases, both transistors will be cut off (the AND function). When cut off occurs for both transistors, the collector voltage (output) decreases negatively. This signal inversion is the NOT function.

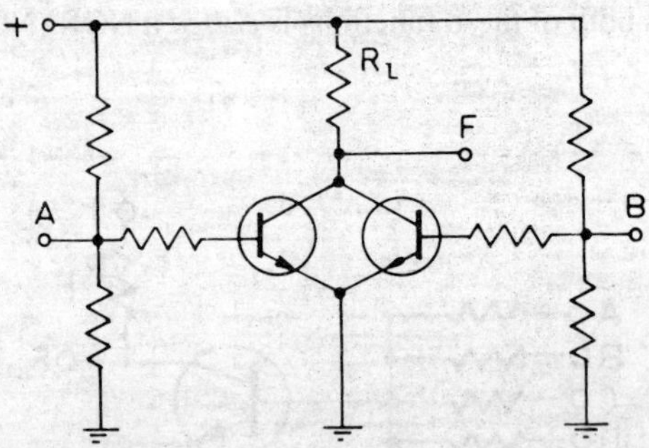


**Fig. 3.29:** *NAND* circuit, npn transistors.

In Fig. 3.29, the npn transistors are also biased for saturation conduction. When both transistors are cut off by coincident negative signals, the output voltage from the collector increases positively. This is a negative logic *NAND* circuit. This operation can also be performed by other circuits. The circuit in Fig. 3.30 provides the NOT (inversion) function by placing a transformer in the input leg of diode D3.

Diodes D1 and D2 are forward biased allowing them to conduct. This places the output voltage at a low level (binary 0).. Diode D3, however, is biased so that the diode is cut off with no signal applied to input C. With a signal applied to either input A or B, either D1 or D2 will be cut off, however, the other diode will continue to conduct, holding the output at a low potential.

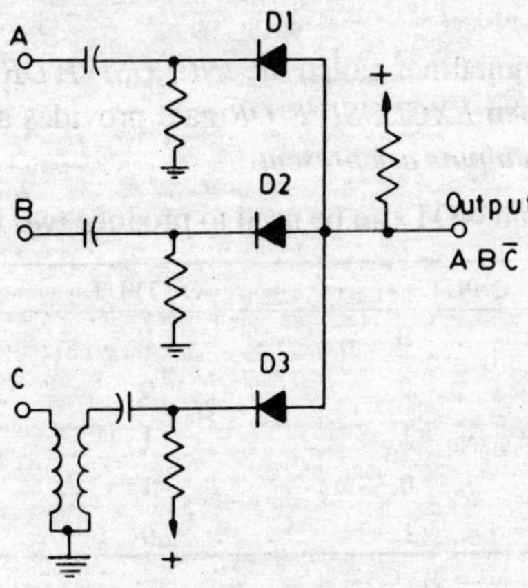


**Fig. 3.30:** INHIBITOR circuit.

When inputs A and B both have a signal applied, with no-signal applied to input C, both D1 and D2 will be cut off and the output will rise to a high positive potential (binary 1).

If a pulse is applied to input C, the pulse will be inverted across the transformer. This will overcome the bias on D3, allowing it to conduct, holding the output to a low level regardless of the state of the other inputs.

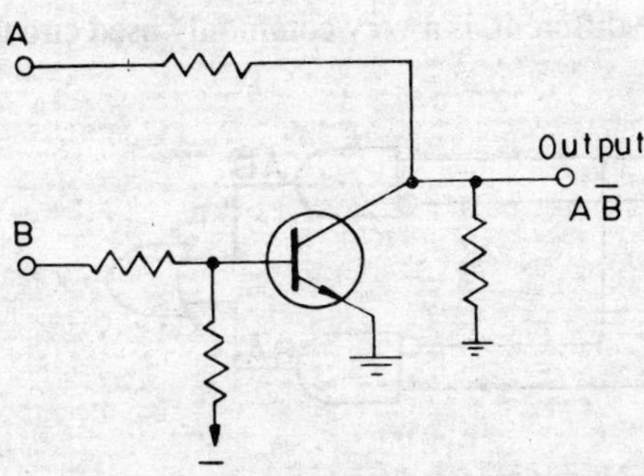


**Fig. 3.31:** INHIBITOR circuit, npn transistor.

Yet another circuit which will perform the inhibit operation is shown in Fig. 3.31. This circuit employs an npn transistor with the emitter grounded. Note that the emitter-base circuit is biased in the reverse direction, therefore, the transistor will be normally cut off.

If a positive pulse is applied to input A, current flow through the load resistor will produce a corresponding signal at the output terminal, when an input is present at input terminal B.

If, however, a positive pulse appears at input B at the same time, there is one present at the input A, this input would overcome the reverse bias and the transistor would conduct to saturation. When at saturation, the transistor is effectively a short circuit and the input pulse at A would be effectively lost; thus it would not appear at the output. In this fashion, a pulse applied to input B, inhibits the appearance of a pulse at the output terminal.

### *EXCLUSIVE OR Gate*

The OR gate discussed so far is sometimes called the *INCLUSIVE OR* gate since it produces an output when both inputs are available. An *EXCLUSIVE OR* gate provides an output when one input or the other is present *but not when both inputs are present.*

Three basic gates, AND, OR and NOT can be used to produce two input EXCLUSIVE OR gate.

| INPUT | | OUTPUT |
|---|---|---|
| A | B | F |
| 0 | 0 | 0 |
| 0 | 1 | 1 |
| 1 | 0 | 1 |
| 1 | 1 | 0 |

**Fig. 3.32:** EXCLUSIVE OR gate using AND-OR-NOT gates.

The circuit which is sometimes called an *inequality comparator, or detector*, since it only produces an output when the two inputs are different, is a very commonly used circuit in the arithmetic section of a computer.

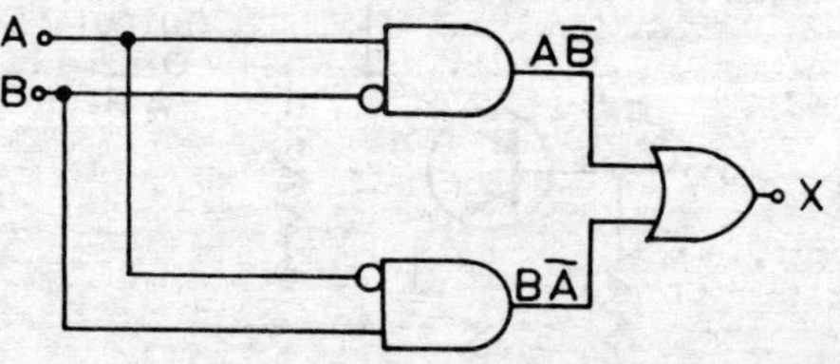


**Fig. 3.33:** Alternative logic diagram showing inverters as circles.

Some manufacturers provide in one 14-lead IC package four such EXCLUSIVE OR gates, called *Quad EXCLUSIVE OR Gate*, such as the Motorola MC 771. Each of the four gates consists of interconnected gates shown in Fig. 3.33. Two input leads and one output lead are associated with each gate and provision is made for the connection of $V_{CC}$ and ground, making a total of 14 leads.

**Fig. 3.34:** Standard symbol for the EXCLUSIVE OR gate.

The area occupied by the four gates is very small compared with the overall package, but the number of gates provided is limited by the number of connecting leads.

### 3.10 FLIP-FLOP Circuits

The flip-flop circuit, *bistable multivibrator,* is used in digital computers to supply an *output and its complement simultaneously* as an OFF/ON trigger and for storage purposes. It forms the basic circuit used in most registers.

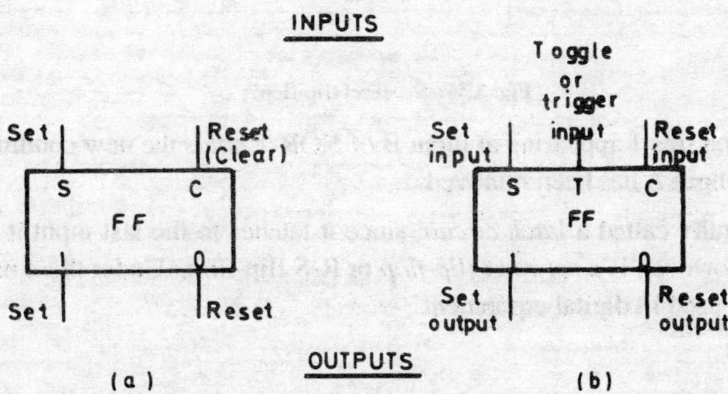


**Fig. 3.35:** Flip-flop symbols.

Like other logic symbols, *the flip-flop logic symbol is functional,* it is the same regardless of internal circuit. Thus any number of multivibrator types may be represented by the same logic symbol. One flip-flop symbol is a rectangle, Fig. 3.35(a), with terminals extending from opposite sides to indicate *set and reset* input terminals and set and reset output terminals. Reset is used interchangeably with *clear.* The OR circuits and the inverter combined in the flip-flop are not shown in the symbol although an understanding of the operation of these circuits in the flip-flop is essential in determining what affects the various types of input pulses will have on the output.

A second flip-flop symbol, Fig. 3.35(b), uses a *toggle or trigger* input termnal(T). An input signal in this terminal is applied to both halves of the flip-flop and causes both sides to change their conducting condition, from 0 to 1 or from 1 to 0.

Fig. 3.36 shows one example of interconnection between two NOR circuits. To determine just what happens here let's assume first that NOR-1 is producing a1 output (both its inputs are at 0 level). The 1 output from the NOR circuit is applied to input C of NOR-2, forcing the output of that NOR circuit to be a 0, without regard to the signal applied to input D. The 0 present at the output of NOR-2 is applied back to input B or NOR-1. If neither input A nor input D, receives a 1 input, the circuit will *maintain* its condition indefinitely.

Lets apply a 1 to input A for a moment, though, and then remove it. As soon as we apply the 1 to input A, the output of this NOR circuit drops to 0. That 0 at input C of NOR-2 allows the output of

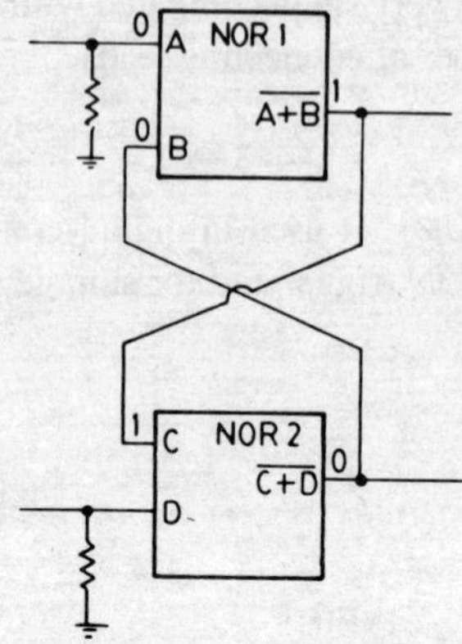


**Fig. 3.36:** Set-reset flip-flop.

NOR-2 to rise to a 1, and this 1 appearing at input B of NOR-1 *holds* the new condition rather than the old-even after the 1 at input A has been removed.

This circuit is generally called a *latch circuit*, since it latches to the last input it receives. A more common name for it, however, is a *set-reset flip-flop* or R-S flip-flop. Under these names the circuit is one of the most widely used in digital equipment.

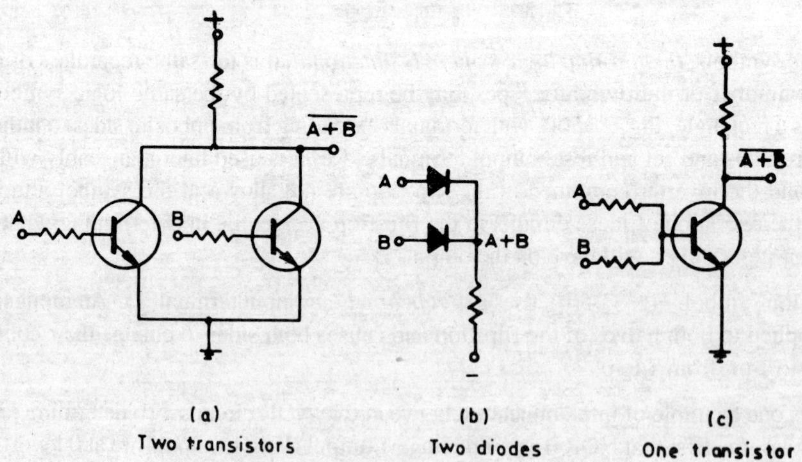


**Fig. 3.37:** OR function circuit.

The R-S version, however, is not the only type of flip-flop in use, nor is the specific circuit in Fig. 3.36 the most common way of wiring a flip-flop. Before going into circuit details of various types of flip-flops, let's look at the several methods of obtaining the OR function.

The resistor-transistor logic (RTL) OR circuit, shown in Fig. 3.37(c), operates similarly to the circuit in Fig. 3.37(a), except that it has only one transistor instead of two. Base current for this transistor may be supplied from either input regardless of the source of the base current, it will turn the transistor on.

Actually both the circuits in Fig. 3.37(a) and 3.37(c) are NOR circuits rather than OR circuits, since the transistor (s) invert the OR function. If we simply replace the original NOR circuits in the R-S flip of Fig. 3.36, with the circuit in Fig. 3.37(c), we obtain the basic flip-flop circuit shown in Fig. 3.38. *This circuit is typical of discrete component circuit practice.*

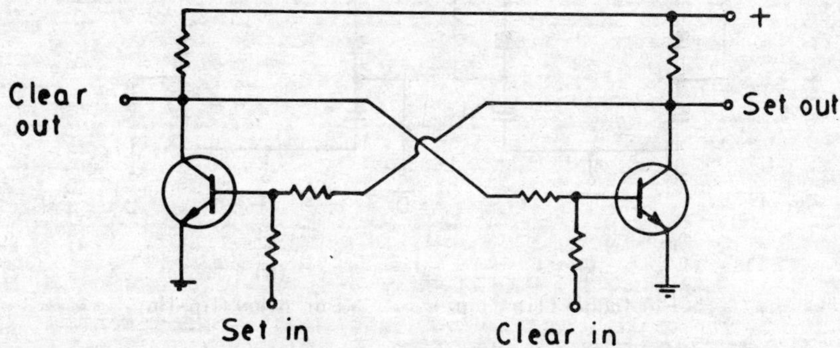


**Fig. 3.38:** Set-reset transistor flip-flop.

Frequently, designers want a flip-flop to act as a counter, or as a storage element, but not necessarily as a latch. For these purposes the flip-flop input circuits are modified. Each modification has its own name. Most (but not all) of the names were originally bestowed by engineers of the International Business Machines Corporation, but they are now virtually universal.

The R-S flip-flop latches to its most recent input signal. However, if both its inputs are made positive at precisely the same time the resulting state of the flip-flop cannot be predicted.

To overcome this, and to provide the possibility of counting, the J-K flip-flop was developed. Functionally, it differs from the R-S version only in that *it will change state if both inputs are made true at*

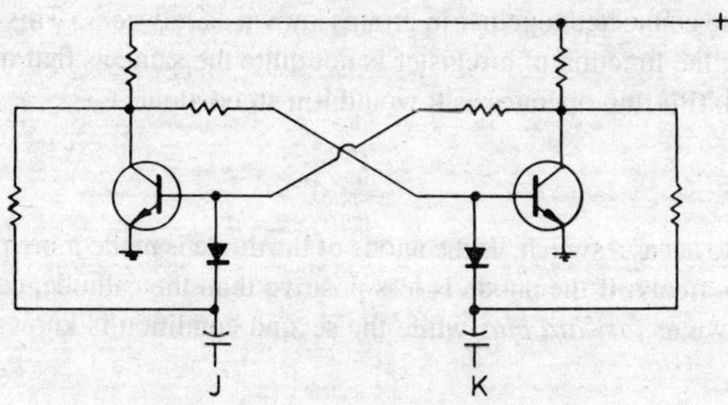


**Fig. 3.39:** J-K flip-flop circuit.

*the same time*. Fig. 3.39 shows the change made in the input circuit. The diodes act as gates to *steer* the true input signal only to the side of the flip-flop which was previously false.

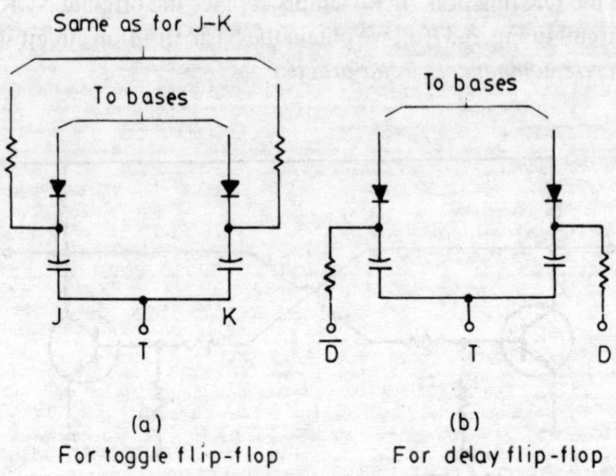


**Fig. 3.40:** Modifications of J-K input circuit.

Because of its versatility the JK flip-flop is virtually the only type available in integrated circuits. In discrete component circuits though, the other special types are often found. One is the T(toggle) flip-flop which consists only of a JK flip-flop with both inputs connected together and is used for binary counting, Fig. 3.40(a). The other is the D(delay) flip-flop, which is similar to the T type but has input resistors brought out to the D and $\overline{D}$ input terminals, Fig. 3.40(b). When a logic signal is applied to D and its complement to $\overline{D}$, and the flip-flop is subsequently triggered, the state of the logic signal will be stored.

Flip-flops may have several types of gating built into them so that some specific combination of input signals is required either to set or clear the flip-flop. However, complexity of this sort consists mainly of combinations of circuits already discussed.

Frequently, flip-flops are connected together in groups known as *registers*. This adds apparent complexity of a different type; the function of a register is not quite the same as that of a simple flip-flop. Electrically, however, each flip-flop operates as it would if it stood alone.

### 3.11 Diode Logic

A diode can also be made to act as a switch. If the anode of the diode is made more positive with respect to the cathode, it conducts freely; if the anode is less positive than the cathode, current is almost nill. The first condition is known as *forward bias* while the second condition is known as *cut off*. This is illustrated in Fig. 3.41.

This characteristic of the diode has been exploited in digital industry. Most digital devices in service to day contain at least several hundred diodes. Many devices contain thousands.

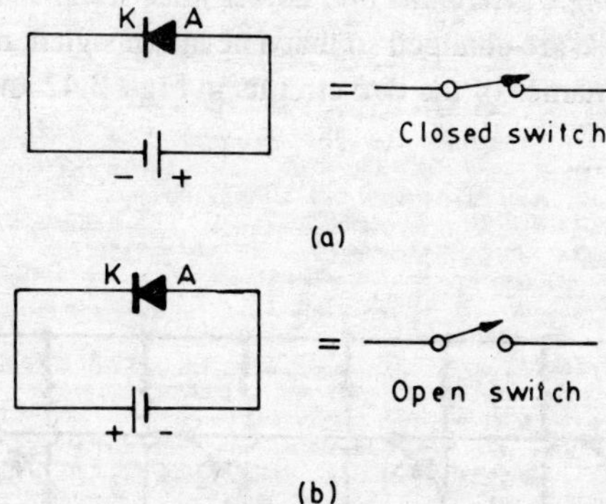


**Fig. 3.41:** Switching action of a diode.

Diode switches are used most often to peform the *AND* and *OR* functions. Fig. 3.42 shows a diode AND circuit for use where a true condition is represented by a positive voltage and a false condition is represented by ground or a negative voltage.

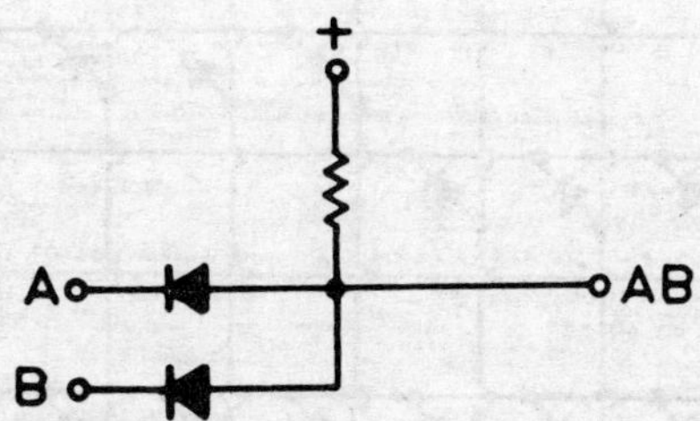


**Fig. 3.42:** Diode AND circuit.

If either input A or input B in Fig. 3.42 is false (at ground potential) its associated diode will be forward biased by the voltage (called *pull-up voltage*) applied through the resistor and all the current will be switched to the grounded input terminal. If both A and B are true (positive) the pull-up voltage will have no possible route to ground and so will appear at the output.

Fig. 3.43 illustrates the diode OR circuit for the same true and false conditions. If either input A or input B is true (positive), its diode will be forward biased and the input signal will appear at the output terminal. If both A and B are false (negative), the pull-up voltage will make the output false also.

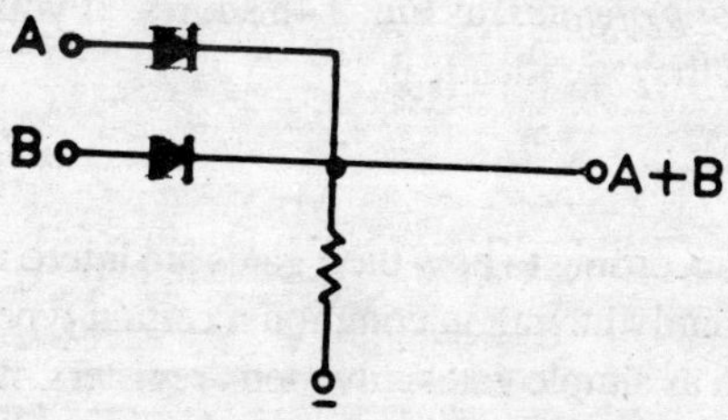


**Fig. 3.43:** Diode OR circuit.

*The similarity between ANDing true signals and ORing false signals is still present with diode logic.* If the definitions for true and false are changed so that a negative signal represents a true condition and a positive one a false signal the names of the two circuits in Figs. 3.42 and 3.43 reverse also.

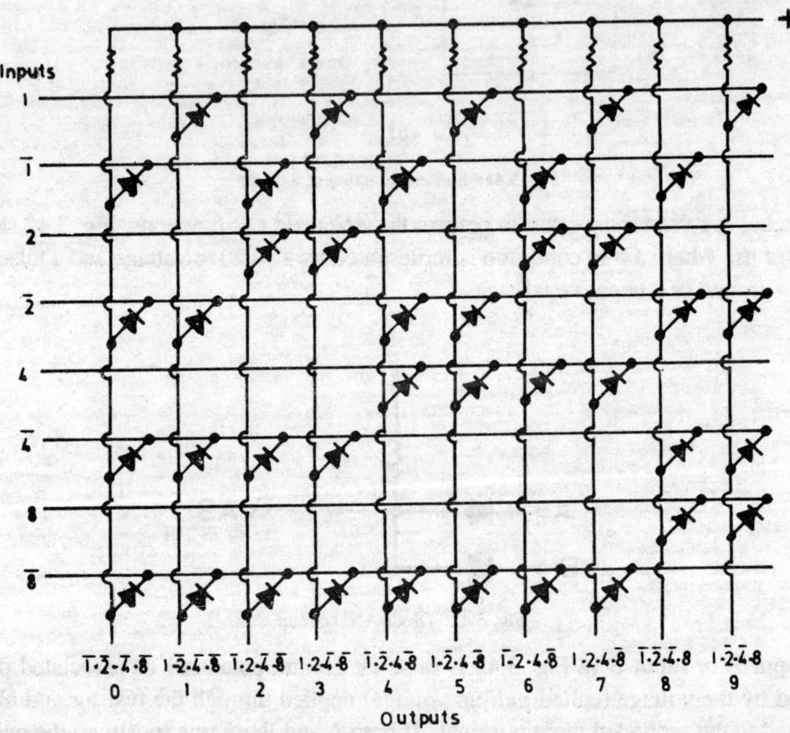


**Fig. 3.44:** Diode matrix circuit.

Not all diode logic circuits used in digital devices are simple two input gates. One frequently used circuit is the *matrix* in Fig. 3.44. This particular matrix circuit translates four-bit binary numbers into the ten decimal digits; other matrix circuits perform many other functions. For all its apparent complexity, *the matrix is only a group of gates*, as Fig. 3.45 shows. If you trace the circuits of Figs. 3.44 and Fig. 3.45, you will realise that they are identical.

## 3.12 Logic Families

Digital ICs may be classified according to how their gates are interconnected. They may thus be placed in logic families, each logic family having in common a certain type of *interconnection* and consisting of *similar logic functions* such as simple gates, inverters, registers, flip-flops etc.

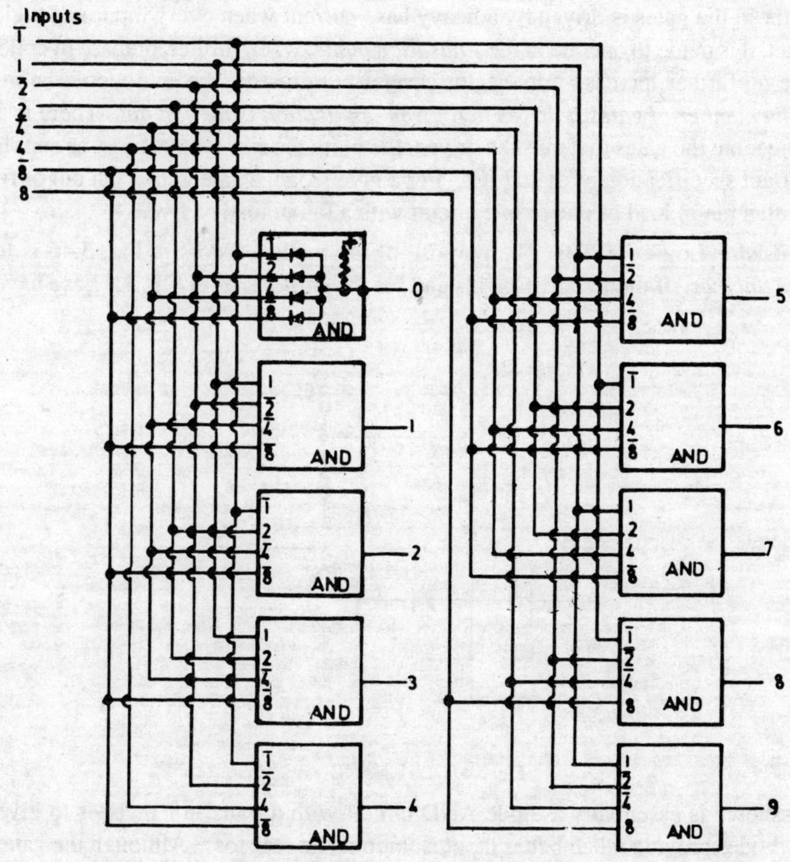


**Fig. 3.45:** Matrix as group of gates.

**1. Resistor/Transistor Logic (RTL):** A resistor/transistor logic NOR gate is shown in Fig. 3.46. The logic element is a transistor and resistors connect it to preceding gates. A positive input at A, B or C will

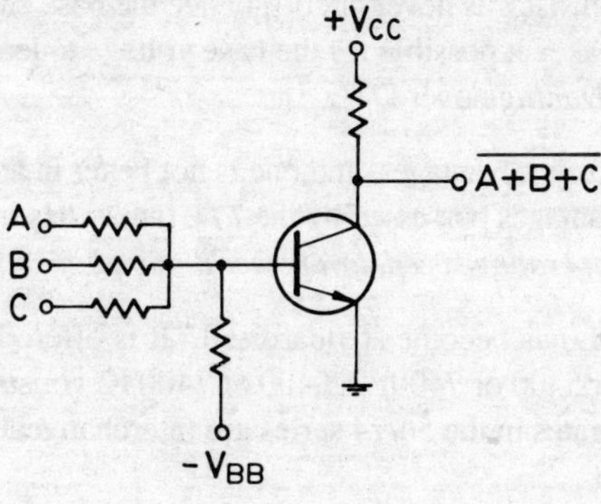


**Fig. 3.46:** RTL NOR gate.

drive the transistor into conduction, increasing the voltage drop across the collector resistor and decreasing the positive output voltage.

The transistor in the gates is driven by a heavy base current when every input is a logic 1 (positive voltage). In fact, it is made to operate in the *saturation mode*, when further increase in collector voltage would produce no further increase in collector current. Ordinarily, the gate would be made to drive other gates. *The number of gates it drives in a given application is the fan out.* There is a limit to the fan out since the more the gates that must be driven, the more the current that must be supplied. The fan out is an important specification of digital ICs. For a given load, a gate with a fan out of five can drive five circuits with a fan in load of one or one circuit with a fan in load of five.

**2. Diode/Transistor Logic (DTL):** The purpose of the input resistors in Fig. 3.46 is to *isolate the inputs from one another*. If diodes are used instead for the purpose, as in Fig. 3.47, we have diode/transistor logic.

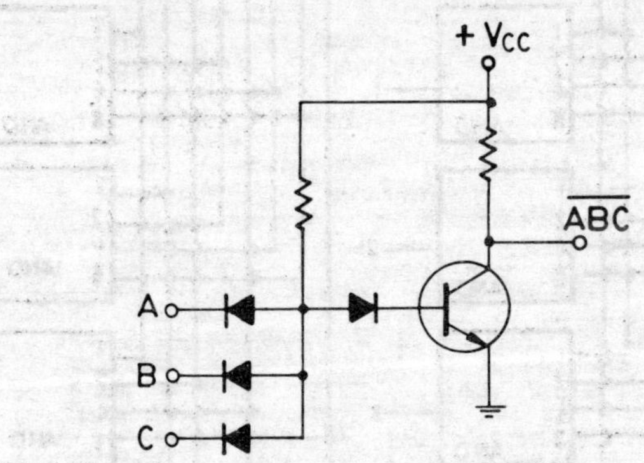


**Fig. 3.47:** DTL NAND gate.

The circuit shown is essentially a diode AND circuit with a transistor inverter to give the NAND function. The diodes provide much better input isolation than resistors. Although the gates in the DTL family can switch faster than the gates in the DTL family, the diode places a limit on the *switching speed* due to the charge stored in their junction.

**3. Transistor/Transistor Logic (TTL):** A transistor/transistor logic NAND gate is shown in Fig. 3.48. The multiple diode arrangement of the DTL circuit is replaced by a *multi-emitter transistor*. The base-collector junction of this input transistor is never cut off, hence the base current of the next stage always has a low resistance path. This makes it possible for the base voltage to leak off quicker than in the DTL circuit and gives TTL a *speed advantage* over DTL.

All of the logic families have their advantages, and one is not better in an overall sense than the others. However, the combination of advantages possessed by the *TTL* family has made it the most popular so far. *It offers a high fan out of 10, fast operating speed, low power dissipation and good immunity to noise.*

Because of its popularity, TTL has become standardised. It is offered by many manufacturers in a series beginning with the number 5400 or 7400. A 5400 or 7400 IC consists of four input NAND gates, no matter, who makes it. Logic gates in the 54/74 series are interchangeable with other ICs of the same number.

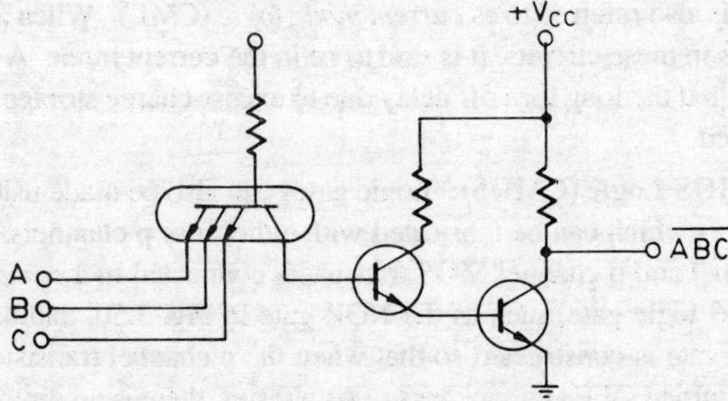


**Fig. 3.48:** TTL NAND gate.

An improved version of TTL is called *Schottky TTL*. It differs from conventional TTL in that Schottky barrier diodes with metal-semiconductor junctions are used in its construction. Logic gates of this family are designated as 54S/74S series (beginning with 54S00 and 74S00) and offer *much faster switching speeds*. Outwardly, there is little difference between the 54/74 series and the 54S/74S series since they have identical packaging and pin connections.

Low power version of both conventional and Schottky TTL gates are also available, these are designated as the 54L/74L series and the 54LS/74LS series, respectively. Reducing the power consumption of gates has advantages in reducing the size of power supplies, but it also slows down the switching speed of logic gates.

**4. Emitter Coupled Logic (ECL):** Emitter coupled logic is the *fastest of all logic families*. Its application is limited primarily to large computers, where extreme speed is required. Another advantage is that a *complementary output is available*, as shown in Fig. 3.49.

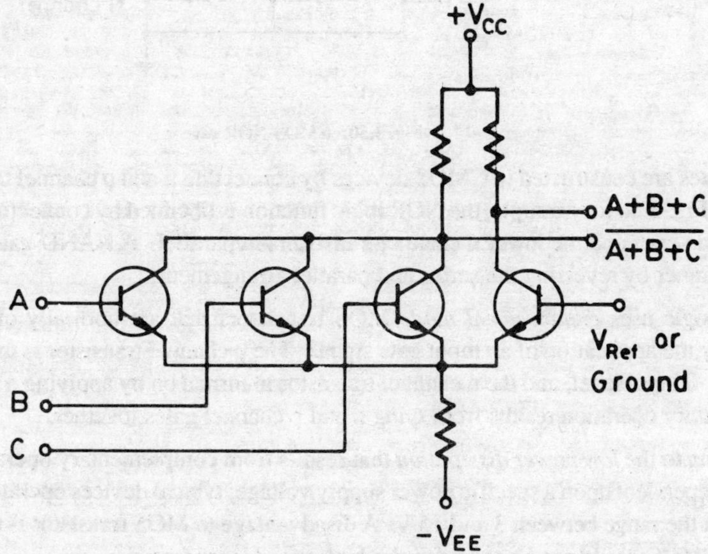


**Fig. 3.49:** ECL gate showing complementary OR/NOR outputs.

This logic function is also referred to as *current mode logic* (CML). When a transistor operates in its *nonsaturating mode*, as in these circuits, it is said to be in the current mode. A result of operating in the nonsaturating mode is that the long turn off delay due to excess charge storage is avoided. This is what gives ECL its high speed.

**5. Complementary MOS Logic (CMOS):** Logic gates can also be made using metal oxide semiconductor (MOS) transistors which can be fabricated with either n or p channels. Of particular interest is the use of both n channel and p channel MOS transistors connected in a *complementary* arrangement. A complementary MOS logic gate, such as the NOR gate in Fig. 3.50, exhibits extremely low power dissipation. The logic gate is constructed so that when the p channel transistors are turned ON, the n channel transistors are turned OFF and vice versa. As a result, there is no direct path for current to flow from the power supply to ground except for very small leakage current through OFF transistors.

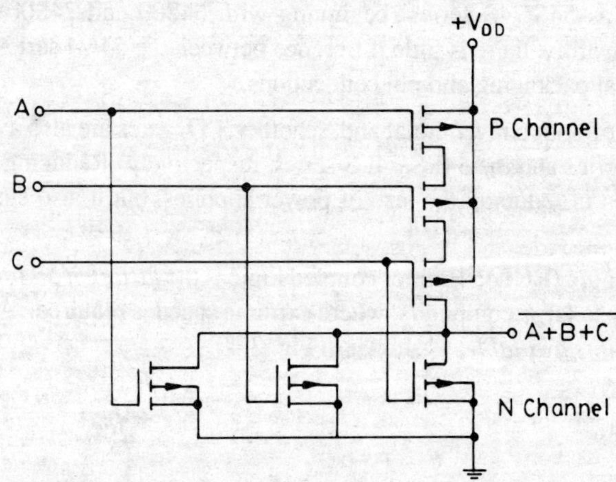


**Fig. 3.50:** CMOS NOR gate.

Logic gates are constructed in CMOS devices by connecting n and p channel transistors in series and parallel. In Fig. 3.50 for example, the NOR logic function is obtained by connecting the upper p channel transistors in series and the lower n channel transistors in parallel. A NAND gate could be obtained in a similar manner by reversing the series and parallel arrangement.

CMOS logic uses *enhancement mode* MOS transistors that are normally off and which must be turned on by the application of an input gate signal. The p channel transistor is turned on by applying a low or logic-0 input level, and the n channel transistor in turned on by applying a high or logic-1 signal. Complementary operation results from tying n and p channel gates together.

In addition to the *low power dissipation* that results from complementary operation, the CMOS logic gate is not dependent upon a specific power supply voltage; typical devices operate with supply voltages anywhere in the range between 3 and 15 V. A disadvantage to MOS transistor is that they switch rather slowly, so *CMOS would not be considered a high speed logic family*.

## 3.13 Counters

*A flip-flop can divide the frequency of an input pulse train by a factor of two.* This property can be put to good use in digital counters. A single toggle flip-flop is indeed a one bit binary counter. On the first input pulse the 0 output will go high and on the second pulse the 0 output will go low again. The flip-flop is therefore counting from 0 to 1. Several flip-flops can be combined to form counters that will count to any desired number.

### *Reset and Set*

When power is first turned on in a computer, the triggers in it may assume either the ON or the OFF state, depending on which transistors are conducting first. For this reason *a reset pulse is sent to all triggers befor the start of an actual calculation.* This pulse *clears* the triggers and *prepares* them for their storage function. In some circumstances, it may be necessary to set all the triggers before the start of a computation. In that event, a pulse is sent on the *set line* to place all the triggers in the required 1 state.

### *Negative and Positive Shift Input*

A trigger may be flipped by a dc or ac input. The dc input, as long as it remains on, exhibits a *holding action.* This means that as long as the input is on, the trigger cannot be flipped into its other status by another input. An ac input does not exhibit this holding action. Fig. 3.51 shows a negative shift input. The negative facing diode blocks a positive shift but allows a negative shift applied to a transistor base

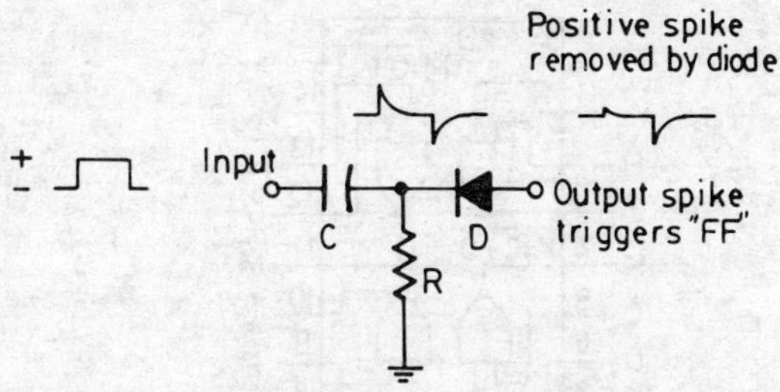


**Fig. 3.51:** Negative shift input.

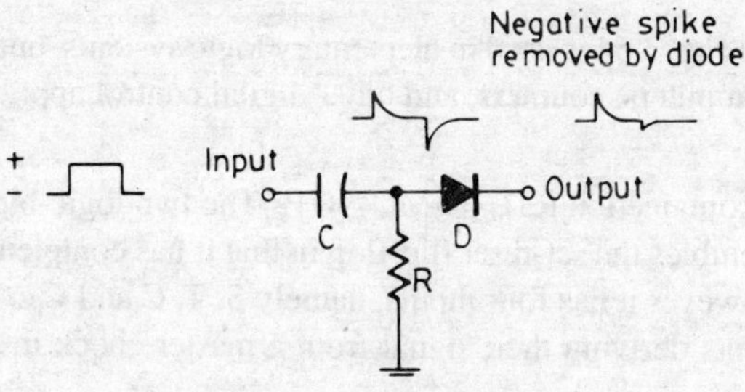


**Fig. 3.52:** Positive shift input.

to cut off that transistor, if it is npn and if it was on when the pulse was applied. Because of the RC circuit involved, *recovery time* of the input circuit is necessary. This means that once the trigger is flipped by the negative shift input circuit, a short period of time must be allowed for the RC circuit to recover (recharge) before another pulse is placed on the input.

Fig. 3.52 shows a positive shift input circuit. The positive facing diode blocks a negative shift but allows a positive shift applied to a transistor ba.:e to turn off that transistor, if it is pnp and it was on when the pulse arrived.

### *Binary or T input*

A binary or T input allows the flip-flop to be flipped regardless of its pervious state. One way of accomplishing this is to feed the same signal to both transistors, as in Fig. 3.53. Thus a pulse applied to the binary input turns the flip-flop on if it was off, or turns it off if it was on.

### *IC Flip-Flop*

The CDA 4001A IC, Fig. 3.53, housing four two input NOR gates, may be connected to give two two-input flip-flops. The NOR gate whose two inputs and outputs are, respectively, terminals 1, 2 and 3 is cross-coupled to the NOR gate whose two input terminals and output are, respectively, terminals 5, 6 and 4. Cross-coupling is achieved by connecting terminal 3 to 6 and terminal 4 to 2. The inputs are then terminals 1 and 5, and the outputs are terminals 3 and 4. One such flip-flop is shown in Fig. 3.53.

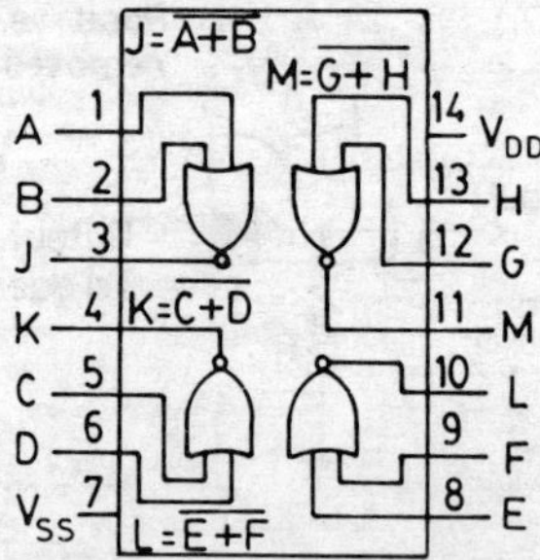


**Fig. 3.53:** Block diagram of IC CDA 4001A quad two-input NOR gates, showing terminal connections.

The set reset flip-flop in Fig. 3.54 is used in elementary logic systems, but more soplisticated circuits are employed in digital computers, counters, and other digital control applications. One such flip-flop is the JK type

Dual JK flip-flops are contained in RTL IC-MC790 P. The two logic-block symbols for this IC are shown in Fig. 3.55. It resembles the set-reset flip-flop in that it has complementary outputs Q and $\overline{Q}$ or binary 1 and binary 0. However it has four inputs, namely S, T, C and $C_D$. Of these, the first three are *clocked inputs*, that is inputs deriving their signal from a master chock in a computer, while $C_D$ is a *direct 'clear' input.*

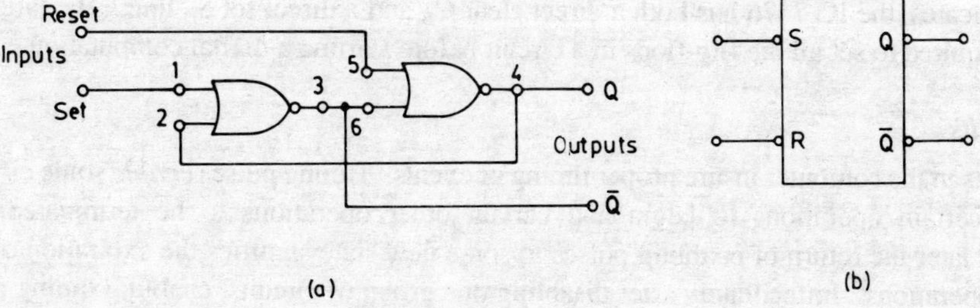


**Fig. 3.54:** Set-reset flip-flop.

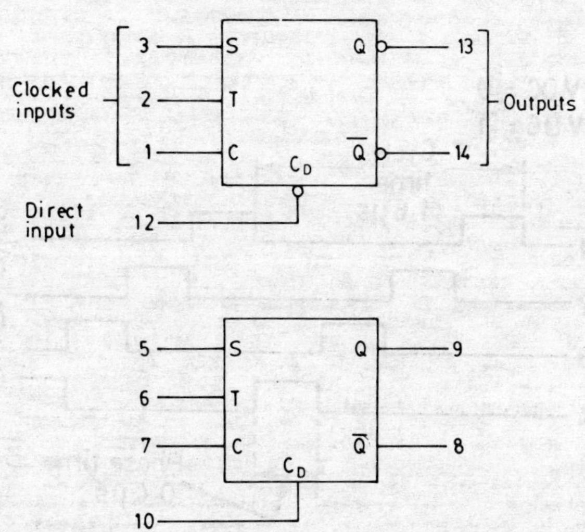


**Fig. 3.55:** MC-T90 P, dual JK flip-flop logic symbol, RTL logic.

The $C_D$ input is used to clear each flip-flop when that is required before starting an operation. The clocked inputs supply the logic information for the system. The T input is a binary, while the S and C are 'set' and 'clear' inputs.

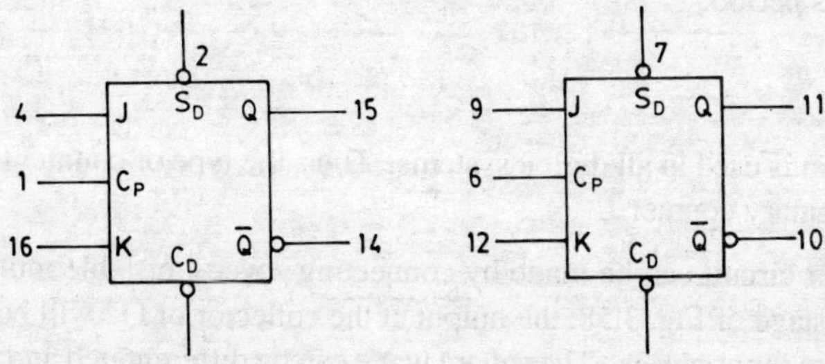


**Fig. 3.56:** IC-C7476, dual JK flip-flop with direct clear ($C_D$) and direct set ($S_D$). Clocked inputs are JK, $C_P$. Outputs are Q and $\overline{Q}$. TTL logic.

Fig. 3.56 shows yet another IC, the 7476, housing dual JK flip-flop. The clocked input to these flip-flops J, $C_P$ and K, correspond respectively to S, T and C in Fig. 3.55. The outputs are Q and $\overline{Q}$. As the figure indicates, the IC 7476 has both a direct clear $C_D$ and a direct set $S_D$ line. The latter may be used if it is required to set all the flip-flops in a circuit before starting a digital computation.

***Timing Circuits***

Timing circuits in the computer insure proper timing of events. Timing pulses *enable* some circuits, and they permit certain operations to begin and certain other operations to be terminated. A few microseconds later the return of enabling pulses to some new state signifies the expiration of time for the selected operations. Immediately after disabling one group of circuits, enabling timing pulses are applied to the others, and another set of operations is performed. *This process of enabling some circuits, and disabling others in sequence is repeated over and over again until the program has been completely executed.*

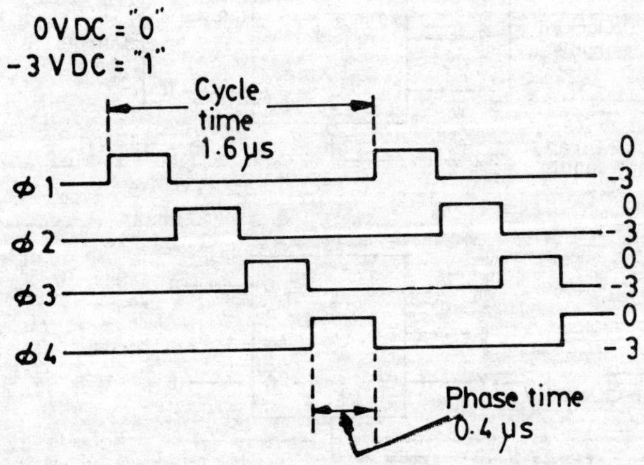


**Fig. 3.57:** Timing sequence for 4-pulse master clock.

A circuit arrengement called a master clock provides the main timing signals with-in a digital system. The master clock system, Fig 3.57, utilises a 4-phase output with a complete cycle time of 1.6 μs. The four clock phases are actually the master clock output pulses,arranged in groups of four. The four phases are obtained from an arrangement of gates and multivibrators in such a way that only one phase is high in any given 0.4μs period.

## 3.14 Binary Counters

The binary number system is used in all digital systems. Thus, the type of counter used most frequently in digital systems is the binary counter.

A simple basic counter circuit can be made by connecting several bistable multivibrator circuits in series. Taking the basic stage of Fig. 3.58, the output at the collector of Q2 will be a square wave with a frequency half that of the input pulses. The square wave can be differentated and positive pulses from the ON step used to trigger the next stage and so on.

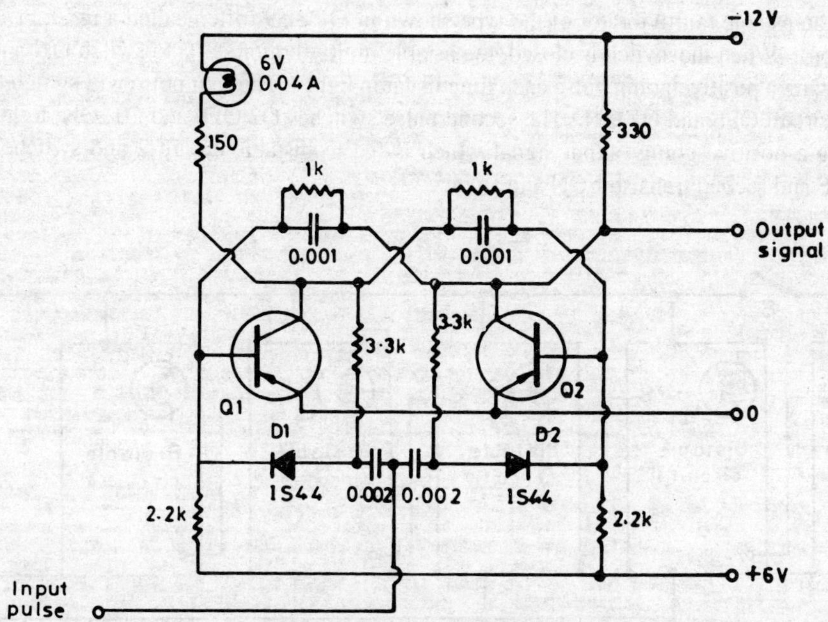


Fig. 3.58: Counting with bistable multivibrator.

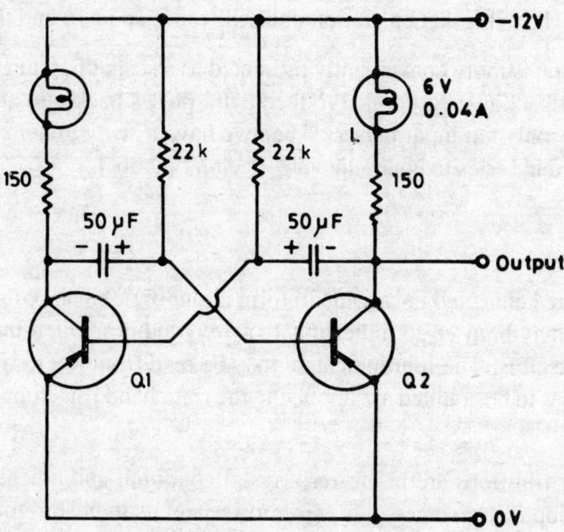


Fig. 3.59: Counting with astable multivibrator.

To demonstrate this counting arrangement four binary circuits can be arranged, as shown in Fig. 3.60 and fed from an astable multivibrator of the type shown in Fig 3.59 (often called a *pulse generator* in this application). When the switch is closed, the astable multivibrator starts to switch ON and OFF at about 1Hz, giving a positive going pulse each time its lamp lights. The first pulse will switch Q1 of the first bistable circuit OFF and Q2 ON. The second pulse switches Q2 OFF and Q1, ON, lighting lamp L1 and giving a positive going output signal which is fed to bistable circuit 2 and switches its first transistor OFF and second transistor ON and so on.

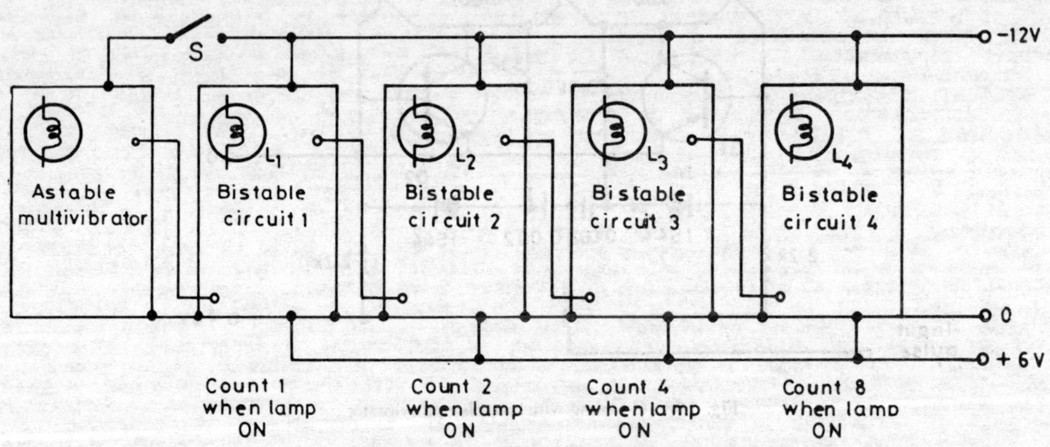


**Fig. 3.60:** Demonstration counting unit.

The sequence of operations will pass through the chain, the number of input pulses upto any moment being indicated by which lamps are ON, adding 1 when L1 is ON, 2 when L2 is ON, 4 when L3 is ON and 8 when L4 is ON. Thus, this demonstration unit will count up to 16 and then start again.

The output information is more conveniently presented as a scale of 10 to give decimal information. To accomplish this, feedback is incorporated to feed extra pulses back into an earlier stage so that the four stages count 16 for only ten input pulses. Then we have a *scale of ten counter*, and several such circuits can be combined in series to count upto any decimal number.

## 3.15 Counting

Four flip-flop circuits are connected in cascade to form a four digit binary counter, as illustrated in Fig. 3.61. Indicator lamps may be used to indicate a 1 or *set* condition when they are ON, or 0 or *reset* condition when they are OFF. The four indicators may be read from left to right as a four digit binary number. The input pulses to be counted are applied to the right hand flip-flop which represents the least significant digit.

Assume that all four flip-flops are in the zero (reset) condition and a negative input appears at the input, Fig. 3.61. Flip-flop FF1 changes state (reset to set)and its indicator goes ON, but FF1 produces no output and the other three flip-flops continue to indicate logic 0. The four indicators may be read as a binary number, 0001. This is shown in Fig. 3.61(a).

With the second input, FF1 returns to 0, and produces a negative going input to FF2. Flip-flop FF2 changes to the logic 1 state, but produces no output. The binary indication now is 0010 (the binary equivalent of binary 2) indicating that two inputs have been counted.

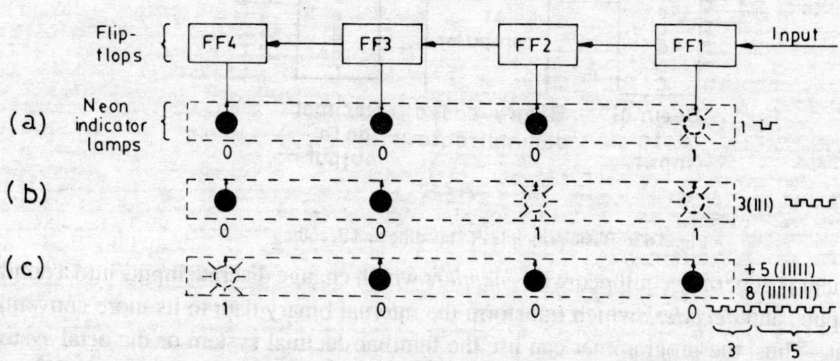


**Fig. 3.61:** Adding function of the flip-flop.

With the third inputs, Fig. 3.61(b), FF1 changes to the logic 1 (sets), but it produces no output. FF2, therefore, remains in the logic state 1, and the binary indication is now 0011 (decimal 3).

Now let five more pulses go into the flip-flop. The first pulse of this group of five produces outputs from FF1 and FF2 since they are ON. It also switches FF3 ON, but in going through it turns FF1 and FF2 OFF. The second pulse of the five merely turns FF1 ON, so that the indication is 0101, which is 5 in binary notation. You can follow the same sequence upto eight, at which point the flip-flops will indicate 1000, as shown in Fig. 3.61(c).

The counter returns to its original state (0000) after 16 inputs. If another flip-flop were added the counter would count upto 1111 (decimal 31), and return to the original state in the 32 nd input, and so on.

This example illustrates the counting function possible with a series of flip-flops. With modifications and elaboration of this, and the addition of other circuits, it is possible to perform the fundamental arithmetic operations.

The flip-flops are capable of *storing information*-in other words, they have *memory*. The unit which stores a number, then adds another number to the stored number and then the sum of the two, is called an *accumulator*.

### 3.16 Encoders and Decoders

Digital circuits are designed to operate on data that is in the binary form, that is either 0 or 1. In most computers it is not practicable to enter numbers and other information in binary form, as this would require the programmer to spend too much time in the detailed effort of accurately representing large numbers of complicated alphabets and symbols. Likewise, it is not usually desirable to present the final output of a digital circuit in binary form as this requires too much time in reading and interpretting. Thus it becomes necessary to perform suitable *conversion* on both input and output information.

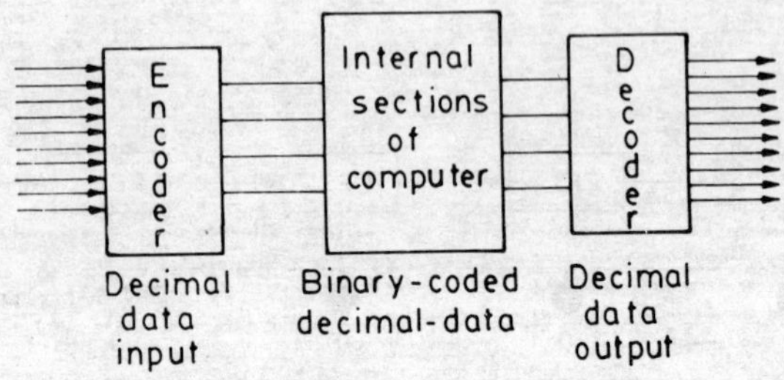


**Fig. 3.62:** Basic principle of encoding and decoding.

Digital equipment is generally equipped with *encoders* which change discrete inputs into a combination of coded outputs, and *decoders* which transform the internal binary data to its more conventional form at the output. Thus, the programmer can use the familiar decimal system or the octal system to programe computers, depending on the encoder used to convert the data to the binary form as required internally by the computer. The output of the internal circuits of the computer is then changed by the decoder back to the familiar decimal form. The basic principle involved is illustrated in Fig. 3.62.

### *Encoders*

The simple *decimal-to-binary* encoder, Fig. 3.63, receives any one of ten decimal numbers at its input and produces a binary coded decimal output. This action can be seen by a study of the circuit using OR logic. There are more complex types of encoders.

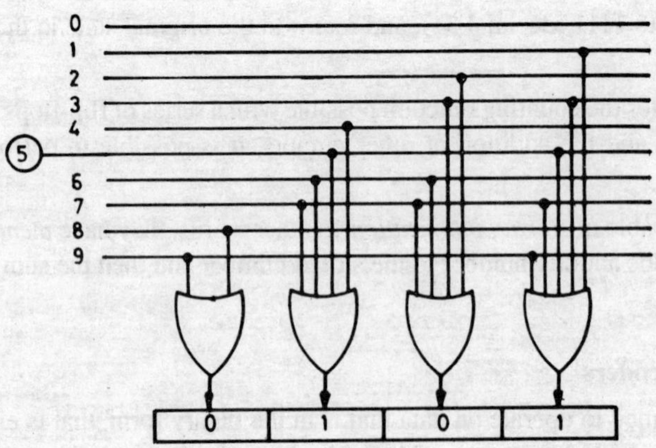


**Fig. 3.63:** Decimal-to-binary encoder.

The basic principle of *binary-to-decimal* decoding, Fig. 3.64, uses flip-flops to provide the input to the AND element, which in turn, produces the decimal output. This type of circuit produces an output only when the right combination of flip-flop outputs is applied to the inputs of the AND element. Appropriate use of either inverters or flip-flop output connections makes possible the decoding of any binary-coded decimal digit.

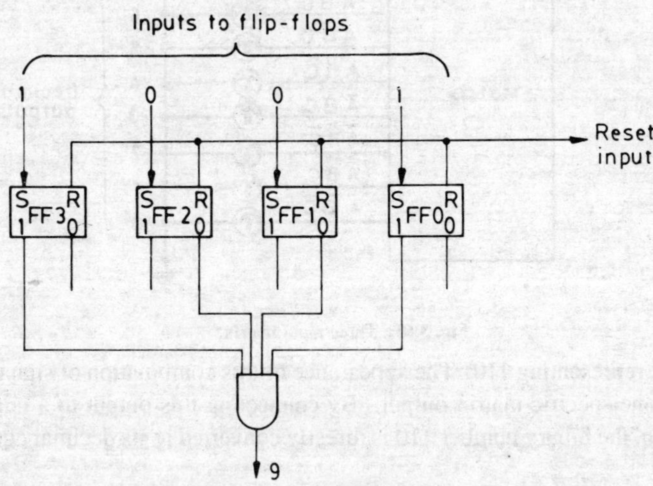


**Fig. 3.64:** Binary-to-decimal decoder.

Encoding and decoding uses a *matrix* to translate between number systems. A matrix is an array of circuit components such as wires, diodes, magnetic cores and relays. The matrix performs a specific function, such as converting from one numercial system to another. In mathematics, matrix means a set of terms which operate on one type of number to produce a second type-in other words a *number translator*.

For any particular logic circuit, a single output is produced if some specific combination of inputs exists. Further, there is a fixed number of possible input combinations (depending on the number of inputs used) each one of which is called a *minterm*.

A circuit with two inputs has ($2^2$) or four minterms, AB, $\overline{A}B$, $A\overline{B}$ and $\overline{AB}$. A circuit with three inputs has ($2^3$) or eight minterms, ABC, $AB\overline{C}$, $A\overline{B}C$, $A\overline{BC}$, $\overline{A}BC$, $\overline{A}B\overline{C}$, $\overline{AB}C$ and $\overline{ABC}$. In general, for n inputs, there are $2^n$ minterms. *Each minterm indicates one of the possible input combinations that can occur.*

The logic matrix differs from other circuits, discussed so far in this chapter, in one respect that *a matrix with n inputs has $2^n$ outputs-one output for each minterm. It produces a signal at one specific output terminal for any combination of inputs.*

Consider the simplified illustration in Fig. 3.65. The matrix has three inputs (A, B and C) each of which can be in either the 0 or 1 state. Each combination of digits represents a binary number so that ABC is interpreted as 111, $\overline{A}\,\overline{B}\,\overline{C}$ as 000, $\overline{A}\,B\,\overline{C}$ as 010, and so on. In the illustration, a signal exists at

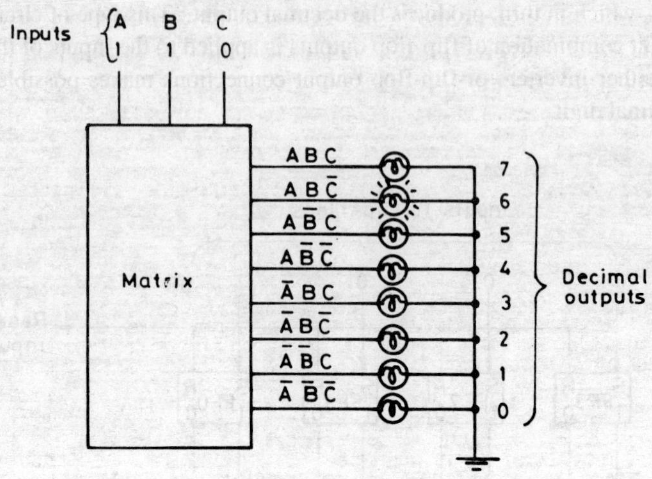


**Fig. 3.65:** Three input matrix.

A and B, but not at C, representing 110. The appearance of this combination of signals at the input will produce a signal on one specific matrix output. By connecting this output to a numbered indicating device, such as a lamp, the binary number 110 is directly converted to its decimal equivalent 6.

## 3.17 Decade Counters

Decade counters consist of a set of *decade counting units* (DCUs), one for each figure displayed in the read out. Each DCU is made of flip-flops cascaded in such a way that when one changes all the others will be affected. Each flip-flop is a binary counter that is worth a 1 or 0 according to its state. If we start out with our flip-flops all in the 0 state and supply a triggering pulse to the first flip-flop, it will change to the 1 state. A second pulse will return it to the 0 state but in doing so will trigger the second flip-flop to the 1 state, and so on with succeeding pulses.

Since each binary coded decimal (BCD) counts from zero through nine, you can use them to count any size of number, with one DCU for each digit in the read out, as shown in Fig. 3.66. As the tenth

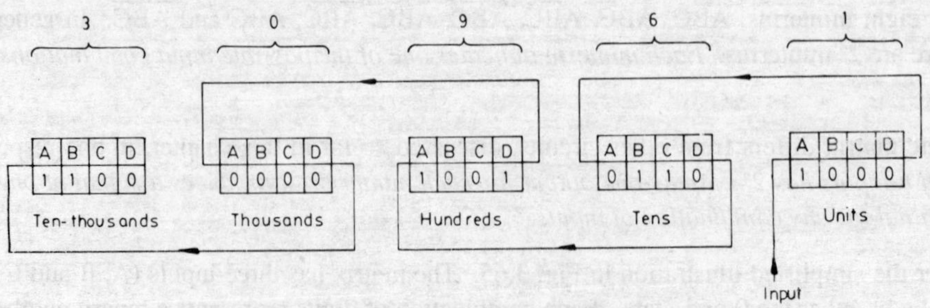


**Fig. 3.66:** Decade counting units.

pulse causes the flip-flop in the units DCU to return to the 0 state, it causes the flip-flop in the tens DCU to change state to 1. In this way the tens DCU counts 1 for every ten counts of the units DCU. A1 is carried forward to the next DCU to the left each time the DCU to the right counts to ten, just as in simple addition.

In Fig. 3.66, the state of each DCU is shown in this binary form with its decimal equivalent above. The latter is what we want the read out to display.

The circuit of a four digit decade counter is given in Fig. 3.67. There is a set of ICs for each DCU and an associated display unit. One of these ICs is for the decade counter and the other IC operates as the decoder-cum-driver. The decade counter ICs are of the "divide by ten" type.

An incoming negative pulse changes the state of the first decade counter IC1. After every ten negative pulses, a carry pulse is generated by IC1 to change the state of the second decade counter IC2. After every hundred pulses a carry pulse is generated by IC2 to change the state of the third decade counter IC3, and so on. These changes, brought about by *flipping and flopping* are fed to decoders/drivers which decode and drive the light emitting diodes in the display unit.

The ICs operate most satisfactorily at voltages between 1.75 to 5.25 V. Although the circuit is for four digits only, it can be extended further to read the count of any number of digits.

As the counter functions on a negative going pulse, the input is left open and is considered to be at a high level. The counter continues with each negative pulse of about 0.8V. The counter triggers only with a square wave pulse. Any other wave should first be converted to a square wave pattern.

The counter can also be operated by a relay (IC-9). The IC is labelled and its connections are also given. The common contact of the relay is grounded and the other two contacts are connected to the input terminals (pins 2 and 5). The operating dc voltage for the IC is fed to pin 14 and negative going pulses are available at pins 3 and 4 which are shorted. The relay contacts can also be connected directly to the counter (without IC-9) but that may cause bouncing of the relay to give a faulty count.

The counter can also be attached to the relay in a light operated switch. When the light beam is interrupted the counter operates.

The counter can be reset to 0000 by applying a high pulse to pins 2 and 3 of ICs 1, 2, 3 and 4. As these pins (2 and 3 of ICs 1 through 4) are connected in parallel all the decade counters are reset to 0000 by disconnectng S2 momentarily. This is done before each count. A reset 9999 is also provided. When this is opened all the four ICs are reset to the 9 state.

List of components

| | | | |
|---|---|---|---|
| ICs 1 to 4 | IC-7490 | C1 | 500MF/25V |
| ICs 5 to 8 | IC 7447 | C2 | 500MF/12V |
| IC 9 | IC 7400 | Z1 | Zener diode 4.6V/400 mW |
| D1 to D4 (Displays) | FND 507 (common mode) | TR1 | 2N3055 with heat sink 4" × 4", 2 mm thick aluminium sheet |
| T1 | Primary 220V Secondary 9-0-9V | | |
| D1 and D2 | 1N 4001 | $R_1$ $R_2$ to $R_{29}$ | 1K/ 1/2W 180Ω/1/4W |

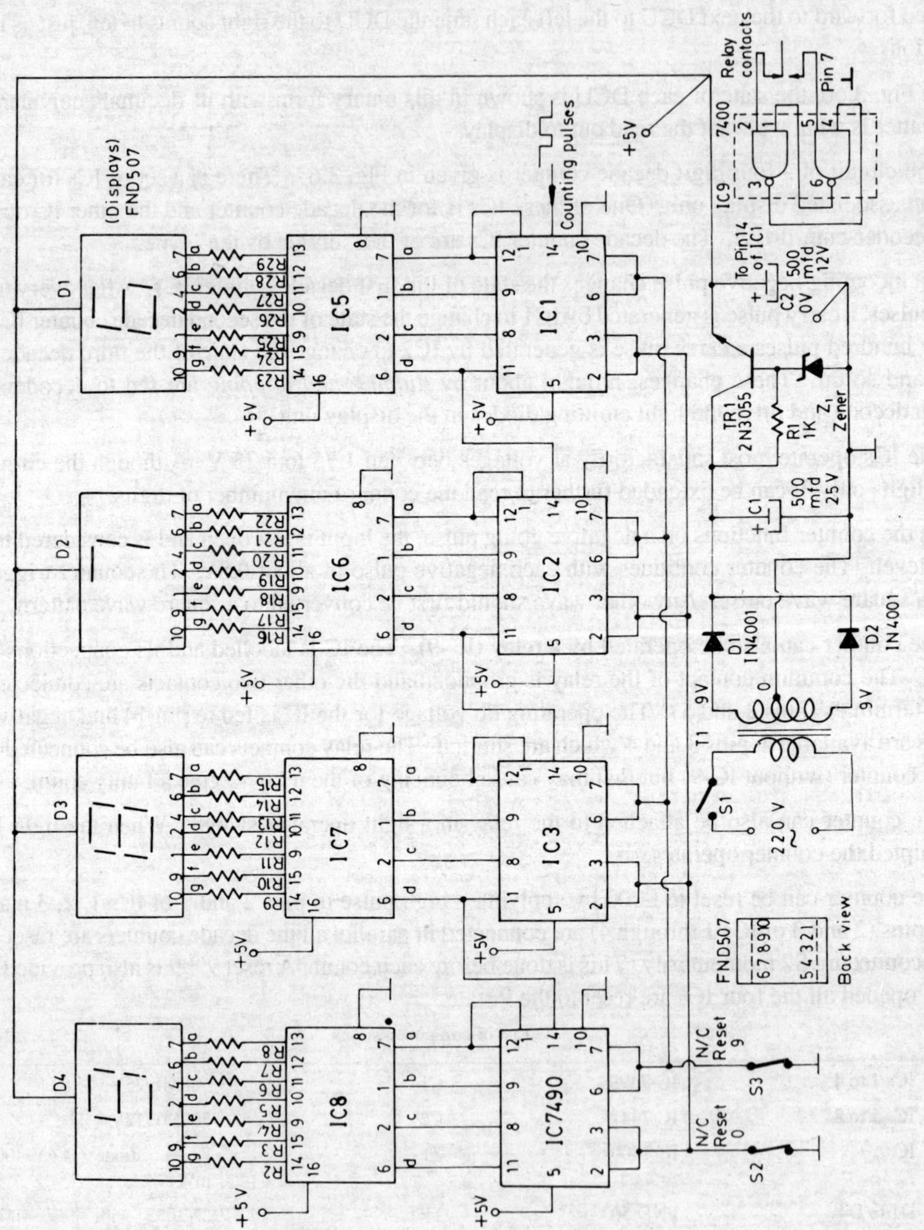


**Fig. 3.67:** A four digit decade counting unit (DCU).

## 3.18 Frequency Dividers

Frequency dividers, as their name implies, perform the function of frequency division. This is generally accomplished by impressing a single frequency at the input of the divider. *Frequency dividers always divide by a whole number*, that is, the output frequency is one-half, one-third, one-fourth, one-fifth, or a still smaller integral fraction of the input frequency. Viewing the operation in reverse, it can be said that *the input frequency is always related harmonically to the output frequency.* The output frequency is said to be a *subharmonic* of the input signal.

Harmonic generation and modulation are the basic principles involved in frequency multipliers and frequency translators. Although these principles also have relevance in the performance of frequency dividers, a large family of frequency dividers is more directly based on the concept of synchronisation.

### *Synchronisation*

Numerous circuits generate relaxation oscillations for which a common approach may be used to attain synchronisation. The process of synchronisation leads directly to frequency division for *division by*

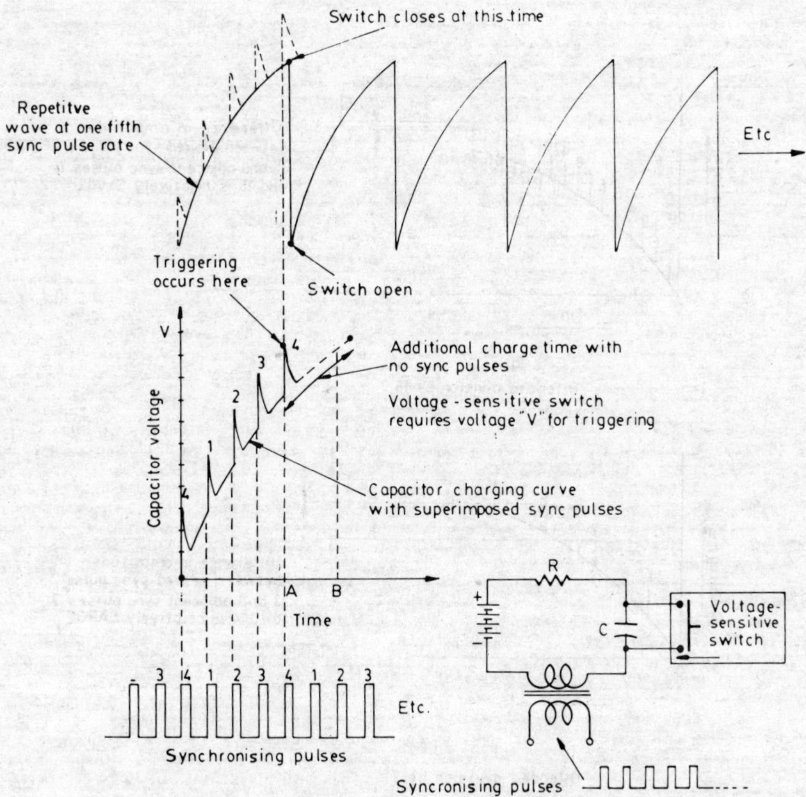


**Fig. 3.68:** Arrangement for simulating synchronisation mechanism in a large number of electronic devices.

*factor* **n** *is brought about by causing the relaxation cycle to repeat every* **n***th pulse of the synchronising signal.*

Free running oscillators oscillate on their own accord when the synchronising signal is removed. Many, but not all, relaxation osciallators utilise some type of *voltage sensitive switch* which is impressed with the voltage from a capacitor being charged. When the capacitor voltage reaches the *firing potential* of such a switch, a more or less effective discharge path is placed across the capacitor. The stored voltage is *depleted* until the voltage sensitive switch opens again. Then a new charge cycle commences. This process is repetitive, and its consequence is a near sawtooth waveform. *Synchronisation consists essentially of speeding up the charge cycle so that the nth sync pulse suffices to fire the voltage sensitive switch. The repetition rate of the sawtooth waves then is forced to be 1/n times the rate of sync pulses.*

Fig. 3.68 shows an arrangement simulating such synchronisation. Without synchronisation B time units are required for the capacitor voltage to reach firing potential (V) of the switch. With synchronisation, V is reached in the *shorter* time period of A. The "hurry up" process due to the syncpulses superimposed on the capacitor charging curve, then repeats itself so that one relaxation cycle is always

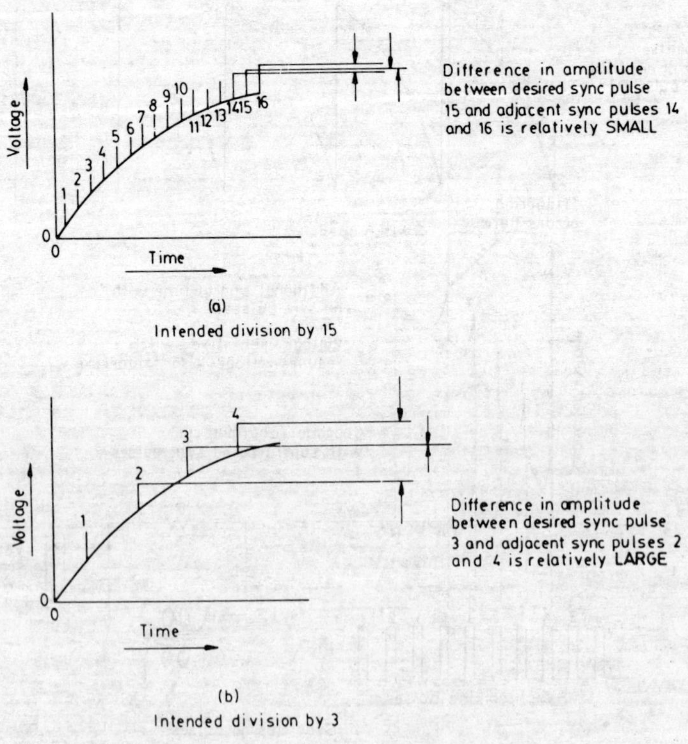


**Fig. 3.69:** Capacitor charging curves with superimposed sync pulses showing why sync tends to be more stable for small dividing networks.

terminated at the end of n sync pulses. As illustrated in Fig. 3.68, n is 4 and the synchronised oscillator then performs as a 4-to-1 frequency divider.

The greater the division ratio involved in synchronisation, the more the difficulty experienced in maintaining synchronisation by factor n rather than (n+1) or (n–1).

A greater amplitude difference exists between adjacent sync pulses on the charge wave when a few pulses are involved rather than when there are many pulses (see Fig. 3.69). *Sync stability tends to be better when the firing pulse rides on the steeper portion of the charging curve.* If synchronisation occurs late in the charging cycle, say after the elapse of several time constants, there will not be much amplitude difference between adjacent pulses, even for low division ratio.

## 3.19 UJT As Frequency Divider

Fig. 3.70 illustrates the characteristcis of the unijunction transistor (relevant to its use in frequency divider circuits), with emitter current as a function of the voltage applied between the emitter and base 1. This representation is more practically related to frequency changing than are the curves frequently encountered in technical literature; these curves portray voltage as a function of current.

In virtually all frequency divider circuits utilising UJT, a rising capacitor voltage is, from a circuit view point, the *cause* of operation; current change from this viewpoint appears as an *effect*.

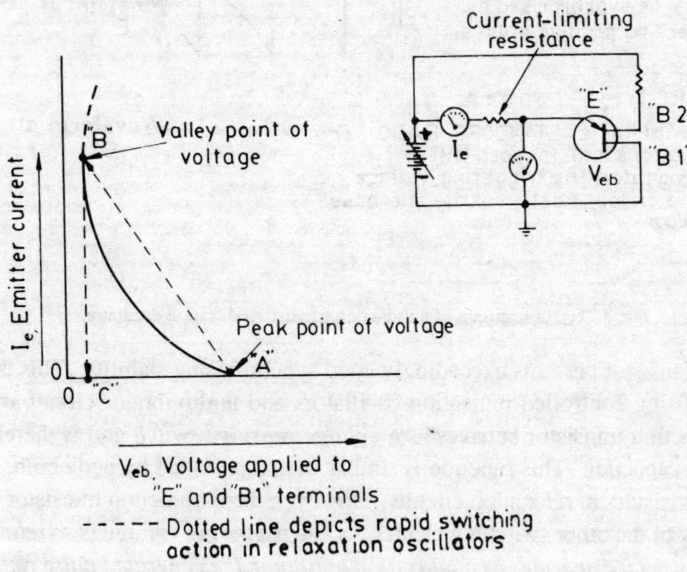


**Fig. 3.70:** Static characteristics of emitter-base 1 section of a unijunction transistor.

In relaxation oscillators whenever emitter limiting resistance is used, the switching from the peak point A (Fig. 3.70) to the valley point B occurs very rapidly. This is likewise true for the switching transition between points B and C. The characteristic between A and B has a *negative slope*. This represents negative resistance and accounts for the *instability* of the device when net positive resistance does not exceed the negative resistance between emitter and base1. Although instability often denotes undesirable properteis it is, in this case, the very characteristic responsible for abrupt switching between points A and B. The second switching transition occurs when the applied voltage between emitter and base 1 is reduced between the voltage valley point B. This change of conductive state is analogous to deionisation of neon lamp.

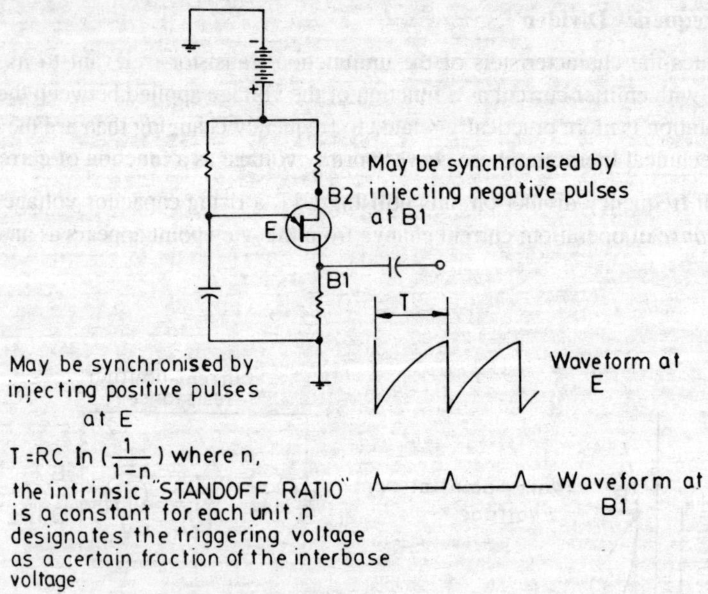


**Fig. 3.71:** Basic circuit of unijunction transistor relaxation oscillator.

The unjunction transistor persents exceedingly good synchronising stability. This is attained in frequency dividers utilising controlled relaxation oscillators and multivibrator circuit arrangements. In such circuits, unijunction transistor behaves as a *voltage sensitive switch* and is thereby used to periodically discharge a capacitor. This function is similar to that provided by neon bulb, thyratron tubes, and conventional transistors in relaxation circuits. However, the unijunction transistor possesses a feature not found in any of the other switching elements. *The triggering voltage is so related to the supply voltage that when the device operates as a relaxation oscillator, the generated pulse rate is substantially independent of the supply voltage.* This is quite contrary to the situation prevailing in relaxation oscillators and multivibrators employing other switching elements. This is perhaps one factor which helps acheive reliable synchronisation in unijunction transistor frequency dividers. Other factors are, *the relative freedom from effects of radiant energy, temperature, mechanical vibration and aging.*

## 3.20 Tunnel Diode Frequency Dividers

The tunnel diode is an excellent element for accomplishing frequency division by means of various switching circuits and relaxation oscillators. Fig. 3.72 shows the equivalent circuit of a tunnel diode amplifier. The relationship between the various parameters determines whether or not the circuit is stable. The frequency divider applications require operation in *unstable mode.*

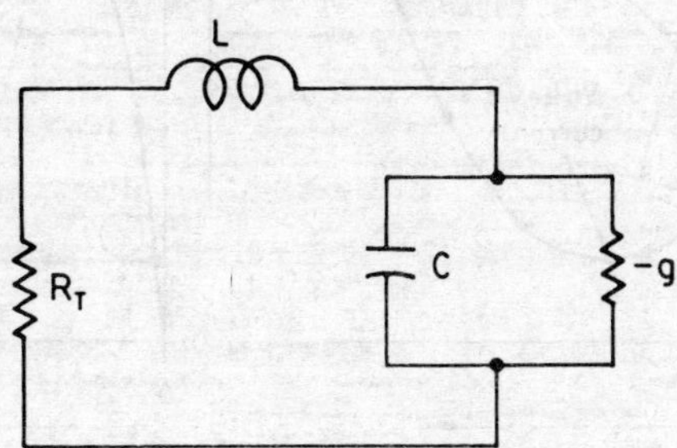


**Fig. 3.72:** Equivalent circuit of a tunnel diode amplitier or oscillator.

The criterion for stability is given by the following inequality:

$$Lg/C < R_T < 1/g \quad 3.1$$

where

L is the total series inductance carrying current to the tunnel diodes;

C is the total capacitance appearing across the tunnel diodes;

$R_T$ is the total series resistance carrying current to the tunnel diode, and

-g is the negative conductance.

The important parameter (g) is generally given in the manufacturer's specifications. For example, the General Electric 1N2939 and 1N2939A are popular tunnel diodes having a value of g equal to- 0.0066 mho. The negative resistance is the reciprocal of g or –1/0.0066 = –151.5 ohms.

When no physical inductor is connected in the tunnel diode amplifier circuit, the two characteristics depicted in Fig. 3.73(a) and (b) are possible. The characteristic shown in Fig. 3.73 (a) is obtainable when the total series resistance carrying current to the tunnel diode is *less than* the numerical value of 1/g. This is the classic N curve which includes the negative resistance region as stable loci.

The characteristic shown in Fig. 3.73(b) results where the total series resistance carrying current to the tunned diode is *greater than* 1/g. This is the *operational mode* most suitable for frequency dividers. Note the abrupt switching transition at points 2 and 4. The negative resistance region is skipped in the cycle of events 1, 2, 3 and 4. If the tunnel diode "sees" *a voltage-regulated power supply*, the characteristic of Fig. 3.73(a) is produced. If the tunnel diode "sees" *a current regulated supply*, the switching cycle of Fig. 3.73(b) may be obtained. Ideally, the first condition corresponds to $R_T$, with a value of infinity.

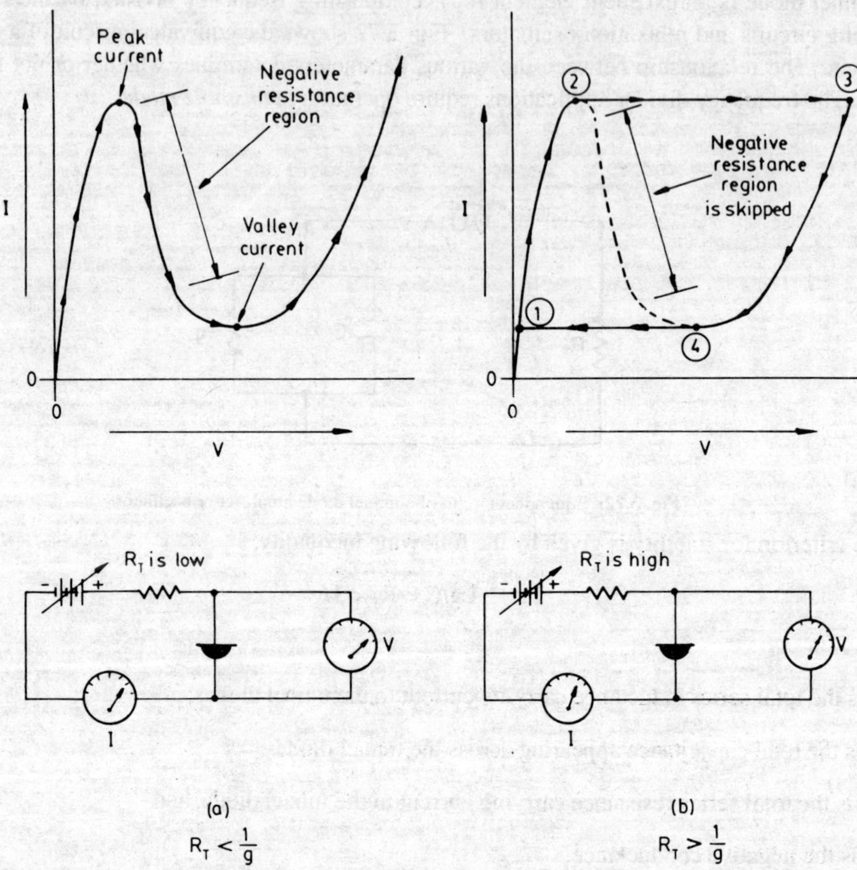


**Fig. 3.73:** Tunnel diode characteristic curves.

In practical circuits, the first condition prevails as long as $R_T$ is less than the numerical value of the negative resistance; the second condition pervails as long as $R_T$ is more than the numerical value of the negative resistance. The facts described above apply when no inductance is added to the circuit.

## 3.21 Ring Counter Frequency Dividers

The ring counter type of frequency dividers consist of *a closed series of ON/OFF switching elements.* There are many elements which can be used. Included in this category are transistors, tunnel diodes, neon bulbs, thyratrons, four layer diodes and unijuction transistors. There are also many variations in circuit operation. However, the block diagram of Fig. 3.74 will serve to illustrate the basic concept of this family of frequency dividers. In this illustration, *all switching elements are triggered in parallel. Only one of these elements is in the ON, or conducting state. All the other elements are in the OFF, or nonconducting state.*

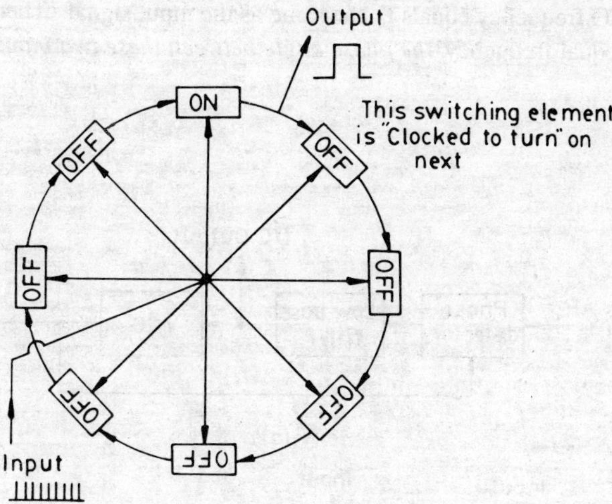


**Fig. 3.74:** Basic concept of ring counters.

The circuit connections are such that the ON element always drives the next (in this case) clockwise element to the threshold of its triggering potential. Due to this feature, each input pulse triggers one element at a time in clockwise sequence. As each new element is turned ON it turns off the preceding ON element. Thus a series of output pulses results in a progressive clockwise rotation of the ON state from element to element. It follows that *an output taken from any element provides frequency division by n, where n is the number of elements connected in the ring.* Such a frequency divider stage is aperiodic and delivers no output in the absence of input.

Increased reliability and greatly extended high frequency performance result from the use of *solid-state switching elements* in frequency dividers based on the ring counter. There are many semiconductor devices capable of performing the requisite *bistable function.*

## 3.22 Phase Locked Loop (PLL)

A phase locked loop is basically a *closed loop feedback system.* The action of the PLL is to lock or synchronise the frequency of a controlled oscillator to that of an incoming signal. Phase locking is not a new prinicple. The implementation of *PLL with discrete components* involves circuits of considerable cost and complexity. For this reason, the use of PLL in the past has been limited to specialised measurements. The development of *integrated PLL circuits* now makes it highly economical as well as reliable. Sophisticated phase lock techniques are currently available in a wide variety of signal detection and signal processing applications.

Fig. 3.75(a) illustrates the block diagram of a phase locked loop. An input signal with a frequency $f_x$ is one of the inputs to a phase detector. The other input comes from a voltage controlled oscillator (VCO). The output of the phase detector is filtered by a low pass filter. This removes the original frequencies, their harmonics, and the sum frequency. Only the *difference frequency* (ac voltage) comes out of the low pass filter. The dc voltage then controls the frequency of the VCO.

The feedback system locks the VCO frequency on to the input frequency. When the system is working correctly, the VCO frequency equals $f_x$, the same as the input signal. Therefore, the phase detector has two inputs with equal frequency, *the phase angle between these two inputs determines the amount of dc output.*

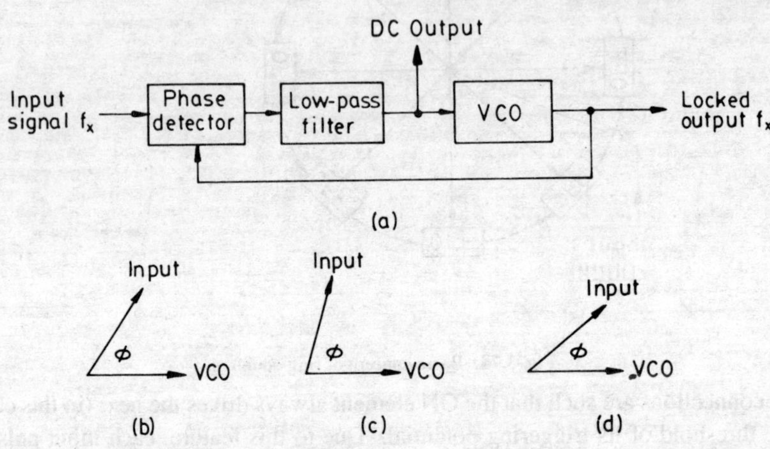


**Fig. 3.75:** Phase locked Loop.

Fig. 3.75(b) shows the phasor for the input signal and the VCO. *If the input frequency changes, the VCO frequency will track it.* For instance, if the input frequency $f_x$ increases slightly, its phasor *rotates faster* and the phase angle increases as shown in Fig. 3.75(c). This means that less dc voltage will come out of the phase detector and force the VCO frequency to increase until it equals $f_x$. On the other hand, if the input frequency decreases, its phasor *slows down,* as shown in Fig. 3.75(d). Now more dc voltage will come out of the phase detector and force the VCO frequency to decrease until it equals the input frequency $f_x$. In other words *the PLL automatically corrects the VCO frequency and phase angle.*

The *lock range* is the range of frequencies the VCO can produce and is given by:

$$B_L = f_{max} - f_{min} \tag{3.2}$$

where $f_{max}$ and $f_{min}$ are the maximum and minimum VCO frequencies.

The PLL can lock on to the input frequency if it lies within the *capture range* $B_C$, a band of frequencies centred on the free running frequency given by

$$B_C = f_2 - f_1 \tag{3.3}$$

where $f_2$ and $f_1$ are the highest and lowest frequencies the PLL can lock into. The capture range is always less than, or equal to, lock range and is related to the cut off frequency of the low pass filter. *The lower the cut off frequency of the low pass filter, the smaller the capture range.*

$$B_C \le B_L \tag{3.4}$$

One use for the locked output $f_x$ of a PLL is to synchronise the horizontal and vertical oscillators of a tv receiver to the incoming sync pulses. PLLs can also automatically tune each tv channel by locking on to the channel frequency.

In general, the locked output is a signal with the same frequency as the input signal. Even though the input signal may drift over a rather large frequency range, the output frequency will remain locked on; *this eliminates the need to tune a resonant circuit for maximum output.*

## 3.23 Voltage Controlled Oscillator (VCO)

Many designs are possible for a voltage controlled oscillator. One approach is to use an LC oscillator with a *voltage controlled capacitor* (varactor). By varying the dc voltage applied to the varactor, we can change the capacitance and control the resonant frequency.

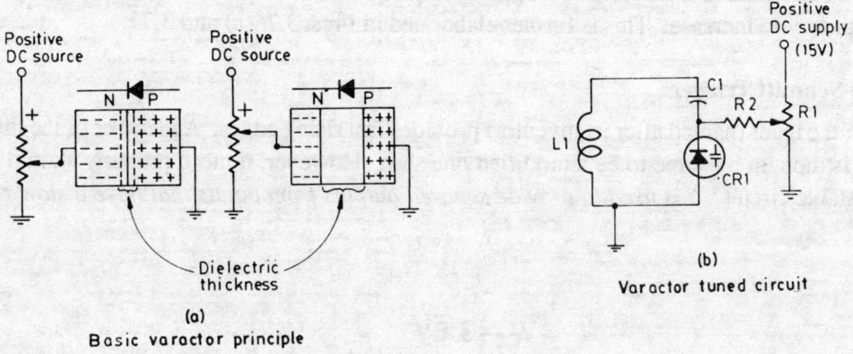


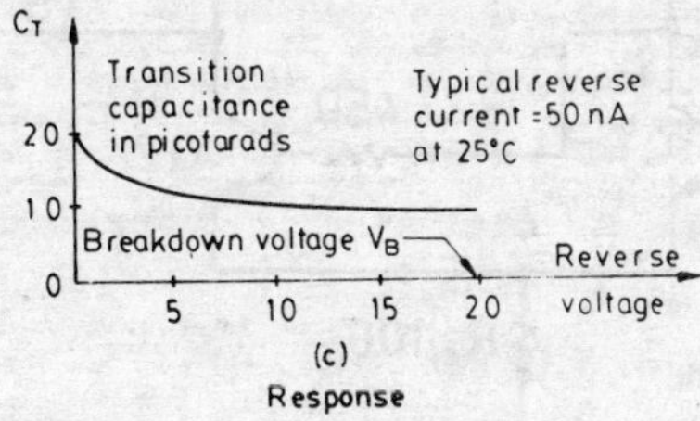


**Fig. 3.76:** Varactor tuned circuit.

Fig. 3.76 illustrates a typical circuit. The capacitance of the varactor diode CR -1(V56) is varied from 32 to 145 pF by varying the bias from 0 to 15 V. The 0.01 μF dc blocking capacitor C1 prevents coil L1 from short circuiting dc bias. The reactance of the varactor is so low that the varactor is the tuning element.

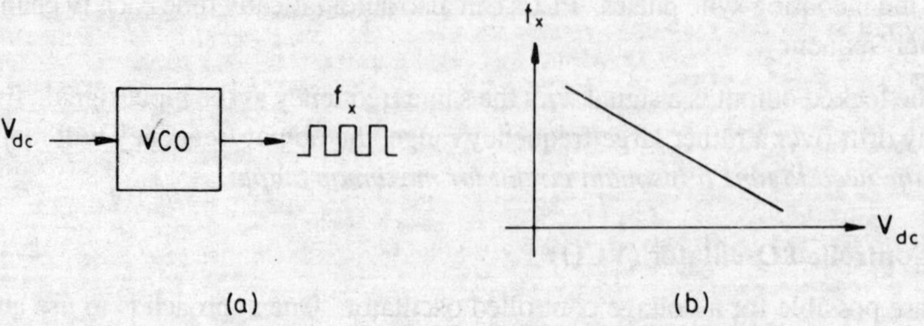


Fig. 3.77: Voltage controlled oscillator.

The important thing to note about any VCO is that *an input voltage controls the output frequency*. In Fig. 3.76(a), an increase in dc control voltage causes the varactor capacitance to decrease and hence the VCO frequency to increase. This is further elaborated in Figs. 3.76(c) and 3.77.

## 3.24 The Schmitt Trigger

The Schmitt trigger (named after its inventor) provides fast rising edges. A member of the multivibrator family, it is thought by some to be a modified one shot. However, it more properly should be referred to as a bistable circuit. *It is used to provide squared outputs from inputs that have a slow rise and fall time.*

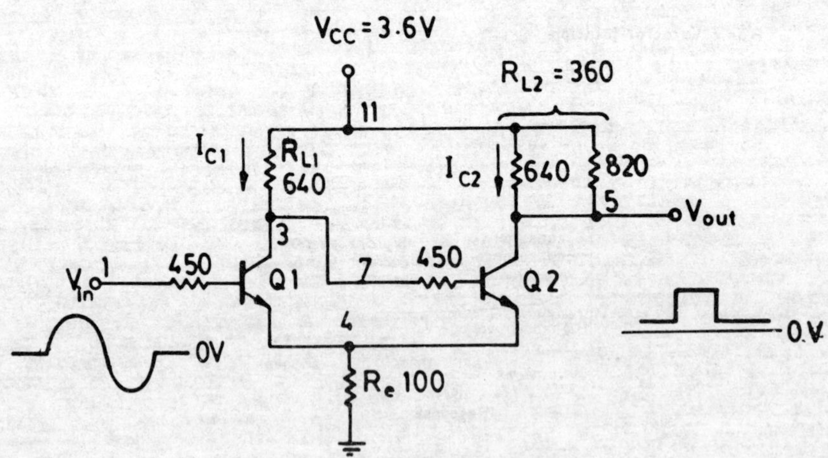


Fig. 3.78: The Schmitt trigger.

The one shot multivibrator circuit is *driven*, not free running. Unlike the bistable trigger, the monostable multivibrator has one stable state to which it returns after an external driving pulse has caused it to execute its cycle. The one shot multivibrator is often used to widen a pulse or delay its trailing edge.

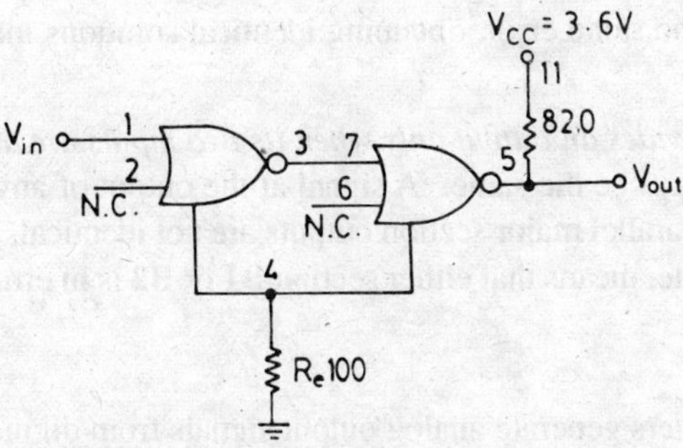


**Fig. 3.79:** A simple way of constructing a Schmitt trigger.

Fig. 3.79 illustrates a simple wav of constructing a Schmitt Trigger from two NOR gates.

## 3.25 Digital Comparator

Unlike the radio receiver, a digital system generally exhibits no *natural symptoms*-such as hum, distortion or erratic volume-to warn the operator that something has gone wrong. Instead, the system simply provides the wrong answer. Furthermore, componenet failure is not the only way by which information can be made erroneous or be lost. There is some distortion each time a pulse passes from one circuit to another. Added to this is the distortion caused by the inductive effects from nearby circuits. Thus, it should be understood that *a pulse itself can sometimes become so distorted that it may produce an error*.

One method of detecting errors, which may be called a *redundancy method*, involves running the problem through the computer twice and noting whether the solutions are identical. However, a computer's major function is to save time, and this type of check *doubles* the amount of time that is requried to solve a problem. In addition, this check is useless if a component has failed, for *the faulty component will probably distort both solutions in the same manner*.

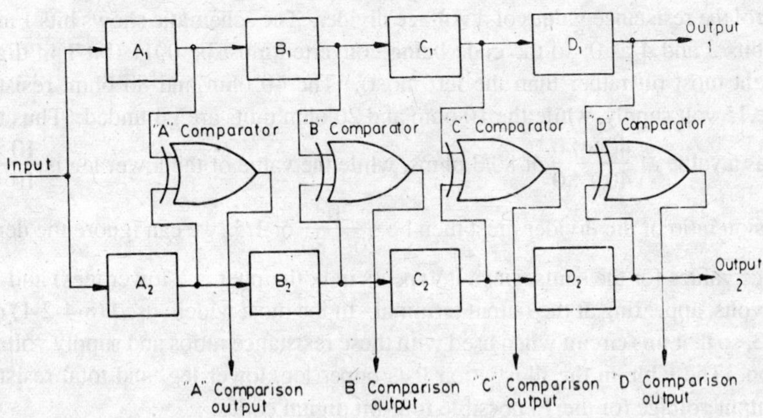


**Fig. 3.80:** Exclusive-OR circuits used as comparator in error detection.

A better solution is to provide two sets of identical circuitry, as shown in Fig. 3.80. When this is done, the same problem is run simultaneosly through both circuits. This is called *parallel operation*. Unless both circuits commit exactly the same error, obtaining identical solutions indicates that the answer is correct.

*An exclusive-OR circuit provides an output only when its two inputs are different.* When using this method, the inputs should always be the same. A signal at the output of anyone of the exclusive-OR cirucits indicates that the two parallel major section outputs are not identical, and that one of them is in error. A signal at B, for example, means that either section B1 or B2 is in error.

### 3.26 D/A-A/D Converters

Digital to analog (D/A) converters generate analog output signals from digital input signals. Together with the opposite equipement-analog to digital (A/D) converters-such devices make possible direct digital or numerical control of manufacturing machines. Both converters may be lumped into a single group under the name of *digital transducers*.

While many types of *digital to analog converters* exist, one of the simplest for converting a digital code to an analog voltage level is shown in Fig. 3.81. In practice, however, logic switches shown in the illustratration as single SPDT switches would be composed of flip-flops or relay contacts.

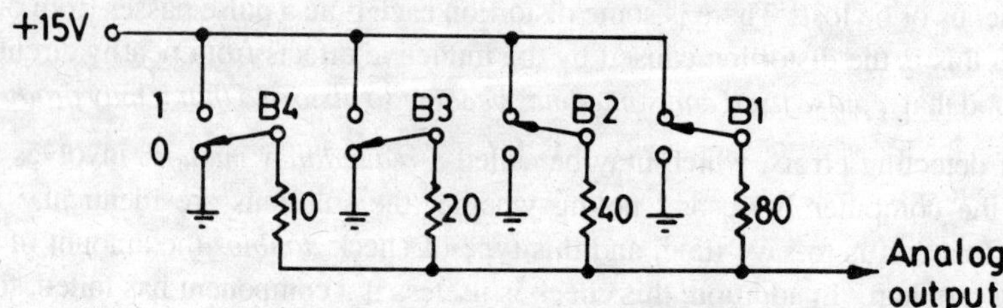


**Fig. 3.81:** Digital to analog converter.

The circuit changes a digital code into the corresponsding voltage level by using the binary digits of the code to control the resistance values of a voltage divider. The schematic shows bits 1 and 2 as being of 1 value and bits 3 and 4 as 0, so the code being converted must be 0011 (bit 1 in digital codes is normally the right most bit rather than the left most). The 40 ohm and 80 ohm resistors are both connected to the 15 volt supply, while the 10 ohm and 20 ohm units are grounded. Thus the upper leg of the divider has a value of $\frac{40 \times 80}{40 + 80}$ or 80/3 ohms, while the value of the lower leg is $\frac{10 \times 20}{10 + 20}$ or 20/3 ohms. The division ratio of the divider must then be $\frac{20}{20 + 80}$ or 1/5 (we can ignore the denominator of 3 in the resistance values for the sums, since it appears in both upper and lower legs) and 1/5 of the 15 volt supply is 3 volts, appearing at the output terminal. In the most widely used (8-4-2-1) digital colde, 0011 stands for 3, so that this circuit when used with these resistance ratios and supply voltage, provides direct conversion. The table in the illustration lists upper leg, lower leg, and total resistance values, together with output voltage for the 16 possible four-bit digital codes.

| DIGITAL CODE | UPPER LEG R | LOWER LEG R | TOTAL RESISTANCE | OUTPUT VOLTAGE |
|---|---|---|---|---|
| 0000 | INFINITE | | | ZERO |
| 0001 | 80 | 40/7 | 600/7 | 1 |
| 0010 | 40 | 80/13 | 600/13 | 2 |
| 0011 | 80/3 | 20/3 | 100/3 | 3 |
| 0100 | 20 | 80/11 | 300/11 | 4 |
| 0101 | 16 | 8 | 24 | 5 |
| 0110 | 40/3 | 80/9 | 200/9 | 6 |
| 0111 | 80/7 | 10 | 150/7 | 7 |
| 1000 | 10 | 80/7 | 150/7 | 8 |
| 1001 | 80/9 | 40/3 | 200/9 | 9 |
| 1010 | 8 | 16 | 24 | 10 |
| 1011 | 80/11 | 20 | 300/11 | 11 |
| 1100 | 20/3 | 80/3 | 100/3 | 12 |
| 1101 | 80/13 | 40 | 600/13 | 13 |
| 1110 | 40/7 | 80 | 600/7 | 14 |
| 1111 | 16/3 | INFINITE | | 15 (SOURCE) |

*Analog to digital converters* often make use of an internal digital to analog converter together with digital code generators and analog comparators to accomplish their function. A typical method is shown in simplified form in Fig. 3.82.

The digital code generator consists merely of a binary counter counting cycles of an oscillator's output, in order to generate quickly and systematically all possible digital codes within the number of bits to be used. Each bit of the counter is supplied to a digital-to-analog converter, essentially the same as shown in Fig. 3.82, to generate its corresponding analog voltage (which is changing at the rate deter-

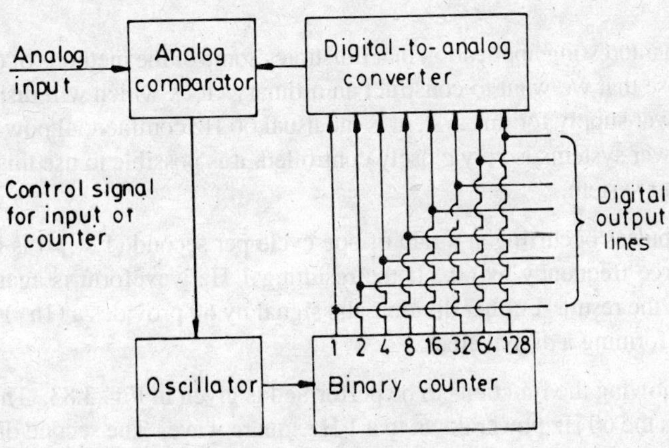


**Fig. 3.82:** Analog to digital converter.

mined by the frequency of the oscillator) and the external analog voltage which is to be encoded are both applied to a difference amplifier. The output of the difference amplifier is of one polarity if the external voltage is the larger, is zero when both the input voltages are equal, and changes to the opposite polarity as the internal voltage exceeds the external.

The "external voltage larger" output polarity of the difference amplifier is used to control the digital code generator so that as long as the external voltage exceeds the internally generated voltage the digital code generator will continue to generate codes for larger voltages. When the two voltages are equal, or if the internal voltage becames larger, the digital code generator is stopped by opening its input gate. At this point the bits of the binary counter represent the smallest coding level which is not smaller than the analog voltage to be converted (since the generator always starts from zero voltage, and is stopped instantly when its code produces a voltage greater than the analog input). A simple test for equality is not practical since the encoded voltage must be a series of steps and an analog input might (in fact frequently does) have a level which is between the encoding steps. If the steps are sufficiently small this effect is negligible.

Response time of such a circuit cannot be instantaneous. If the input voltage range of the device is from 0 to 10 volts and the interval between steps is to be 0.010 V, then it is possible to generate 1000 current digital codes before reaching the correct one. With 1mv step levels and a 10 V range, upto 10,000 attempts may be necessary. Digital circuitry is sufficiently fast so that the oscillator may run at speeds of 100 kHz or more, but this still requires 1/10 second for the correct code to be generated.

If the input range is exceeded with the basic instrument shown in Fig. 3.82 no correct output will ever be generated. Most practical applications include circuitry to detect an *overrange condition*, many include *automatic range selection* so that the over range signal causes automatic insertion of attenuation ahead of the actual converter, thus bringing the signal back into the normal operating range.

Such a device may be useful as a digital volt-meter by detecting the coded output and converting it back into decimal digits for display. However, the most frequent use of D/A and A/D converters is in process control.

### 3.27 Digital Clocks

The digital clock is an interesting application that illustrates some of the methods of employing counters and decoders. Suppose that we want to construct an ordinary clock which will display hours, minutes and seconds. The power supply for this system is the usual 60 Hz commercial power. Since the 60 Hz frequency of most power systems is very closely controlled, it is possible to use this signal as the basic clock frequency for our system.

In order to obtain pulses occurring at a rate of one cycle per second ($1\ s^{-1}$) it is necessary to divide the 60 Hz power source frequency by 60. If the resulting 1 Hz waveform is again divided by 60, a ($1\ min^{-1}$) waveform is the result. Further dividing the signal by 60 provides a ($1h^{-1}$) waveform. This is the basic idea used in forming a digital clock.

A block diagram showing the functions to be performed is given in Fig. 3.83. The first divide by 60 counter simply divides the 60 Hz power down to a 1 Hz square wave. The second divide by 60 counter changes state once in each second and has 60 discrete states and can, therefore, be decoded to provide signals to displays seconds. This counter is then referred to as the *seconds counter*. The third divide by

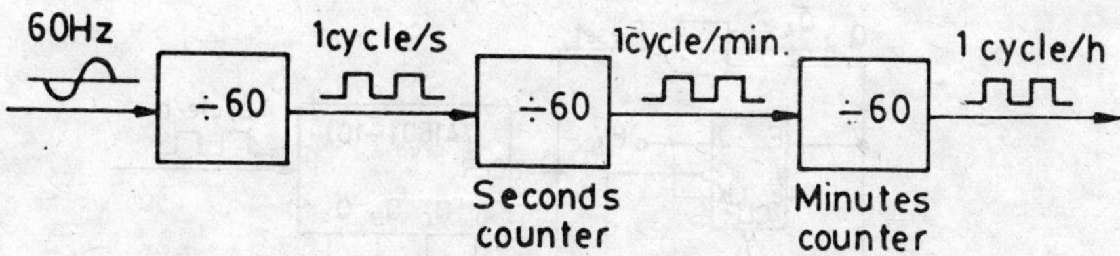


**Fig. 3.83:** Block diagram of a digital clock.

60 counter changes state once each minute and has 60 discrete states. It can thus be decoded to provide the necessary signals to display minutes. This counter is then the *minutes counter*. The last counter changes state once each 60 minutes (once per hour). Thus if it is a divide by 60 counter it will have 12 states that can be decoded to provide signals to display the correct hours. This then is the *hours counter.*

There are a number of ways to implement a counter. The first counter must divide by 60, and it need not be decoded. Therefore, it should be constructed in the easiest manner with the minimum number of flip-flops. The second counter in the system also divides by 60 and could be implemented in the same way. However, the seconds counter must be decoded. We are interested in decoding this counter to represent each of the 60 seconds in 1 minute. This can most easily be accomplished *by connecting a mod-10 counter in series with a mod-6 counter for the divide by 60 counter.* The mod-10 counter could then be decoded to represent the tens digits of seconds. *The modulus defines the number of states through which a counter can progress.*

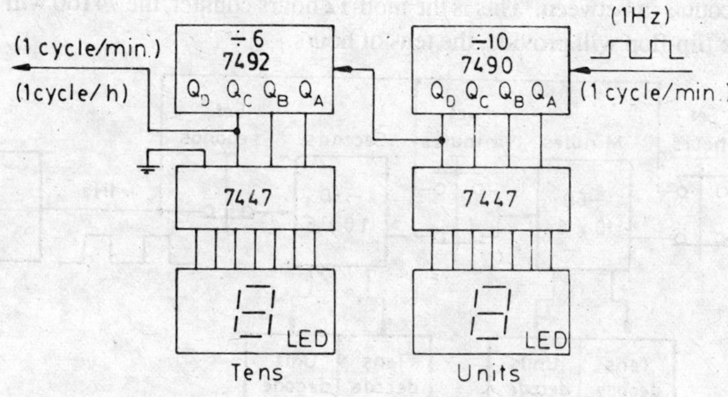


**Fig. 3.84:** A 10 × 6 mod-60 counter with units and tens decoding.

The minutes counter is usually the same as the seconds counter, except that it is driven by the 1min$^{-1}$ square wave from the output of the seconds counter, and its output is 1 h$^{-1}$ square wave as shown in Fig. 3.84. The divide by 12 hours counter must be decoded into 12 states to display hours. This can be accomplished by connecting a mod 10-(54/74160) decade counter in series with a single flip-flop E as

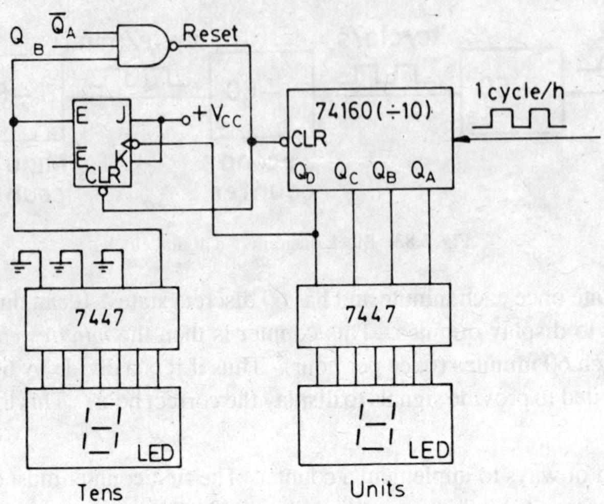


**Fig. 3.85:** Mod-12 hours counter.

shown in Fig. 3.85. This forms a divide by 20 (10 × 2 = 20) counter. Feedback is then used to form a mod-12 counter.

The hours counter must count through states 00, 01, 02,...10 and back to 00. The NAND gate in Fig. 3.85, will go low as the counter progresses from count 11 to count 12 and this will immediately clear the 74160 to 0000 and reset the flip-flop E to 0. The counter actually skips from count 11 to count 00 omitting the eight counts in between. This is the mod-12 hours counter, the 74160 will provide the units of hours, while the flip-flop will provide, the tens of hours.

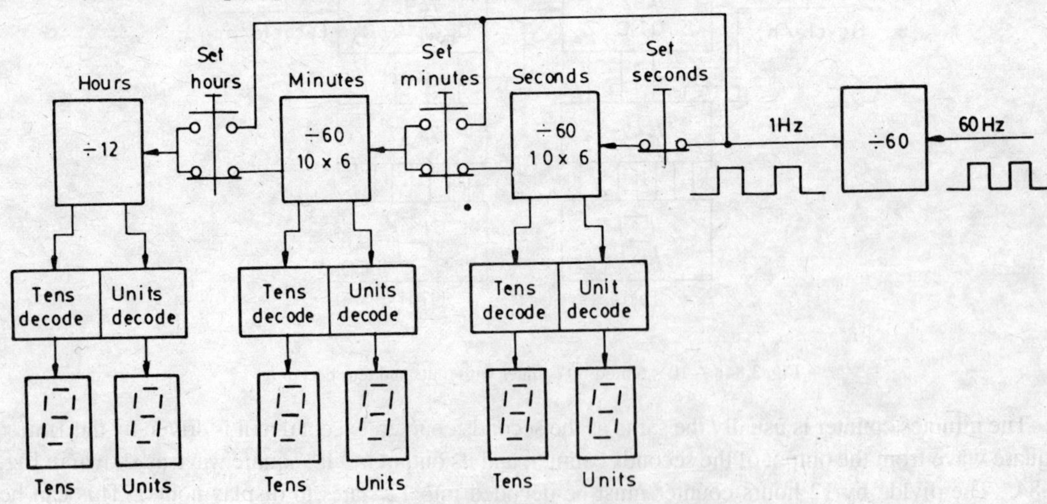


**Fig. 3.86:** Digital clock.

Some means must be provided to set the clock because the flip-flops will assume random states when the power is turned off and then turned back on again. Setting the clock can be quite easily accomplished by means of the SET push button shown in Fig. 3.86. Depressing the SET HOURS button causes the hours counter to advance at 1-$s^{-1}$ rate, and thus this counter can be set to the desired hour. The minutes counter can similarly be set by depression of the SET MINUTES button. Depression of the SET SECONDS button removes the signal from the seconds counter and the clock can thus be brought into synchronisation.

*Chapter 4*

# CONTROL CIRCUITS

---

*The form of a control circuit depends very much on the application for which it is required. The control itself can be mechanical or electronic. Further, the control can be exercised locally or from a remote location. Electromechanical control with movable parts as switching elements is incompatible with the miniaturised equipment being commercially marketed. Electronic control exploits the switching capability of diodes, bipolar and unipolar devices. These devices and their practical application in control circuits form the subject of study in this chapter.*

---

## 4.1 Mechanical Relays

Relays are electromechanical devices that *switch* output circuits ON and OFF in accordance with the *status* of the input signals. They are easy to handle and hence are used extensively for *control purposes*. Relays are composed of two parts: an electromagnet and a contact spring. The contact spring has a contact attached to it and this is the *moving part* that opens and closes the electrical circuit. Electromagnets find specific use in electric relays. Maintenance of electric current in a coil of wire energises its core which then acts as an electromagnet. The movement of the core controls a number of *contacts*, a few of which may *close* and control the performance of a specific circuit, whereas the remaining contacts may *open* to control the performance of some other circuits. When the flow of electric current is stopped, the electromagnet becomes de-energised and the contacts which were previously closed now *open*, whereas the contacts which were previously open now *close*. This is exactly opposite to the sequence of events which took place when the magnet was energised. The electric symbols connected with relays are given in Fig. 4.1.

Mechanical relays can further be classified as *normally open* (NO) relays, or *normally closed* (NC) relays. *In normally open relays, the relay contacts normally remain open* i.e., in de-energised state of the relay, as shown in Fig. 4.2. *In normally closed relays, Fig. 4.3, the contacts normally remain closed* i.e., in de-energised state of the relay, as shown in Fig 4.3.

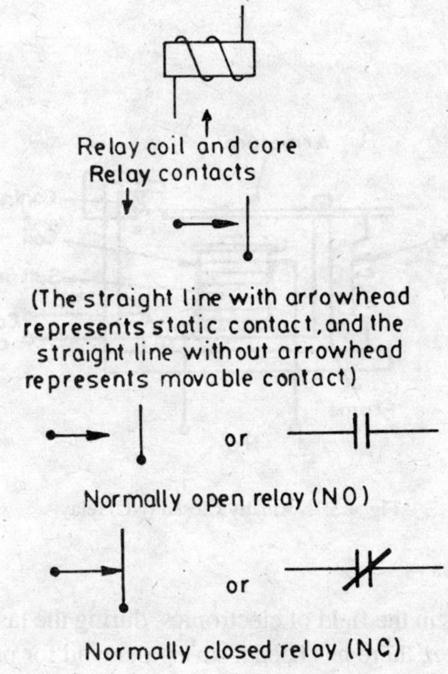


**Fig. 4.1:** Electric symbols connected with relays

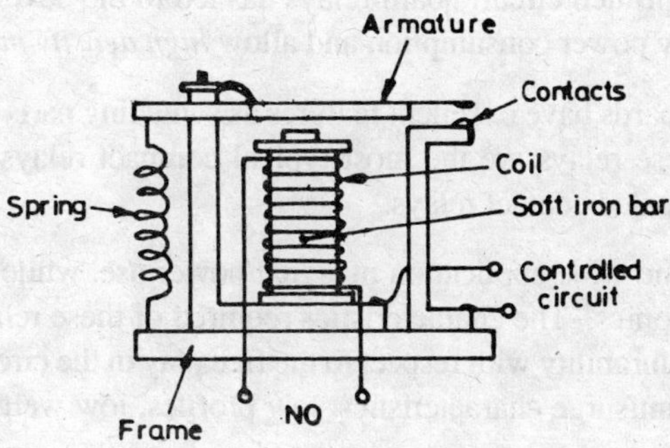


**Fig. 4.2:** Normally Open (NO) relay

Mechanical relays are *sluggish* (due to the inertia of the moving contacts), *introduce noise* (due to chattering of relay contacts), and hence are not suitable for use in high frequency and critical applications. Relay contacts *wear down* because of prolonged usage. Special *contact materials* have been developed to increase the wear resistance of relay contacts.

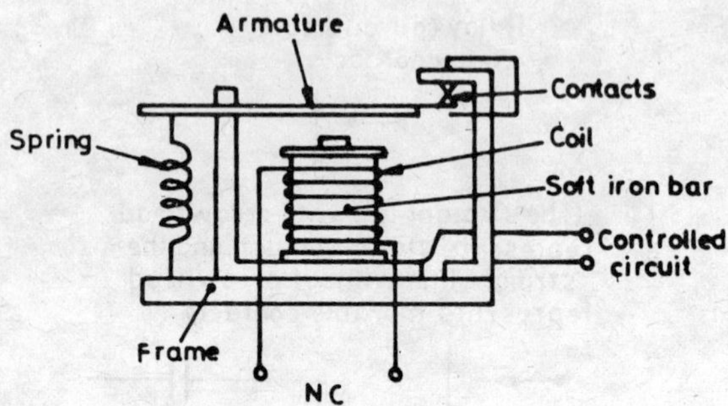


**Fig. 4.3**: Normally Closed (NC) relay

## 4.2 Printed Circuit Board Relays

With an unprecedented progress in the field of electronics, during the last three decades and an increasing trend towards *miniaturisation*, there has been a strong demand for printed circuit board relays used in input and output interface parts.

A common boundary between automatic data processing systems or parts of a single system is called an *interface*. In communication and data systems it may involve code, format, speed, or other changes as required. It is also a *common boundary* between two systems or two devices.

The increasing demand of printed circuit board relays has led to the development of relays that are compact, low profile, have low power consumption and allow *high density mounting*.

Relays for printed circuit boards have excellent features as switching parts and can also be used with semiconductor switches. These relays are the most typical compact relays and can be classified as compact solenoid relays and sealed contact relays.

Compact *solenoid relays* find wide application in signal/power use, while the *sealed contact relays* are used as relays in car electronics. The characteristics required of these relays are contact reliability, long term stability, long term durability with respect to the free play in the circuit cables, coil wires with low power consumption and antisurge characteristics, low profiles, low weight and small surface area etc.

The latest trend is towards combined mounting of electronic circuits and relays on printed circuit boards so that more compact units are realised.

## 4.3 Input/Output Relay Terminals

Input/Output (I/O) relay terminals combines *an I/O relay and a terminal*, and that can be connected with a *snap connector* to the I/O units of programmable controllers to allow input and output.

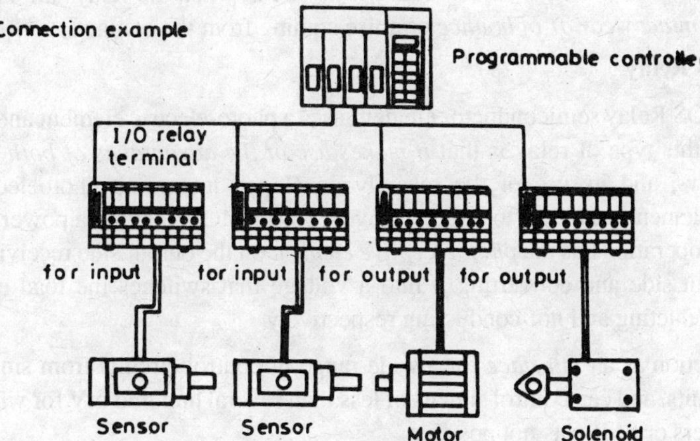


**Fig. 4.4**: Connection example of I/O relay terminals for taking up less space and wiring of control devices

Conventional control boards for the control of input and output signals had to have innumerable relays and sockets built into them, and this led to wiring of great complexity. For example, if 64 single pole relays were to be wired to a programmable controller I/O unit with 64 wiring points, then there would be a total of 256 places where 128 wires would have to be connected. Not only this, the work of attaching the terminal board would become extremely difficult as the number of attachments increases.

I/O relay terminals are the answer to these needs. I/O relay terminals can be connected by a snap connector to the I/O units of the programmable controllers, and through this it becomes possible to have fewer points inside the control point to eliminate wiring, to greatly reduce the number of attachment and wiring processes, and to reduce the overall space. A typical connection example is illustrated in Fig. 4.4.

## 4.4 Solid-State Relays

Its high sensitivity, wide frequency response, and generally low power requirements suit the transistor for use in control devices in which *only one active element* must be used. Thus it is possible with relatively simple equipment to obtain reliable *near control* and *remote control*.

Solid-state relays (SSRs) can be mounted on printed circuit boards along with semiconductors and other electronic parts and when sealed with mold resin they form a completely fixed type of semiconductor switch.

For the time that the input signal is impressed, the negative current is held on, and when there is no more input signal, the load power is turned off, so that the action is that of a semiconductor circuit acting as a *non-contact switch*.

SSRs use a photocoupler on the input side so that there is complete *isolation* between the input and output sides. A triac or power transistor is used on the output side so that there is *no contact*. Since SSRs are fully solid-state elements, they are *durable* and have a *high degree of reliability*. They are also *strongly resistant* to noise and it is possible to completely shut out the noise voltage to provide safety against misoperation due to such noise. There is *no sound* when the relays are operating and there is also *no fear of contact wear or of bounce* or noise coming from the contact.

## 4.5 Photo MOS Relay

The Photo MOS Relay semiconductor element uses a photo-electric element and a power MOS. The main feature of this type of relay is that *it makes use of the advantages of both semiconductors and mechanical relays*, and its use of the recently developed integrated photoelectric element as the photoreceptive element enables it to directly drive a power MOS, making a power source unnecessary. The principle of operation has the *photoreceptive element* on the output side receiving the light from the LED on the input side and converting it into a voltage that switches the load on and off when the MOS-FET is conducting and not-conducting respectively.

This will function as an *interface* for a wide range of control circuits from small analog signals to power drive circuits, and can control signals of less than several hundred mV for which control by triacs and photo-couplers or SSR was not possible.

*Optical MOS-FET relay* is a combination of an LED, an optical exciter element and an MOS-FET, and can transmit and perform high-voltage (600 V) control of low level signals without any distortion.

## 4.6 High Frequency Relays

High frequency relays have excellent characteristics in the VHF and UHF bands and are widely used in *image transmission equipment* instead of conventional signal switching elements.

For improved performance at high frequencies, special contact structures are being used. New magnetic circuits with poles and compact high frequency relays with high sensitivity have also been developed.

For high frequency performance, it is necessary to take into account the loss of the transmission system and to make it uniform at between 50 and 75 ohms. The *isolation characteristic* expresses the leakage across the contacts when the contacts are apart and mainly arises due to the contact separation

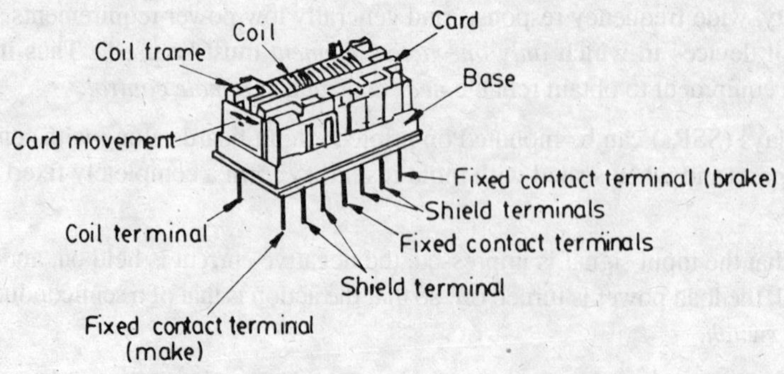


Fig. 4.5: Structure of high frequency relay

capacitance. Because of this, the leakage increases at high frequencies. The *insertion loss characteristic* expresses the insertion loss of the signals when the contact is closed and corresponds to the contact resistance for dc in the low frequency region. The *return loss characteristic* expresses the return loss for the signal when the contact is closed.

Special contact structures of high frequency relays, see Fig. 4.5, give a minimum isolation characteristic of 60 db and a maximum isolation characteristic of 1 db, at 900 MHz.

## 4.7 Surface Mounted Relays

In the modern electronic equipment which is becoming more compact and light weight, surface mounted parts are being used in increasing numbers to achieve *high density mounting* on printed circuit boards.

The diffusion of surface mounted components began with chip parts such as resistors and capacitors used in hybrid ICs giving rise to a growing demand in calculators, radios, cameras, vcrs and other consumer products.

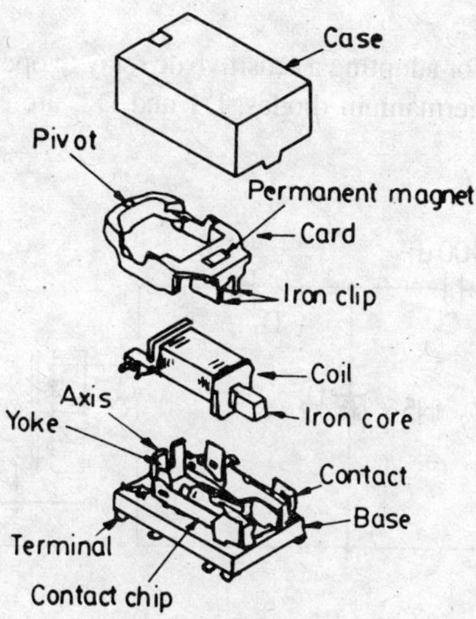


**Fig. 4.6**: Structure of a typical surface mounted relay

Surface mounting has preceded ICs and LSIs to such an extent that it became necessary to have surface mounting for connectors, switches and relays. Surface mounted relays were developed to reduce the total cost through surface mounting and miniaturisation of components used.

The structure of a typical surface mounted relay is given in Fig. 4.6 and its exploded view is given in Fig. 4.7.

When compared to the conventional lead—through relays, surface mounted relay has the following advantages: *effective use* can be made of either side of the printed circuit board, the device is *compact*

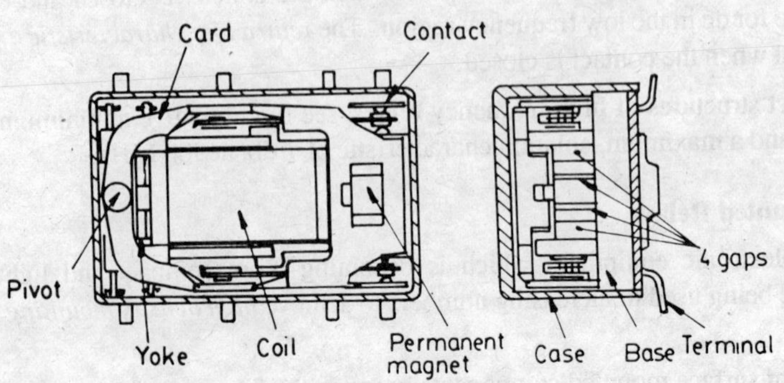


**Fig. 4.7**: Exploded view of surface mounted relay

*and light, automatic mounting* is facilitated, circuit *reliability is improved* and total *mounting cost reduced.*

## 4.8 AF-RF Relay

Fig. 4.8 shows a simple circuit for adopting a sensitive dc relay to operation at AF or RF upto 200 MHz. Two 1N56 high conductance germanium diodes, D1 and D2, are employed in a half-wave rectifier circuit.

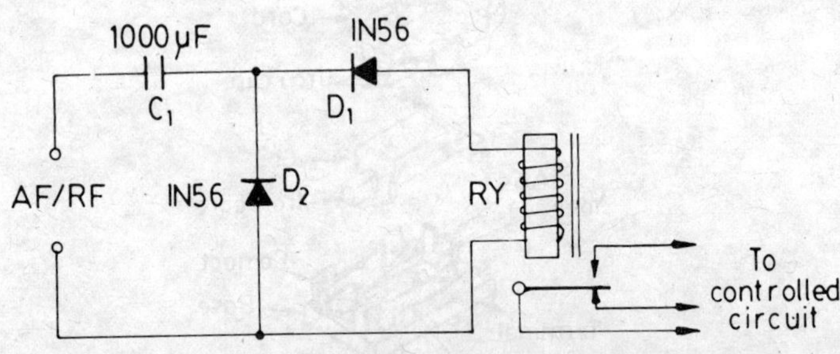


**Fig. 4.8**: AF-RF Relay

Capacitor C1 is not needed unless the signal source has a dc voltage in its output; when such a voltage is present, the diodes must be protected from it by means of a blocking capacitor. At audio frequencies, of 1000 Hz and lower, it may be necessary to connect a 0.1 $\mu F$ for higher capacitor in parallel with the relay coil to prevent *chatter*.

## 4.9 Polarity sensitive DC relay

The circuit of Fig. 4.9 illustrates a dc relay which will operate only when the control voltage has proper polarity. A 1N456 silicon junction diode has been forward connected in series with a sensitive dc relay RY. *The diode acts as a polarity sensitive switch.*

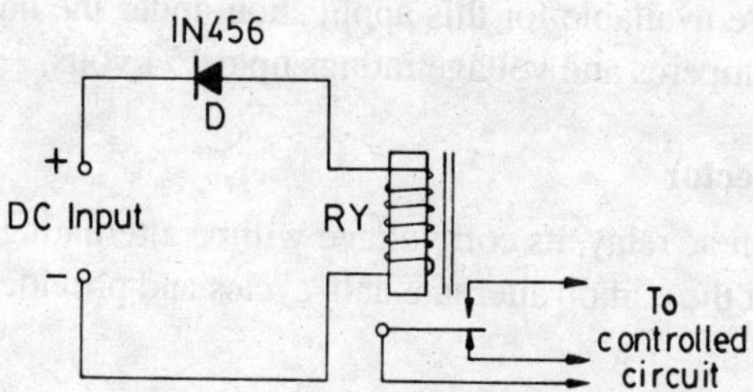


**Fig. 4.9**: Polarity Sensitive dc Relay

When a positive voltage of suitable amplitude is applied to the top input terminal, forward current flows through the diode and *picks up* the relay. But when the polarity of the control voltage is reversed, the diode becomes reverse biased and to all practical intents and purposes, represents an open switch — and the relay remains *dropped out*. This current is suitable for use as an alarm in systems in which a monitored voltage is subject to polarity shifts.

## 4.10 dc Relay Contact Protector

Many circuits use two relays — a *sensitive relay* which responds to the signal and then closes a more *rugged relay,* which, in turn, actuates the controlled equipment. The coil of the second relay constitutes an inductive load on the contacts of the first relay, and the back voltage that it generates on turn-off can destroy the contacts of the sensitive relay.

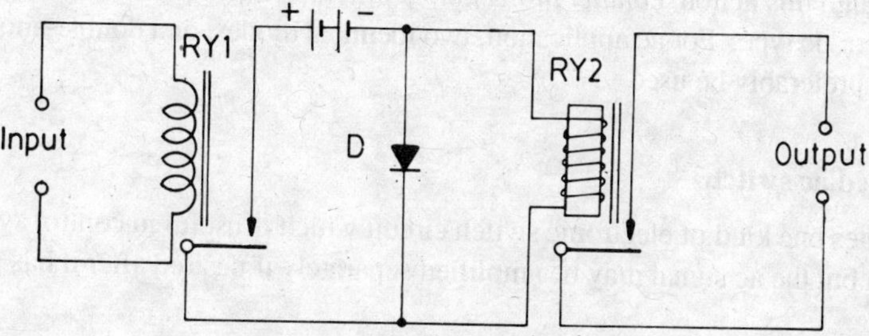


**Fig. 4.10**: dc Relay contact protector

To prevent this damage, a diode (D) may be connected across the coil of the second relay, as shown in Fig. 4.10, to absorb the dangerous transient. The diode is reverse connected with respect to the local supply and therefore does not pass appreciable current, nor does it short circuit relay RY2.

When the contacts of RY 2 open, however, the collapsing magnetic field induces a *transient voltage* across the coil of RY2. The polarity of this voltage is such that the diode is forward biased. Consequently, the diode exhibits a low forward resistance which absorbs the transient and protects the contacts of RY1.

Small selenium diodes are available for this application under the name *contact protection diodes* with current ratings upto 2 amperes and voltage ratings upto 154 volts.

### 4.11 ac Relay contact protector

When RY2, in Fig. 4.10 is an ac relay, its coil voltage will be alternating. If a single diode were used, the diode would short circuit the coil on alternate half-cycles and provide no contact protection half the time.

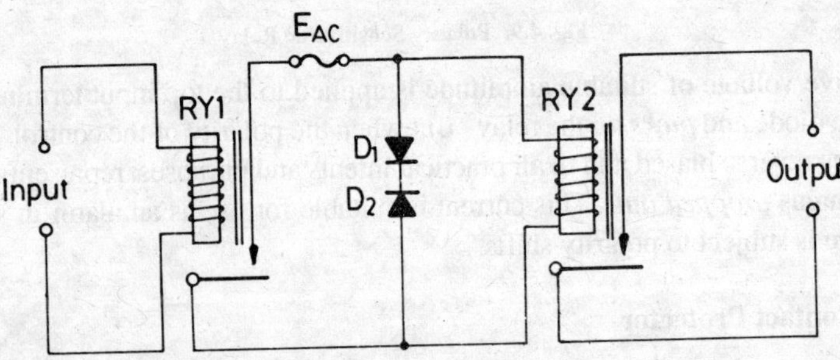


**Fig. 4.11**: ac Relay contact protector

For ac operation, a pair of contact protecting diodes, Fig. 4.11, must be *connected back-to-back* across the ac relay coil. When the upper end of RY2 is positive, D1 conducts and D2 acts as a high resistance. Conversely, when the lower end of RY2 is positive, D2 conducts and D1 acts as a high resistance. Through this action, *contact protection is provided during both the half cycles of ac.* RY2 may be either ac or dc type. For ac application, two identical diodes, or a double-anode contact-protector diode should preferably be used.

### 4.12 dc Operated ac switch

Fig. 4.12 illustrates one kind of electronic switch circuit which is useful in control systems. It provides no amplification but the ac signal may be amplified separately if desired after it has passed through the switch circuit.

D is effectively in series with the ac input and ac output terminals. It is reverse biased by the 1.5 V battery B, and because of its very high resistance in this state, it effectively opens the line between the input and output terminals. Therefore, none of the input signal reaches the output. If a dc voltage of polarity shown is applied to the dc control voltage input terminals, however, the reverse bias will be over-ridden and the diode forward biased. Its low forward resistance then will allow the ac to pass from input to output. A dc voltage of opposite polarity will, of course, have no effect.

An advantage of this circuit is the *common ground* it provides for the ac source, control voltage source, and ac load. This feature reduces stray pick up, interaction, and ground loops.

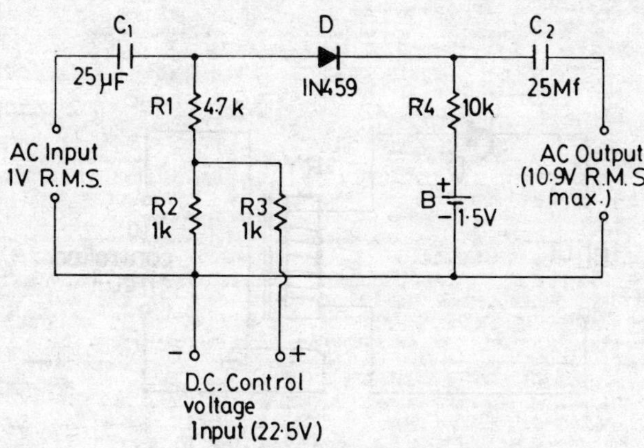


**Fig. 4.12:** dc Operated ac switch

## 4.13 Biased dc relay

When it is desired that a relay close at some particular voltage, this operation is often obtained more easily by suitably biasing the relay than seeking one rated at that specific voltage. What results is a *threshold circuit* in which no current can flow through the relay until the signal voltage exceeds the bias voltage. In this way, a single stock type relay may be used at a number of desired voltages. A diode is required for *isolation*.

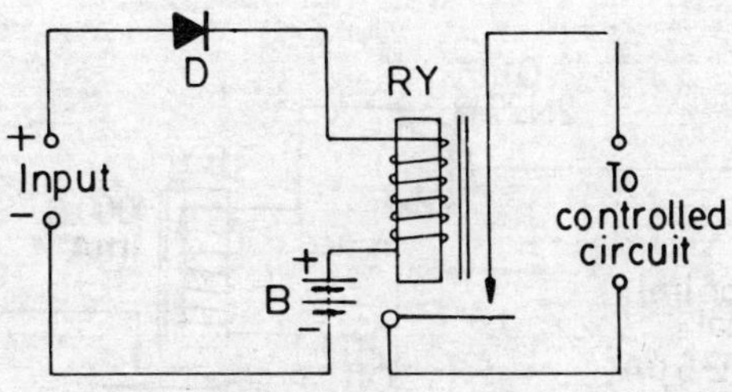


**Fig. 4.13:** Biased dc relay

In Fig. 4.13, the diode is reverse biased by the battery. When the dc input signal voltage exceeds the battery voltage, current flows through the relay coil. A number of such circuits may be operated in parallel across a two-wire transmission line, with each relay biased to close at a selected signal voltage.

## 4.14 Zener Controlled Relay

The biased relay, Fig. 4.13, needs a dc bias source. This difficulty may be resolved in some cases by the use of a zener diode without battery, provided the relay pick up current corresponds to the *breakdown current* of the zener diode and provided also that the *breakdown voltage* is the voltage at which the relay must operate.

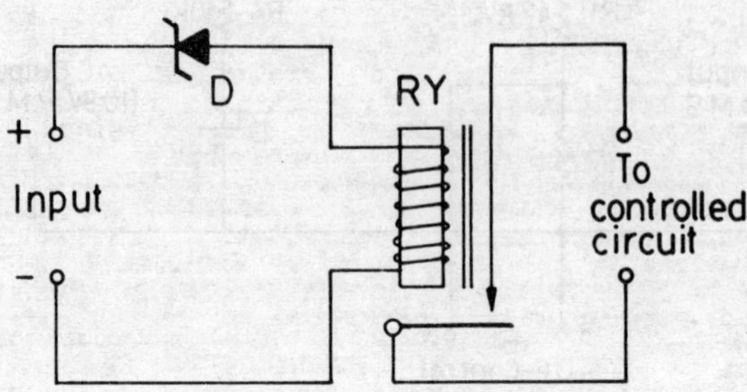


**Fig. 4.14**: Zener controlled relay

The input voltage for closure equals $E_D + E_{RY}$, where $E_D$ is the zener breakdown voltage and $E_{RY}$, the voltage required for closure.

*A number of such circuits may be operated in parallel across a two wire transmission line, with each relay in series with a zener diode chosen to close that relay at a selected voltage.*

## 4.15 dc Relay *npn*

A transistor operated as a dc amplifier in conjunction with a dc relay will perform a number of switching operations. *A milliampere type relay thus may be operated from a control signal of only a few microamperes,* or an ampere type relay may be operated from a control signal of only a few milliamperes.

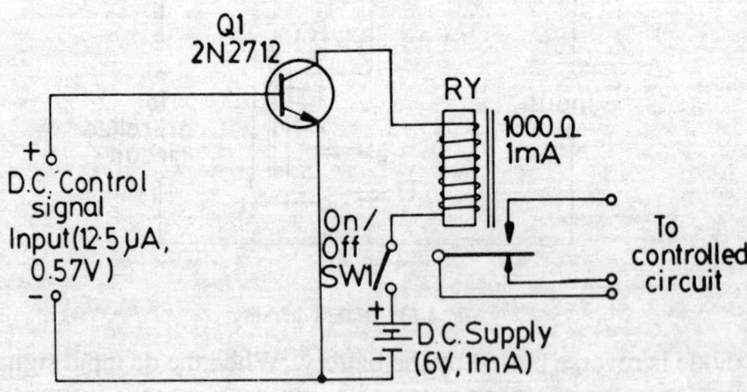


**Fig. 4.15**: Sensitive dc relay npn

Fig. 4.15 illustrates such a circuit employing a 2N2712 npn silicon transistor. The relay (RY) is a 1000 ohm 1 mA unit. A dc control signal of only 12.5 µA at 0.57 V is amplified sufficiently by the transistor to close the relay (i.e. to 1 mA at 1 V). The equivalent input resistance of the circuit is 45.6 k ohm.

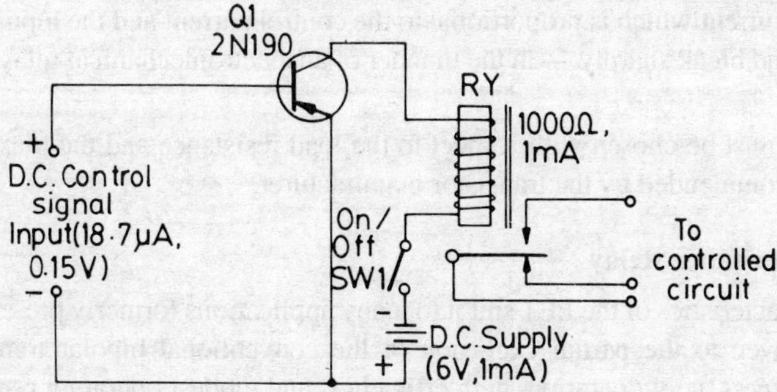


**Fig. 4.16**: Sensitive dc relay (pnp)

Due to the very high collector resistance of 2N2712, the zero signal current through the relay is insignificant hence *no balancing circuit is required.*

The circuit in Fig. 4.16 employs a 2N190 pnp germanium transistor. The chief differences are the negative collector voltage and negative control voltage.

### 4.16 All Solid-State dc relay

When the external load resistance is low with respect to the internal collector resistance of the transistor, the electromechanical relay may be dispensed with entirely and an all solid-state relay circuit is used. *Such a circuit has the obvious advantage of freedom from moving parts.*

In Fig. 4.17, the load is connected directly in the collector circuit. The amount of current amplification obtained with such an arrangement depends upon the load resistance, varying inversely as $R_L$. The maximum dc collector current is 1.5 amperes for the 2 N 301. Higher current operation can be obtained by using suitable power transistors.

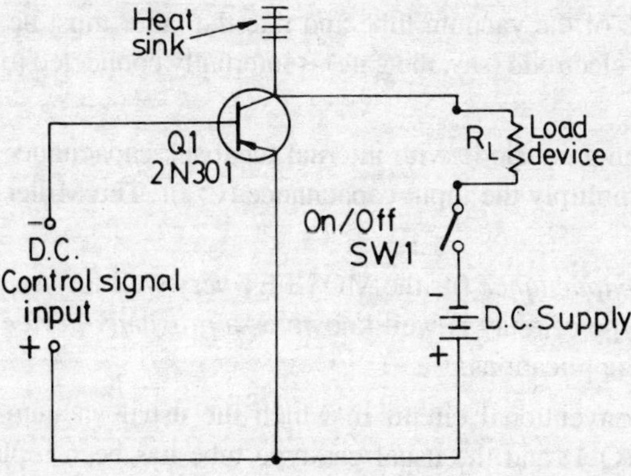


**Fig. 4.17**: All solid-state dc relay

The output load current flows as long as input control signal current flows. This transistor relay provides an input current which is proportional to the control current and the input current, consequently, will not make and break sharply — in the manner of an electromechanical relay — unless the control current also does.

The dc supply must be chosen with respect to the load resistance and the maximum safe collector-emitter voltage recommended by the transistor manufacturer.

## 4.17 INPUT Capacitance Relay

The electrical characteristics of the FET suit it to many applications formerly pre-empted by the vacuum tube, sometimes even to the partial exclusion of the conventional bipolar transistor; small in size, mechanical ruggedness, cool operation, high efficiency, and modest operating power requirements.

The capacitances of an FET are usually so low that they often may be neglected in many critical applications. MOSFET input capacitance is very low. *For operation at VHF and UHF, these capacitances must be considered, since they then become significant components of device impedances.*

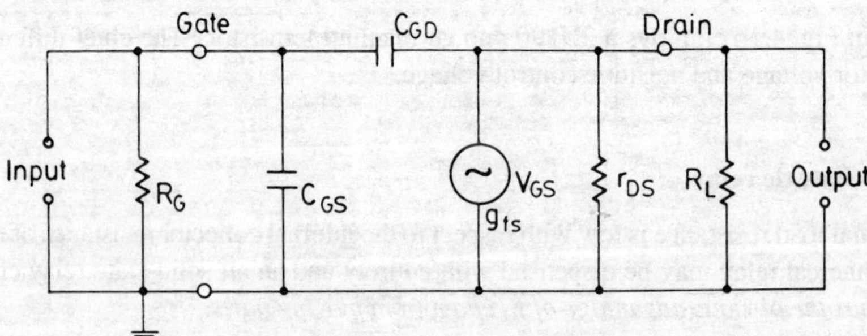


**Fig. 4.18**: Equivalent circuit of a common-source amplifier

In Fig. 4.18, the equivalent circuit of the common source amplifier, the positions of the FET input capacitance ($C_{GS}$) and feedback capacitance ($C_{GD}$) are shown. $R_G$ represents the external gate resistor, $R_L$ the external drain resistor and $r_{DS}$ the internal drain/source resistance ($r_{DS} = V_{DS}/I_D$) of the FET. The equivalent internal generator is represented by its effect $g_{fs} \cdot V_{GS}$. Feedback capacitance $C_{GD}$ is equivalent to the grid/plate capacitance of the vacuum tube and like the latter must be *neutralised* in some FET amplifiers. Sometimes, one electrode (say, the gate) is internally connected to the case, and the relevant capacitance is eliminated.

Because the FET is a high gain triode with internal feedback capacitance ($C_{GD}$), it is subject to the *Miller effect* which acts to multiply the input capacitance ($C_{GS}$). The Miller effect must be taken into account in all applications.

*Its very low gate/source capacitance* fits the MOS-FET very well for use in the oscillator section of a capacitance relay. This type of relay is well known as a *proximity device* because of its industrial, domestic and entertainment applications.

Fig. 4.19 illustrates a conventional circuit in which the usual vacuum tube oscillator has been replaced with a MOSFET (Q 1) and the usual gas type tube has been replaced with an npn bipolar transistor (Q 2). Assembly T1 is a commercial capacitance—relay coil which contains the tapped oscillator inductor L 1, transient capacitor C 2 and RF choke RFC 1. The source resistor R 2 is a wire wound

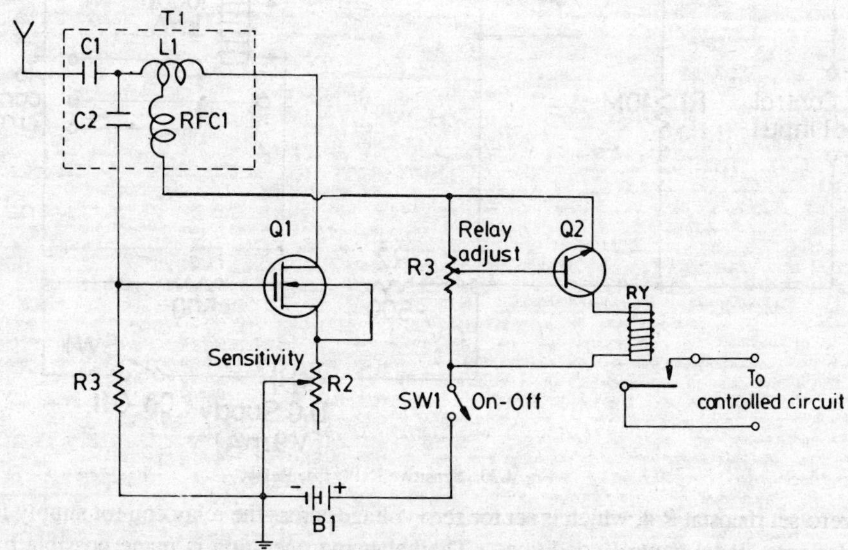


**Fig. 4.19**: MOSFET Capacitance relay

rheostat for controlling the *oscillator sensitivity* and the drain resistor R 3 is a wire-wound potentiometer for adjusting the *pick-up point of the relay.* The relay itself (RY) is a single-pole signal-throw normally open (NO), sensitive dc type (usually 1 or 2 mA).

When on-off switch SW 1 is closed, the Q 1 circuit usually will oscillate if no conductive objects are near the pick up antenna and rheostat R 2 is set near the center of the resistance range. Also, relay RY will probably close and R 3 must be set to the point at which the relay just drops out. If the operator's hand is held close to the antenna, his body capacitance will throw the circuit out of oscillation, the Q 1 drain current will rise, and the voltage drop across R 3 will increase. The base of Q 2 will receive an increased dc voltage and the relay will close. When the hand is withdrawn, the circuit resumes oscillation and the relay again drops out.

By adjustment of the sensitivity control rheostat R 2 and the oscillator trimmer capacitor C 2, the circuit may be made more or less sensitive, as desired, for operation by objects at some desired distance from the antenna.

## 4.18 Sensitive INPUT dc Relay

Employing a 2N3823 FET, the relay circuit in Fig. 4.20 offers 10 Mohm input resistance. If desired, the input resistance may be made still higher by increasing the value of resistor R1. The relay RY is a 1000 ohm, 1mA unit. A dc control signal of 0.09 μA at 0.9 V is amplified sufficiently by the transistor to close the relay (i.e. to 1mA at 1 V).

The zero signal drain current of the transistor is higher than that required to close the relay, hence it must be initially balanced out of the relay coil. Balance is accomplished by means of the 1000 ohm wire

**Fig. 4.20**: Sensitive INPUT dc Relay

wound zero-set rheostat R 4, which is set for zero voltage across the relay coil (or simply for relay drop out) under zero signal control conditions. The balancing operation is made possible by a four arm bridge consisting of resistors R 2, R 3 and R 4 and the internal drain resistance of the transistor. The circuit draws a maximum of 9 mA from the 6V dc supply.

## 4.19 Thyristors

The *silicon controlled rectifier*, also called a thyristor, is a four-layer structure designated as pnpn, and is best described as two *complementary transistors* connected as shown in Fig. 4.21(b). The controlled rectifier is shown in Fig. 4.21(a) as a four-layer structure. This can be broken up into the equivalent of two complementary transistors, as shown in Fig. 4.21(c). The polarities are for a conducting state. A reverse polarity is always blocked.

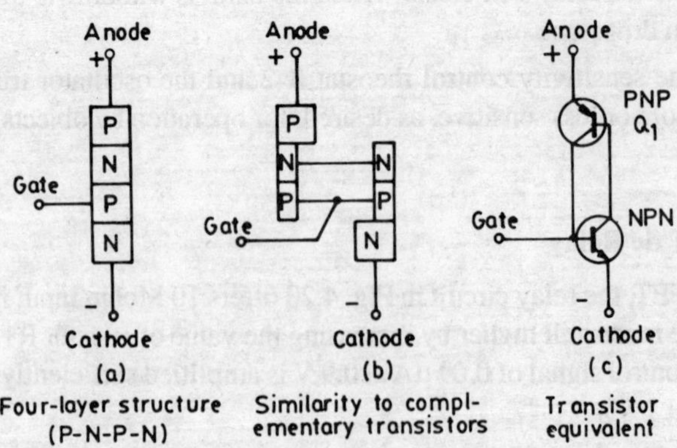


**Fig. 4.21**: Silicon controlled rectifier and transistor analogy

The transistors in Fig. 4.21(c) are connected so that the collector of Q 1 drives the base of Q 2 and the collector of Q 2 drives the base of Q 1. As long as the base-emitter junction of Q 1 remains cut off, the two transistors will remain in the *blocking state*. If a small forward current is produced in the base-emitter junction of Q 2, the transistor will conduct and the collector conduction of Q 2 will produce a current in the base-emitter junction of Q 1, turning on transistor Q 1. Conduction in the collector of transistor Q 1 will drive the base of transistor Q 2, and the conduction of Q 2 will increase. This *positive feedback* loop will cause a rapid rise in the conduction between anode and cathode. As long as voltage is applied between the anode and cathode, the pair will continue to conduct. *Turn off* requires that the power be removed (or reduced below the level needed to sustain conduction), at which time the device will return to a *blocking state*.

The action of an SCR can also be considered as similar to the action of a thyratron (a gas filled triode). The grid can initiate the conduction through the tube, but power must be removed before it can retain a blocking state. A major advantage that the SCR has over the thyratron is that the *voltage drop*

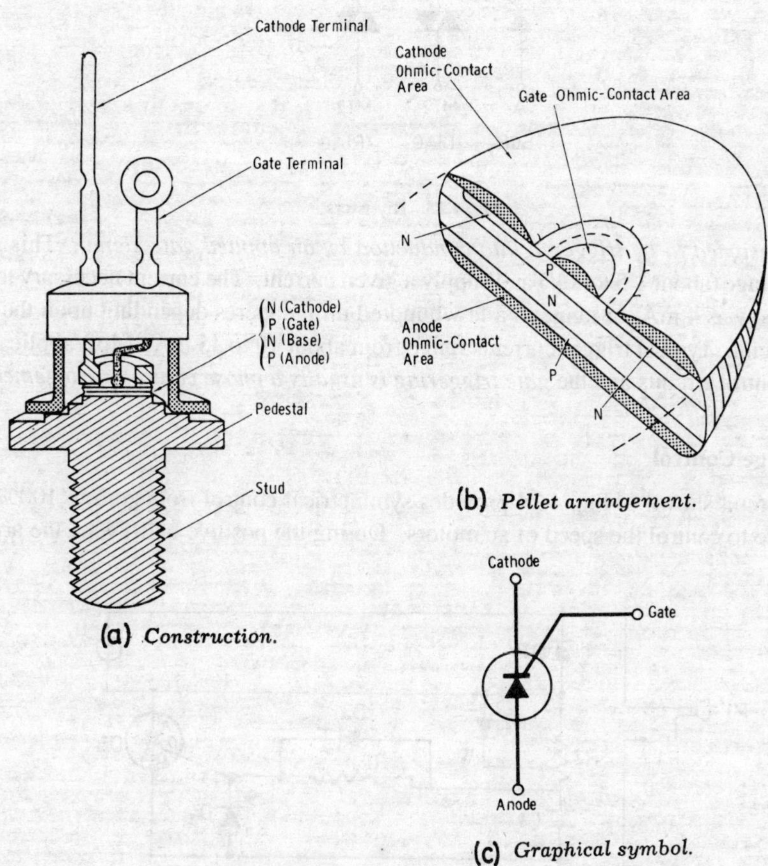


**Fig. 4.22**: Constructional details of a silicon controlled rectifier

*across the unit is very small during conduction*, whereas there is a considerable voltage drop across the thyristor. The constructional details of an SCR are given in Fig. 4.22.

The triac, another member of the thyristor family, is a device which has the property of two SCRs connected back-to-back (*inverse parallel*), Fig. 4.23. *The device conducts in both the directions.* The polarity of the gate voltage for firing will depend on whether MT 1 (main terminal) or MT 2 is positive at the instant of firing. Once the device has fired, it will keep on conducting till the current falls below the *holding current.*

The *diac* is yet another member of the thyristor family, Fig. 4.23. *It is a back-to-back diode, i.e. it conducts in either direction.* However, the conduction in any direction starts only after the average voltage difference between the terminals exceeds the specific voltage.

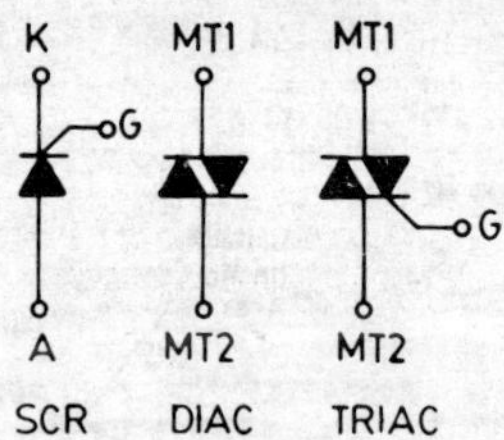


**Fig. 4.23**: Thyristors

*The SCR is designed to be triggered into conduction by an applied gate signal.* This signal must reach a given voltage (about 1.5 to 3 V) and supply a given current. The current necessary to trigger the SCR may be as low as 4 mA, or range to a few hundred milliamperes depending upon the device and the particular circuit. Typical trigger currents range from about 10 to 15 mA. Most applications of the SCR are in ac control circuits and the *gate triggering is usually a phase control arrangement.*

## 4.20 SCR Voltage Control

The full-wave circuit shown in Fig. 4.24 provides symmetrical control from zero to 100% over an ac load. It can serve to control the speed of ac motors. During the positive half-cycle, the arm of poten-

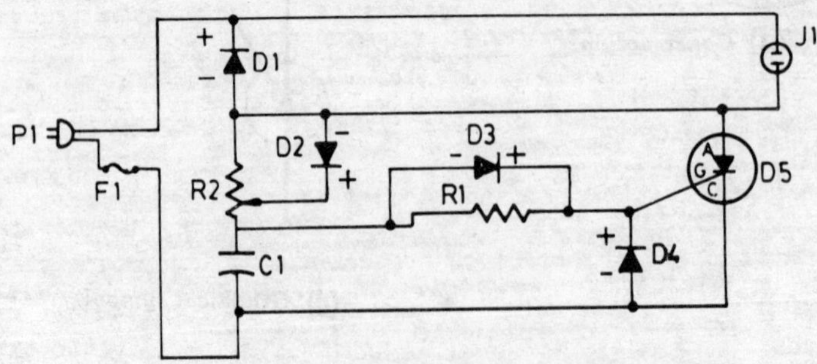


**Fig. 4.24**: SCR voltage control

tiometer samples the supply voltage and compares it with the back voltage of the motor through the gate terminal of the SCR. When the primary voltage rises above that of the rectifier, the SCR is triggered by the diode D 2 which applies the remainder of that half-cycle voltage to the load at the controlled level. D 1 suppresses the *inductive spikes* that could be created when this circuit is used in controlling a motor.

Some additional applications for this unit include light control for projecting slides or home movies. People do not like to be plunged abruptly into darkness from a lighted room and with this unit you can gradually scale down the room lighting as the film gets started. This also gives the projectionist a chance to see that the film loops are properly formed before he plunges into darkness.

## 4.21 SCR Current Control

Fig. 4.25 illustrates a flasher circuit capable of switching 24 watts worth of advertising signs or anything else that needs on and off. Used in an auditory device, it becomes a fine variable speed metronome.

When switch S1 is closed, pulsating dc voltage from KD2103 bridge rectifier is applied across the load and the parallel SCR and resistance capacitance circuit C2, R1 and R8. Since the SCR is in a positive state, the load current is low. As a result, input pulses charge C2 to the neon lamp firing potential. When the lamp fires at approximately 80 volts, about 2 mA of current through R1 provides an input signal to the base of KD2102.

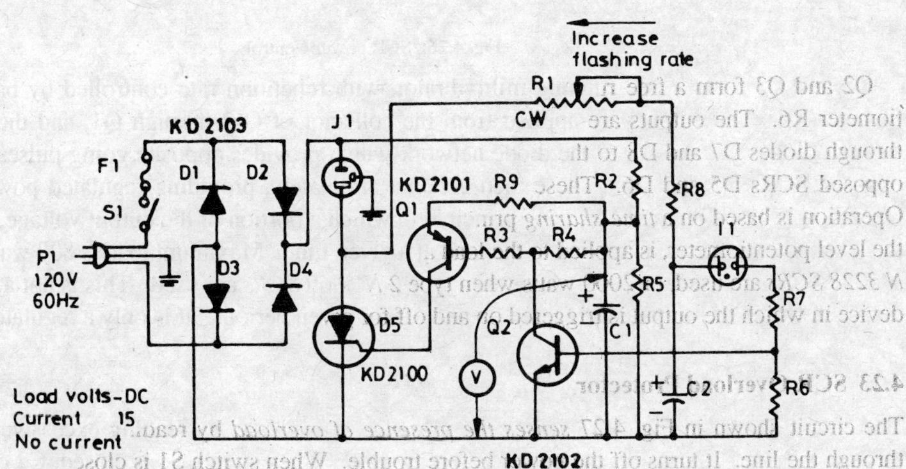


Fig. 4.25: SCR current control

Q1 and Q2 operate as a two stage amplifier rather than a regeneration switch. R6 in the base of Q2 guarantees that the transistor will remain in the off mode when the neon lamp is not conducting. The amplifier output triggers the SCR into conduction and full current appears across the load.

## 4.22 SCR Power Control

This transistorised *integral-cycle ratio power control* can be used in alternating current electric motors — in small hand tools for instance. This circuit could be easily built right into the tool housing, and R6 could be coupled to a spring return trigger. This could result in infinitely variable speed control for the load, with the speed a direct function of trigger position at any given time.

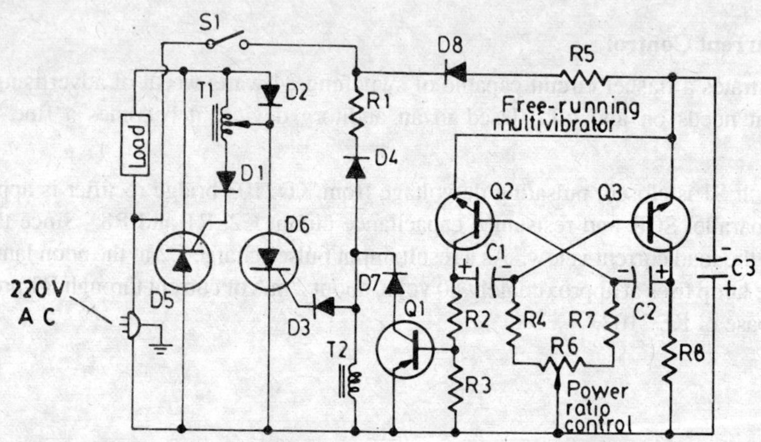


Fig. 4.26: SCR Power Control

Q2 and Q3 form a free running miltivibrator, with repetition rate controlled by balancing potentiometer R6. The outputs are applied from the collector of Q 2 through Q1, and the emitter of Q3 through diodes D7 and D8 to the diode network which provides opposite going pulses to the directly opposed SCRs D5 and D6. These then conduct alternately, providing regulated power to the load. Operation is based on a *time-sharing* principle in which a portion of the output voltage, determined by the level potentiometer, is applied to the load at a given time. Maximum load is 800 watts when type *2 N 3228 SCRs* are used, or 2000 watts when type *2 N 3669 SCRs* are used. This is not a "blinker" type device in which the output is triggered on and off for given periods. It is only a regulator.

## 4.23 SCR Overload Protector

The circuit shown in Fig. 4.27 *senses the presence of overload* by reading excessive current drawn through the line. It turns off the power before trouble. When switch S1 is closed, C2 charges through R7, D2 and D4, when the diodes are forward biased by the power source. On the following half-cycle the charge on C2, triggers the gate of SCR, D7. The voltage placed across the load during this half-cycle is also applied to the combination of D1, R4 and C1 to charge. The residual charge on this capacitor triggers SCR, D6, during the next half-cycle. This process continues to repeat as long as the transistor Q1 is not conducting, which is to say when the overload condition is sensed.

When current drawn by the load causes the regulator switch to conduct, an *overload condition* is present. The potential at which this switch will conduct is set by the potentiometer. R1 in combination

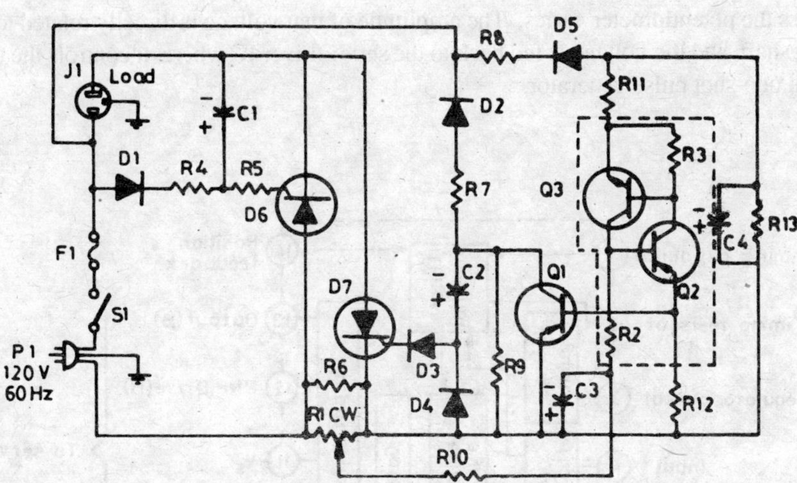


**Fig. 4.27:** Overload protection

with R6 forms a *sensing resistance* that detects increasing current to the load. When this current is high enough to make Q2's base negative with respect to its emitter, the switch conducts. The voltage across the switch is derived from C4, which is charged through R8 and D5 by the power source. When the switch conducts, it turns on Q1 by making its base more negative with respect to the emitter. When Q1 conducts, the charging path for C2 is bypassed.

## 4.24 Servo Control

Radio control (R/C) servos are good for much more than just operating model aircraft or boats. They can be used to keep a signal generator on frequency, steer a small robot, govern your lawn mower, interface a computer to run a strip-chart recorder — in fact, these *miniature servos provide about the ideal electrical-to-mechanical interface. They are simple to use and cost effective, and the circuits needed to drive them are simple, easy to build* — you can even put them together on prototype boards.

*Servos are slow running electric motors that can be stopped and started almost instantaneously. As long as they receive a control signal they run; when it stops so do they.* That makes them well suited for use in proportional control systems, such as used in modern airplanes, boats and cars. The control signal can be sent by radar, or fed directly to the servo control circuits.

R/C servos generally operate from TTL logic level voltages — about five volts — which makes them very convenient to interface with the computer or other TTL circuits. The devices — which may generate several pounds of thrust-draw a maximum of 400 mA depending on the particular type used. *Servo control signals are pulses whose widths vary from one to two milliseconds in length; the servos convert that pulse width modulated signal into a precise mechanical movement.*

Inside the servo housing is an IC that provides all the functions needed to convert the width of a pulse into the signals needed to drive the servo motor so that the output shaft is correctly positioned. Fig. 4.28

shows the block diagram of a typical servo driver IC and its connections to the servo motor. The output shaft of the servo motor is attached to a potentiometer. As the output shaft is driven by the motor, the voltage across the potentiometer varies. The amplitude of that voltage is directly related to the position of the output shaft, and the voltage is fed back to the servo-driver IC where it controls the time duration of an internal one-shot pulse generator.

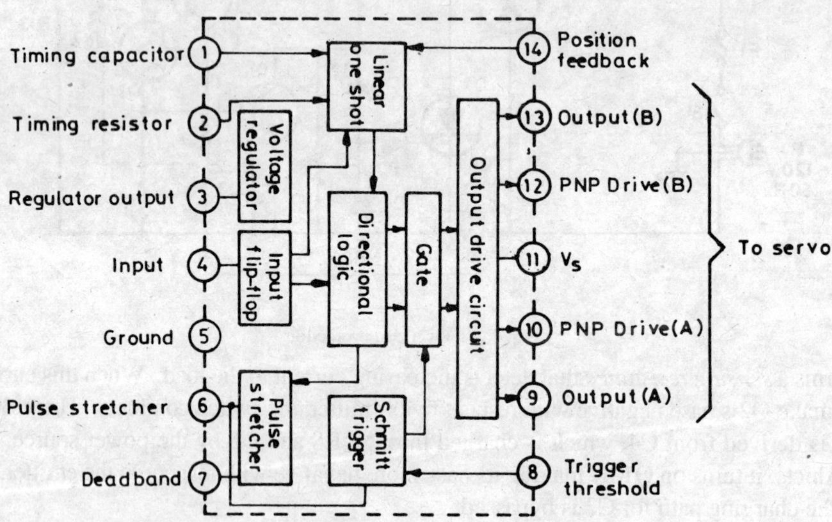


**Fig. 4.28**: Servo-driver IC decodes input pulses and input signals that cause the servo motor to turn in one direction or the other.

When a logic level input pulse reaches the servo-driver IC, two things happen. First, the one-shot pulse generator is *triggered*. The time duration of the one shot output pulse is controlled by the feedback voltage from the potentiometer on the output shaft of the motor. The duration of the output pulse is *compared* to that of the input pulse by two NAND gates. The result of the comparison will trigger one of the two *pulse stretchers*, that in turn will either SET or RESET the directional flip-flop.

The pulse strechers are required because in any servo application — model aircraft, for example — as many as eight channels can be in use at the same time with information being transmitted sequentially to as many as eight servos. Under these circumstances as many as 25 milliseconds may pass before a particular servo receives another input pulse. *The pulse strechers keep the servo motor running for most of the time between input pulses, keeping chatter to a minimum and allowing greater speed and efficiency.*

The output of the flip-flops will cause the servo-motor to be driven by one or the other of the motor-driven circuits. The circuits selected will determine the *direction* in which the motor will turn the output shaft until it reaches a position that causes the input pulse and the pulse from the one-shot to be of equal duration. A one millisecond output pulse will drive the output shaft of the servo motor to one extreme and a two millisecond pulse will drive it to the other.

In most applications you will need to convert the amplitude of a dc voltage into a pulse width modulated signal that can be used to drive the servo-motor. For example, you may have a remote tv camera that you want to be able to aim from four monitoring locations. Your control console could have a joystick for that purpose which, when moved, would move the shaft of a potentiometer. The output of the potentiometer would be a dc voltage related to the position of the joystick and would have to be transferred to a pulse width modulated signal that could be used to control the servo. That signal can be generated by a single quad op amp (one IC containing four separate op amps) like the National Semiconductors LM 324.

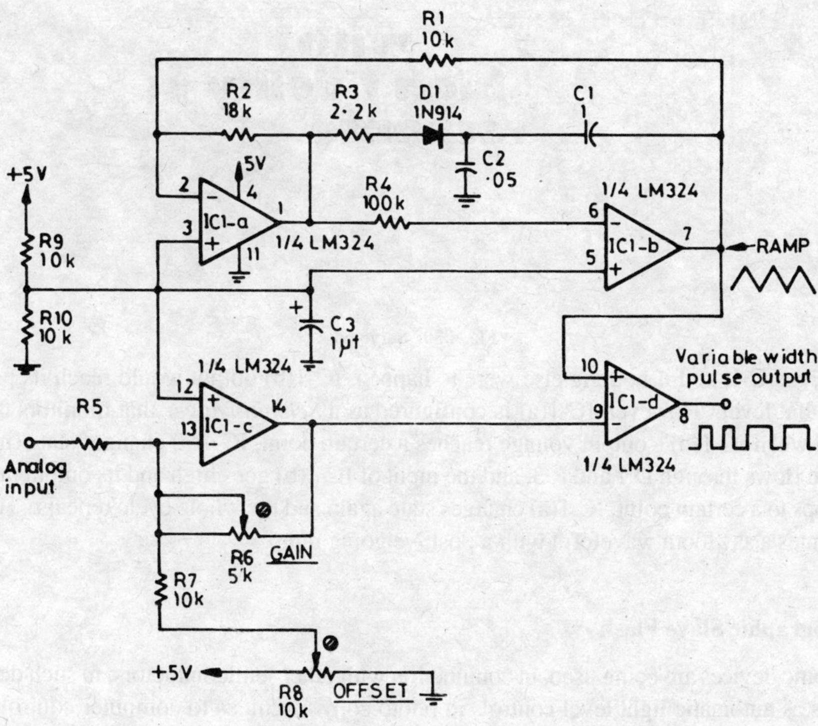


**Fig. 4.29**: Analog-to-pulse width converter senses dc voltage level and outputs a corresponding variable pulse-width string to servo-driver IC.

An easy to build pulse-width modulation (PWM) circuit using that IC is shown in Fig. 4.29. Two sections of IC - 1 function as a *ramp generator* whose output is a sawtooth waveform. op amp IC-1(c) is a *conditioner* that limits the dc level and peak-to-peak swing of the input signal. The last section IC-1(d) compares the output of the signal conditioner with the ramp (triangle) wave-form. *The signal conditioner is adjusted to provide a comparable output that is compatible with the servo's requirements.*

The heart of the ramp generator is IC-1(b). It operates as *integrating amplifier* with the capacitor C 1 as the integrating or feedback element of the circuit. If the output of IC-1(a) is negative, a current will flow through R 4 and try to pull inverting input to IC-1(b) down. In response, the current flow through C 1 continues to equal the current flow through the input resistor R 4.

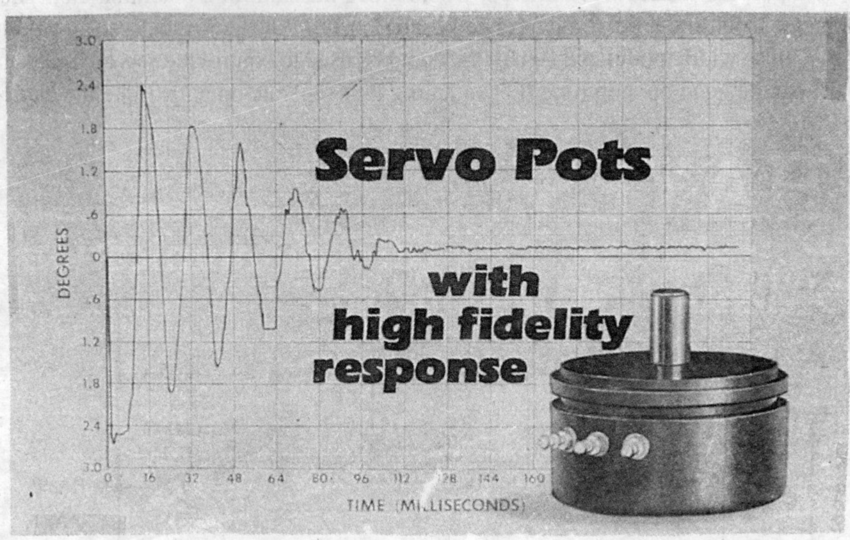


**Fig. 4.30:** Servo pots

This charges C 1, and if nothing else were to happen, IC-1(b) output would reach its positive limit and stay at that level. However, IC-1(a) is configured as a *Schmitt trigger* that monitors the output of IC-1(b) and when IC-1(b)'s output voltage reaches a certain point, IC-1(a) changes state (it goes high). Current then flows through D 1 and R 3, and the input of IC-1(b) goes high and its output drops rapidly. When it drops to a certain point, IC-1(a) changes state again and the whole cycle repeats. The repetitive cycle generates a sawtooth waveform with a positive going ramp.

## 4.25 Photographic Slave Flash

Optoelectronic devices are being used, in conjunction with other semiconductors, in such device control applications as automatic light level controls in photo copy machines to computer control of machine tools.

Included in optoelectronic family are LEDs, photodiodes, phototransistors, light activated SCRs (LAS CRs). optocouplers, optoisolators, and solid-state relays (SSRs).

*Photo electrons convert light into an electric current. Photoemitters convert an electric current into light.* An LED is a diode which emits light from its pn junction when forward biased. An LASCR is a light activated SCR which acts both as a photo detector and an SCR. An IR (infrared) diode emits visible light from the pn junction when it is forward biased. IR diodes serve as the light source for photocouplers (isolators). Optoisolators completely enclose in their housing an IR diode, a photo detector, and the light path between the IR diode and the photo detector. *There is complete isolation between the input and output terminals of an optocoupler.*

Using a single light source in photography produces shadows. However, if more than one flash bulb is used, how do you get them both to fire at the same time? The answer is a *slave*. When the flash bulb in the camera goes off, the photocell causes the slave to fire its lamp - at almost the same time; certainly close enough, so the film will see its light from two sources.

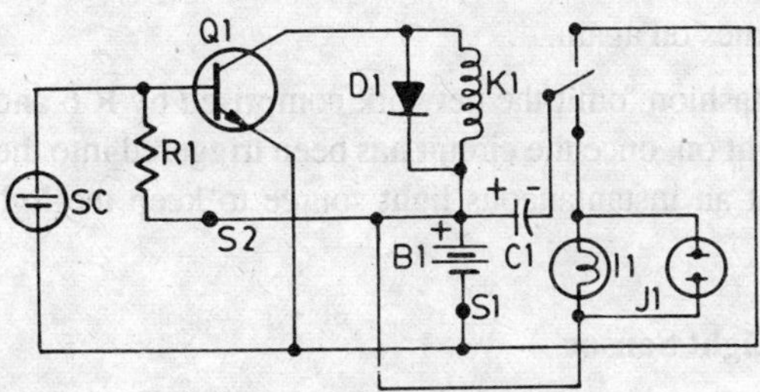


**Fig. 4.31:** Photographic slave flash.

In the circuit shown in Fig. 4.31, the light sensitive photocell (SC) is triggered by the flash from the master unit. This current is then amplified by transistor Q 1 and actuates the relay. The relay then switches connections between B 1 and C 1. The capacitor, which has been collecting a charge from the battery, dumps its voltage across the socket, causing the flash bulb to go off. The slave can be tested by closing I 1. This circuit is more than adequate for use with conventional flash bulbs, but not sufficiently fast for electronic flash equipment.

## 4.26 SCR Light Switch

Q 1 and Q 4 in Fig. 4.32 act as an electronic switch triggered by the conduction of SCR D 5. The sensor which drives the SCR into conduction is the KD 2106 (PC-1), shown here wired remotely and connected

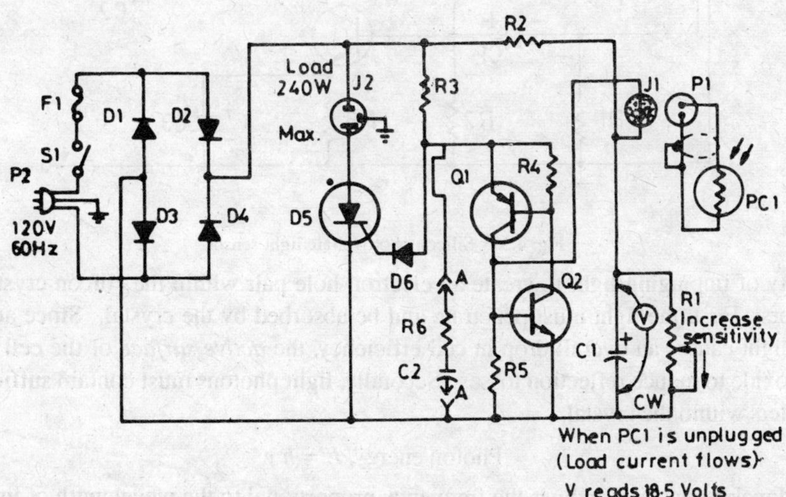


**Fig. 4.32:** SCR Light Switch.

by jack J 1 - plug P 1. *This circuit is designed to operate when ambient light level falls below pre-determined illumination.*

To set up the unit, place the sensor so that it faces a window, and plug a lamp into the load socket. When evening falls, and the ambient light levels are such that you would want the lamp to turn on, adjust potentiometer R 1 until the lamp lights. In the morning, notice that the increased ambient light causes the bulb to turn off. If it does not turn off, readjust R until the lamp goes off and recheck the following night to see that the lamp comes on again.

To use the circuit in this fashion, omit the network comprised by R 6 and D 2, which function in a *latching mode* to keep the light on, once the circuit has been triggered into the on condition. In applications where you would want an instantaneous light source to keep the light off, include these components.

### 4.27 Silicon Photovoltaic Light Sensor

As a light sensor, this circuit is unique. PC1 is highly sensitive to changes in light intensity and Q 1 serves as a pre-amplifier. *The circuit can be used for reading a motion picture optical sound track.*

The photovoltaic cell consists of a large silicon pn junction which is exposed to incident light energy. Impinging electrons having sufficient energy break a covalent bond within the junction and create an electron-hole pair. *The current flow thus generated is directly proportional to the number of electron hole pairs created, the intensity of incident light, and the junction area.* Two electrical contacts are provided so that current may be applied to a load. The voltage developed is a complex function of light intensity.

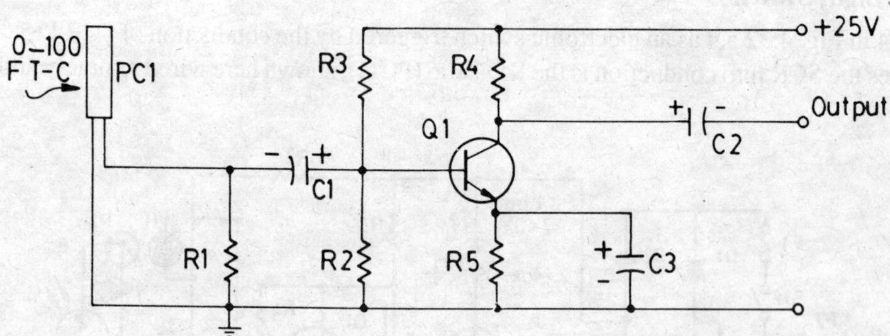


**Fig. 4.33:** Silicon photovoltaic light sensor.

The ability of impinging light to create an electron-hole pair within the silicon crystal depends on several factors. First, the light must penetrate and be absorbed by the crystal. Since any reflected or transmitted light causes an overall drop in cell efficiency, the *active surface* of the cell is coated with silicon-monoxide to reduce reflection losses. Secondly, light photons must contain sufficient energy to liberate carriers within the crystal.

$$\text{Photon energy, } E = h\nu$$

where $h$ is Planck's constant, and $\nu$ is the frequency proportional to the wavelength of incident light.

The number of photons is a function of light intensity. Thus current flow depends on the amount of absorbed light, the intensity, and the wavelength, with a maximum sensitivity specified at a particular wavelength.

## 4.28 Photo-FET Type Photoelectric Relay

The light sensitive device employed in the circuit in Fig. 4.34 is a photo electric field-effect transistor called a *Photo-FET*. This device combines light sensitivity with the amplification and high input impedance of the FET.

A 10 mA current form the 12 V dc supply flows through resistors R 2 and R 3 in series. Consequently a 1.5 V drop occurs across R 2, which is applied as positive bias to the gate of the Photo-FET, and there is a 10.5 V drop across R3, which is applied as negative operating voltage to Photo-FET drain.

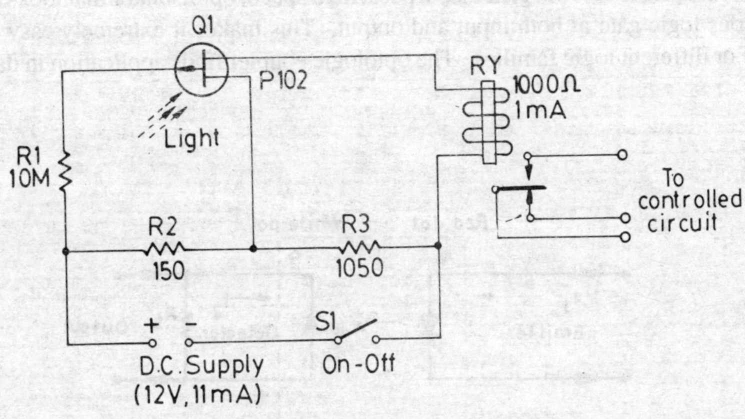


**Fig. 4.34:** PHOTO_FET type relay.

Relay RY is a 1000 ohm, 1 m A unit. A measurable zero signal (dark) current flows through the relay, but it is too small to cause false closure; hence it does not need to be balanced out of the relay coil. Illumination of less than 10 foot-candles will cause the relay to close. The photo-FET has a lens in its nose. The circuit draws a maximum of 11 m A from the 12 V dc supply.

## 4.29 Optocouplers/Optoisolators

Optocouplers are devices that consist basically of a light emitting diode and a phototransistor that are optically coupled within a light excluding case. Optocouplers may be used with analog or digital signals. They are normally specified by their isolation voltage, also called *common mode rejection* (CMR), *speed* (propagation delay), and forward coupling-also called *current transfer ratio* (CTR). A typical good quality optocoupler has a CMR of about 2 kV, a propagation delay of about 5 ns and a CTR, expressed as the ratio of the output current to the input current, of 30%.

$$CTR = I_{out} / I_{in} \tag{4.2}$$

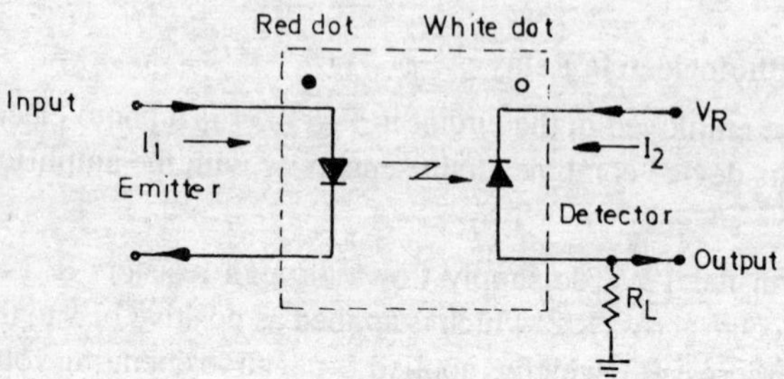


**Fig. 4.35:** Photodiode coupled pair operating in photoconductive mode

The optologic coupler is the first general purpose, high speed, optoisolator that looks exactly like any common 74-series logic gate at both input and output. This makes it extremely easy to interface between the same or different logic families. The optologic coupler finds application in data communication.

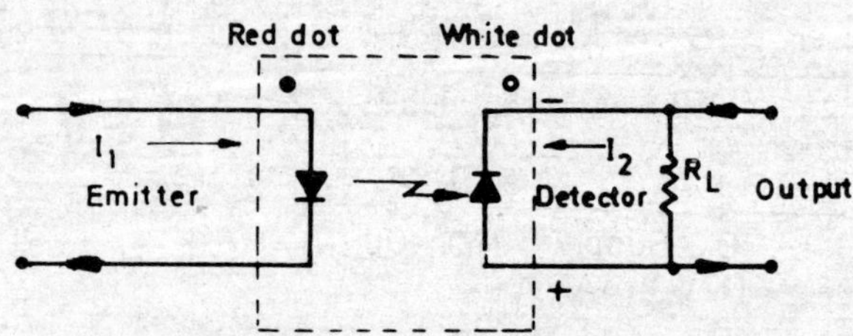


**Fig. 4.36:** Photodiode coupled pair operating in photovoltaic mode.

Fig. 4.37 shows how to make a functional two-way optoisolator by installing two LEDs *face-to-face* in a short length of heat shrinkable tubing. For very high voltage isolation, or for applications in which

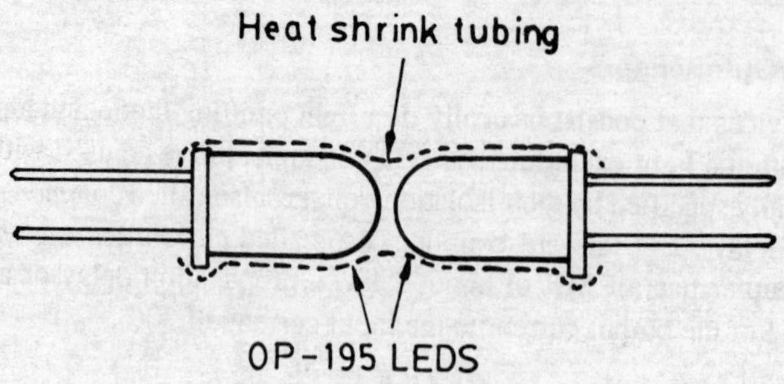


**Fig. 4.37:** How to assemble an LED/LED opto-isolator.

two circuits are some distance apart, the LEDs can be coupled to one another by means of a fiber optic cable.

Though many commercial LEDs can function as both sources and detectors, GaAs, GaAs Si and Al GaAs Si near infrared emitters work better as detectors than do most visible light emitting diodes. Fig. 4.38 shows the current transfer of a pair of TRW Optron OP-195 GaAs-Si near infrared emitters arranged in the configuration shown in Fig. 4.36. Data was obtained by operating the detector LED in an unbiased, photovoltaic mode. If this is done, the resulting current transfer curves are almost identical to those shown in Fig. 4.38.

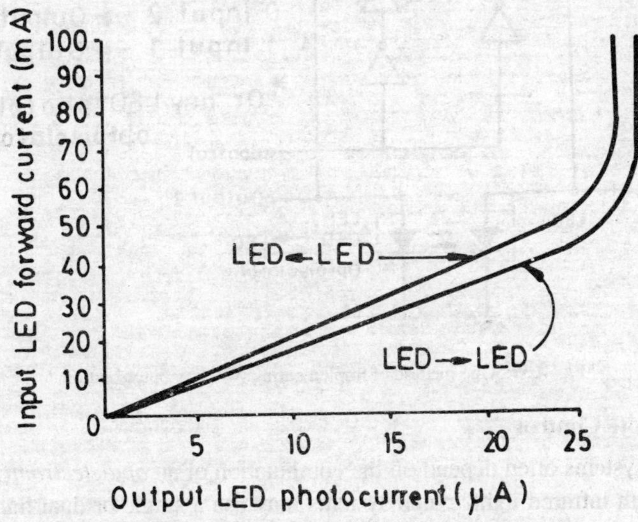


**Fig. 4.38:** LED/LED optoisolator current transfer.

The current transfer ratio, $I_{out}/I_{in}$, for the OP-195 LEDs is only about 0.06% when the input current is 30 mA. While this is much lower than the conventional LED-phototransistor optoisolators, the output current can be easily amplified. The additional circuitry needed can be justified in applications where *two-way optoisolation* is required.

Fig. 4.39 illustrates an experimental circuit to implement, under digital code, two-way optoisolation. The circuit preserves input-output isolation by employing conventional LED-phototransistor optoisolators. In operation, a low or high bit at the control input, forward biases the LED in one of the two conventional optoisolators.

For example, assume that the control bit is high. This causes the LED in optoisolator 1 to be forward biased which, in turn, turns on its phototransistor. Any signal present at the phototransistor's collector can now forward bias LED 1 with LED-LED (two-way) optoisolator. In this case LED 2 functions as a detector. It cannot receive any signal present at the collector of the phototransistor in optoisolator 2 since the transistor is turned off. When the control bit is changed, from high to low, the operating mode is reversed and the LED-LED optoisolator now transmits in the opposite direction.

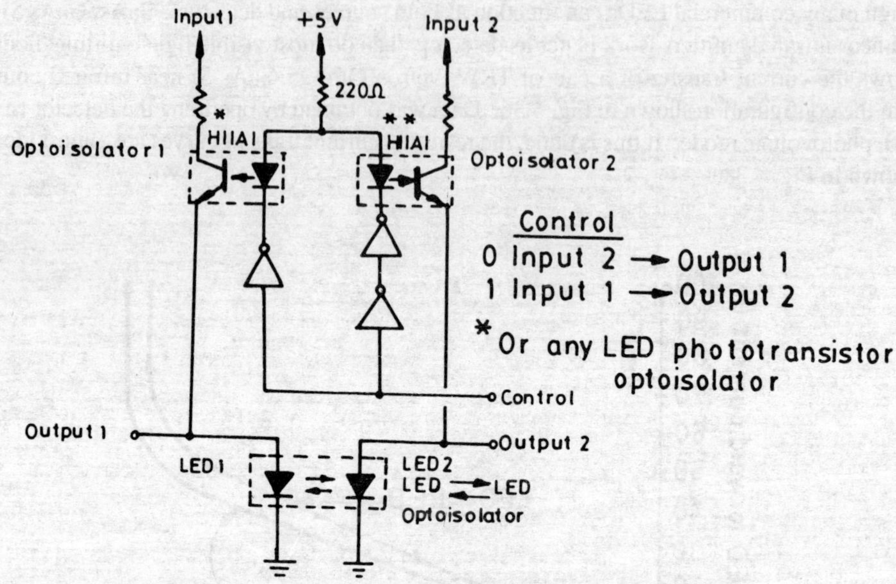


**Fig. 4.39**: One method of implementing two-way optoisolation.

## 4.30 Infrared Remote Control

Alarm and security systems often depend on the combination of an *optoelectronic sensor and emitter*, usually operating with infrared light. Such systems may use a single or dual light beam transmitter-receiver, in which the receiver is switched on to actuate an alarm, when the beam from the transmitter is interrupted by an object. There are also systems that operate by reflecting the light beam back to an integral light emitter sensor with the aid of prismatic mirrors (which simplifies alignment as compared with a plane mirror).

**Fig. 4.40**: Pyroelectric infrared sensor.

Infrared remote control system, Fig. 4.41, for use with tv receivers and audio-systems, also uses a hand held transmitter, and a receiver located in the equipment to be controlled remotely. The transmitter is usually controlled by a five bit (32 codes) or a six bit (64 codes) key pad. The code is transmitted by a number of IR LEDs. The coded signal is received by a photocell and fed to a decoder in the receiver. The decoder usually provides both digital (channel changing, loudspeaker muting) and analog outputs (volume control).

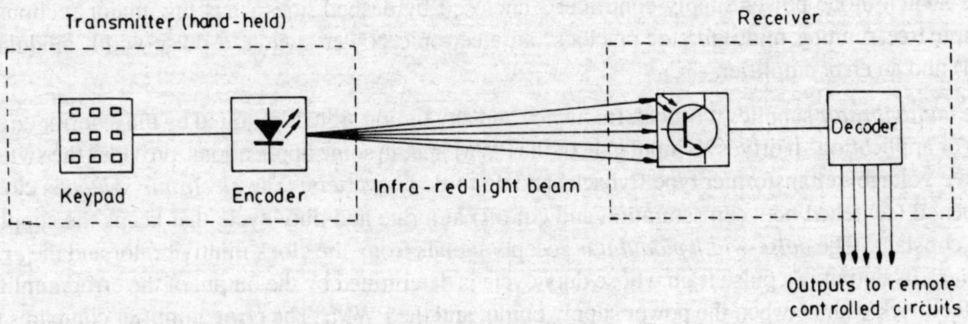


**Fig. 4.41**: Infrared remote control system.

## 4.31 Switch Mode Power Supply Controller

The growing popularity of the switch mode or switching-type power supply has led to the development of more specialised component types than any other electronic circuit since the early days of radio. It

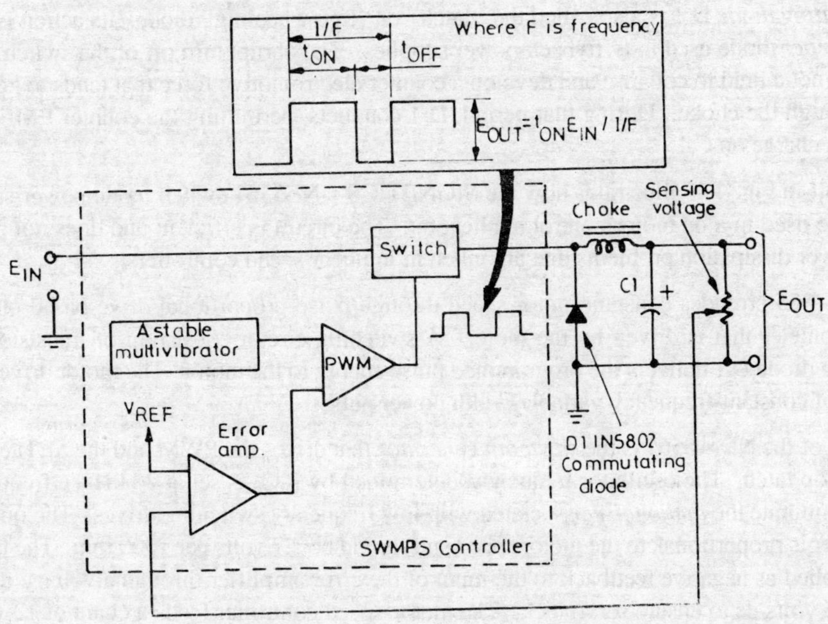


**Fig. 4.42**: Switch mode power supply.

has involved the development of special types of rectifier diodes, inductors and a variety of dedicated ICs.

Basically, a switch mode power supply can be thought of as a type of *series voltage regulator* in which the pass (series) transistor has been replaced by a solid-state switch and a series inductor, Fig. 4.42.

The switch mode power supply controller - enclosed by dashed lines - has few major sections: a precision free running multivibrator or clock, an electronic power switch, a pulse width modulator (PWM) and an error amplifier.

The *multivibrator* supplies the clock frequency and any timing signals required by the exterior circuit design or application. It drives an input gate of the PWM and, in some applications, provides the switching drive voltage in transformer type flybacks and dc-to-dc converters. The *electronic switch* is closed and opened (switched between saturation and cut off) at a rate and duty cycle that keeps the supply's output constant. The *pulse-width modulator* accepts signals from the clock multivibrator and the *error amplifier.* Its output is a pulse train whose duty cycle is determined by the output of the error amplifier in the feedback path between the power supply output and the PWM. The error amplifier compares the voltage tapped from the power supply output with a precision reference-voltage and develops an error voltage based on the difference between the input signals.

The error voltage varies the *switching pulse duty cycle* (the ratio of the on-time to the total time period) as a function of the input voltage and output load conditions. The switch's on time is increased if the sample voltage tapped from the supply output is *low* compared to the reference voltage. If the output sample is *higher* than the reference voltage, the error voltage narrows the switching pulse, decreasing the switch on time.

*Commutating diode* D 1 is also called the "latch" or "free wheeling" diode. Its action is similar to that of a *damper* diode used in tv flyback power supplies. The abrupt turn off of the switch causes the choke's magnetic field to collapse and develop a counter electromotive force that tends to keep current flowing through the choke. During that period, D 1 conducts, permitting the counter EMF current to maintain the charge on C 1.

The circuit in Fig. 4.43 illustrates how the SIGNETICS—NE5561 switch mode power supply controller can be used in a dc motor control application. The circuit is efficient and does not present the heat and power dissipation problems that are inherent in linear speed controllers.

The NE—556 provides constant motor speed through pulse-proportional drive based on feedback from a tachometer that is driven by the motor. A switching circuit consisting of transistor Q 1 and cummutating diode D 1 delivers the programmed pulse energy to the motor. The motor drive-current is in the form of constant-frequency variable-width power pulses.

The heart of the NE—5561 is the *sawtooth generator* that drives the PWM and the SET terminals of an RS flip-flop latch. The oscillator frequency, determined by $R_tC_t$, is set at 20 kHz, a frequency high enough to eliminate the *audionoise* associated with low frequency switching drives. The output of the dc tachometer is proportional to the motor speed and should be 2.7 volts per 1000 rpm. The tachometer output is applied as negative feedback to the input of the error amplifier through a voltage divider that delivers 3.75 volts dc to ensure servo lock. The motor-speed controller feeds a chain of 12 volt pulses to the motor. The duty cycle of the pulses is directly proportional to the load torque demand. The circuit

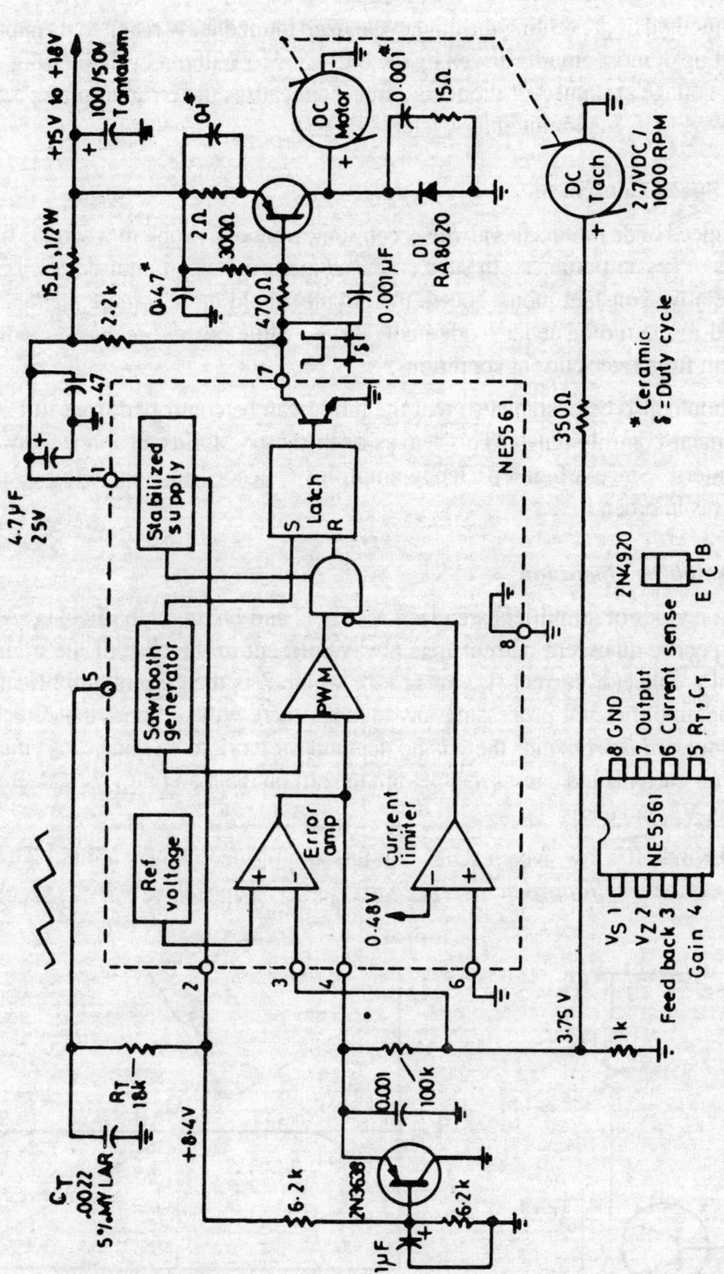


**Fig. 4.43**: The NE—5561 switch mode power supply controller.

can be modified for torque limiting by feeding a derivative of the motor return current (typical current drawn may be 0.3 amp no load and 0.6 amp full load) back into the CURRENT SENSE terminal, pin 6.

The external method of preventing the duty cycle from immediately rising to its maximum is required at the initial start up of most circuits driven by the NE—5561. Without cycle limiting, at initial turn-on the dc feedback voltage at input 3 of the IC is zero. That causes the error amplifier's output to go high and force the PWM to its maximum duty cycle.

## 4.32 dc Motor Speed Control

Controlling the speed of dc motors has always been somewhat of a problem when the operating *efficiency* of a system is of key importance. In such cases, a high efficiency motor drive circuit is imperative. To maintain a relative constant motor speed, the circuit should have *voltage feedback* to increase the current delivered to the motor under loaded conditions, while *current feedback* could be included to protect the system from over-current conditions.

The circuit should also be designed so that the motor can be controlled from full stop to full speed, using a 0-5 V linear control signal. This leaves open the possibility of *interfacing* to other circuits including computers. Since a high efficiency amplifier is needed to control the dc motor, a study of *amplifier classes* is in order.

### *(1) Classes of Amplifier Operation*

The most common types of amplifiers are class A, AB, C and D. In the basic class A amplifier shown in Fig. 4.44, a specific quiescent current $I_B$ is always present in the base of the transistor. This base current results in a collector current ($I_C$), or $I_B \times \beta$, where $\beta$ is the current amplification factor of the device. With the amplifier not processing any signals, there will be some collector current flowing. However, this is *wasted power* since there is no demand for work to be done. Varying the base current varies the collector current, and causes the Q point to shift on the load line. As $I_B$ decreases, the voltage across the device, $V_{ce}$, increases.

According to Kirchoff's law, which states that the sum of the voltages around a loop must be zero, we can see that *voltage not dropped across the load ($V_{CC} - V_{rl}$), must be dropped across the device ($V_{ce}$).*

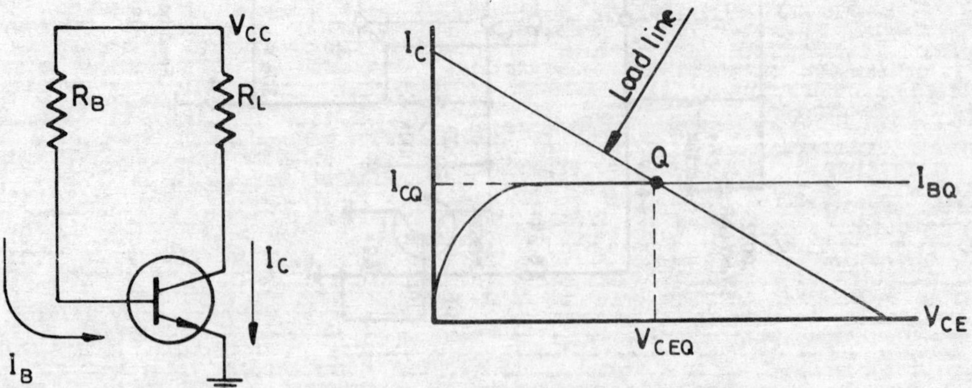


**Fig. 4.44**: Basic class-A amplifier.

Therefore, not only is power being dissipated in the load, but power is also being dissipated as heat ($Vcc \times I_C$) in the device. This mode of operation is very *inefficient*. In fact, the maximum operating efficiency for a class-A amplifier is 50%-unacceptable for the application under discussion.

The class-B amplifier, shown in Fig. 4.45 has greater operating efficiency than a class-A amplifier. In this case, two transistors are biased at cut off, *eliminating the inefficiency of quiescent current*. One-half of the input signal causes one transistor to conduct and the other to be reverse biased. The other-half of the input signal causes the opposite to happen. Unfortunately, any voltage still not dropped across the load is dropped across the device. As in class A operation, *any power not dissipated in the load is wasted as heat*. The maximum efficiency of this amplifier is 78.5% which is better but an unacceptable design.

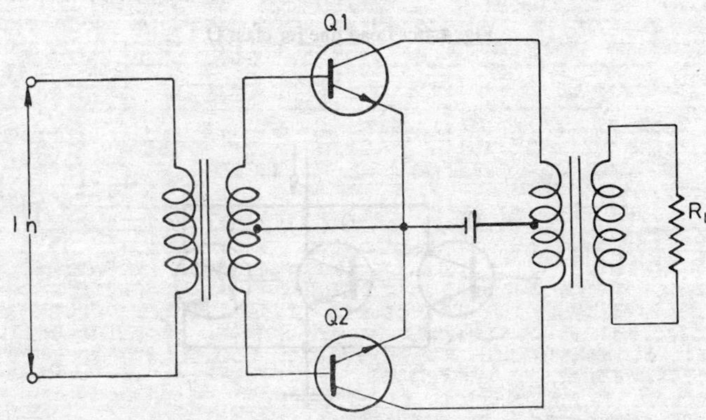


**Fig. 4.45**: Class-B amplifier uses two transistors.

*In any amplifier configuration, inefficiency is present when the difference in power supply voltage and load voltage is dropped across the device (assuming some current flow). Any amplifier configuration in which the device is operating in the linear region cannot be used since maximum efficiency is a mandatory requirement.* Since classes AB and C also operate in this manner, they can be eliminated as possible for this application. Therefore, excluding class A, B, AB and C let's see the class D - the *switching amplifier.*

*The active device in a class D amplifier is never operated in the linear region, as it is either cut off or in saturation*, as shown in Fig. 4.46. If the device is in saturation, the voltage across it is theoretically zero (maximum current) and no power is dissipated. When the device is reverse biased, there is maximum voltage but no current (theoretically) and no power is dissipated. Thus, *the class D configuration can be looked at as a simple on/off switch.*

When the switch is open (cut-off), there is no current flow in the device and as a result, no power dissipation. When the switch is closed (saturation), maximum current flows through the device (determined by $R_L$) and there is zero voltage across the device. Again no power is dissipated in the device.

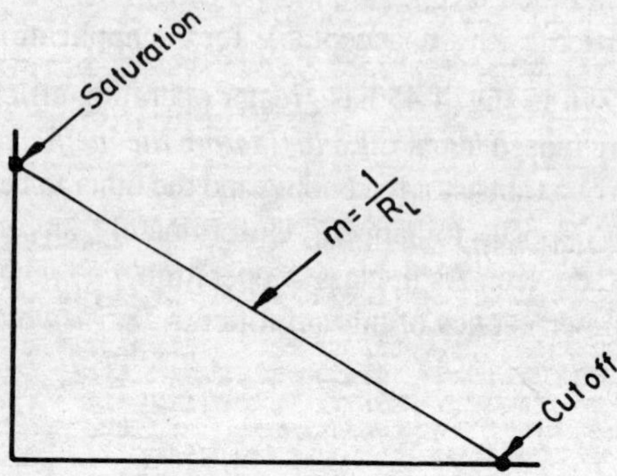


Fig. 4.46: Load line for class D.

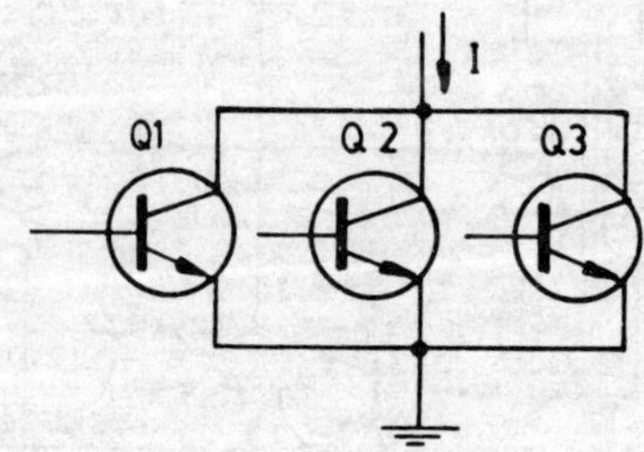


Fig. 4.47: Class D analogy of a switch.

We can see that all the power being dissipated is in the load, and we have achieved a maximum theoretical efficiency of 100%. The actual efficiency is less, of course, due to certain characteristics of the semiconductor switching device. We expect the voltage across a closed switch to be zero. In actuality, there is some small voltage across the device in the saturated state ($V_{sat}$) just as there is some leakage current through the device in the cut-off state. However, this type of amplifier suits this specific need.

### *(2) The Control Voltage*

The next step would be to choose a method of *controlling the switching amplifier* with a 0-to-5 V signal. In this example, a control signal of 0 V corresponds to an open switch, while 5 V corresponds to full power through the load. A question may be raised on how a switching amplifier can promote currents linearly between zero and several amperes. Consider the output of a + 12 V class D amplifier shown in Fig 4.48.

In Fig. 4.48(a), the switch is turned on for 50 ms, and off for 50 ms. Since the *duty cycle* is 50%, the average voltage over 100 ms interval is 50% of 12 V or 6 V. In Fig. 4.48(b), the duty cycle is 25% and the average voltage is $(0.25 \times 12)$ or 3V. *In this manner the average DC voltage is determined by*

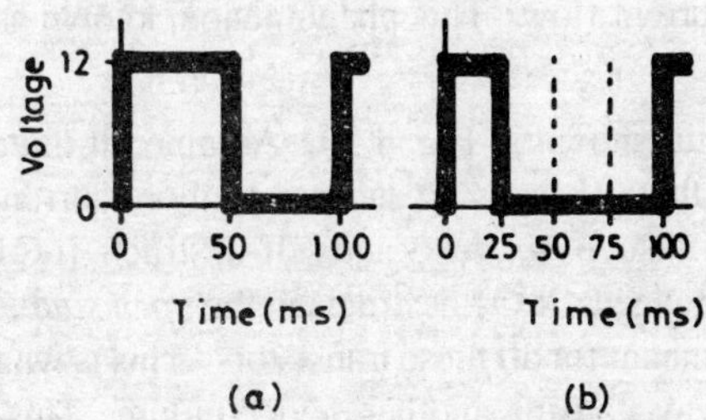


**Fig. 4.48:** Output waveforms of class-D amplifier.

*varying the duration of the pulse width. The speed of a dc motor can, therefore, be controlled by modulating the pulse width.* A typical pulse-width modulator block diagram is shown in Fig. 4.49.

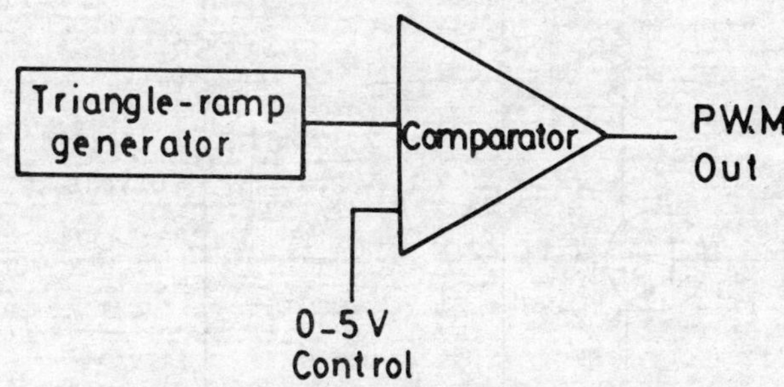


**Fig. 4.49:** Typical pulse-width modulator block diagram.

The output of the pulse-width modulator is fed to the input of the class D switching amplifier. It is important that *the rise and fall times of the switching waveform be as low as possible* since this will result in the switching device being operated in the linear region for only the duration of these times. If the switching times are considerably shorter than the on and off times, then power dissipation in the device will be minimised.

A zero control voltage will result in 0% duty cycle and consequently *no output* from the amplifier. A 5V control signal will result in a 100% duty cycle and *full output* from the amplifier.

### *(3) Devices*

The final step is to investigate the devices that will accomplish the *actual power switching*. The devices are theoretically switches and, unfortunately, semiconductor devices are not ideal. What is needed are diodes having *low on resistance* resulting in low $V_{sat}$ and *low leakage currents*. These two factors reduce the power dissipated in the devices during the on and off cycles. Since it may be necessary to parallel devices to obtain the required constant current capability another important factor in selecting the devices is the *ease with which paralleling can be accomplished.*

Investigating the two main types of devices that would suit this application, *bipolars and MOSFETs*, we find that the main disadvantages of bipolars are their inherent "current hogging" characteristics. The current in a bipolar generates heat. The build up of heat results in a lowering of internal resistance, which leads to an increase in current flow. This phenomenon, known as *thermal run-away*, is characteristic of bipolars in general.

Consider the simplified circuit shown in Fig. 4.50. Assume all three transistors are operating and sharing current I. Now assume that Q1 heats up and passes more current than Q2 and Q3. With more current, it heats still more and will soon *run away* and self-destruct. If Q1 destructs in the *shorted state*, the load circuit will likely be damaged. If Q1 destructs in the *open state*, then Q2 and Q3 will be forced to handle the current originally meant for all these transistors. This is what is meant by *current hogging*. Soon Q2 or Q3 will also run away, causing another device failure. This process will continue until all the devices have self-destructed or the circuit fails.

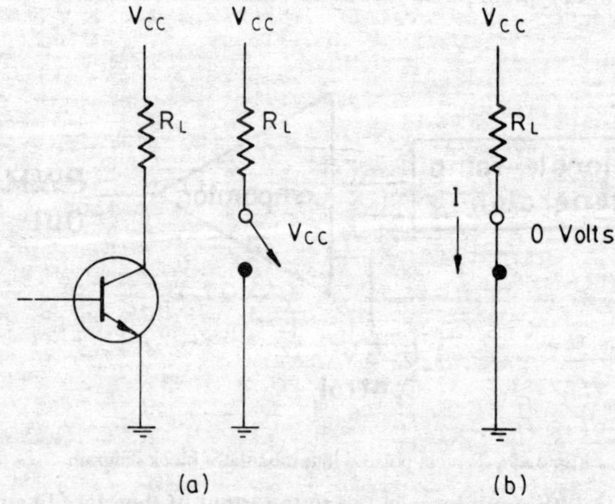


**Fig. 4.50:** Simplified amplifier circuit.

Another circuit could be added to prevent this, but the addition of such circuitry would require extra power, thus reducing the efficiency of the system. Also, bipolars are current operated devices requiring current-drive circuitry. A good engineering rule states that "*the less the component count the more reliable and efficient the system*".

MOSFETs are particularly suitable for this application, especially those known as VFETs and HEX FETs. *The main advantage of using these FETs is their high impedance, voltage operated gate, low on resistance, low leakage current, and negative temperature coefficient.* An amplifier employing VFETs/HEXFETs can be driven directly from the output of the PWM circuit (comparator voltage output) since the high input impedance (>100 kilohms) of these devices eliminates pre-amplifier loading problems. The low on-resistance (typically less than 10 ohms) and low leakage current assure a minimum dissipation in the device during on and off times. Of main importance is their *negative temperature co-efficient*, allowing ease of paralleling. As a VFET/HEX FET heats up it draws less current, forcing the devices to share the current until the first cools. Thus paralleled FETs are *self-stabilising* due to its characteristic. A simplified block diagram of the complete system is shown in Fig. 4.51.

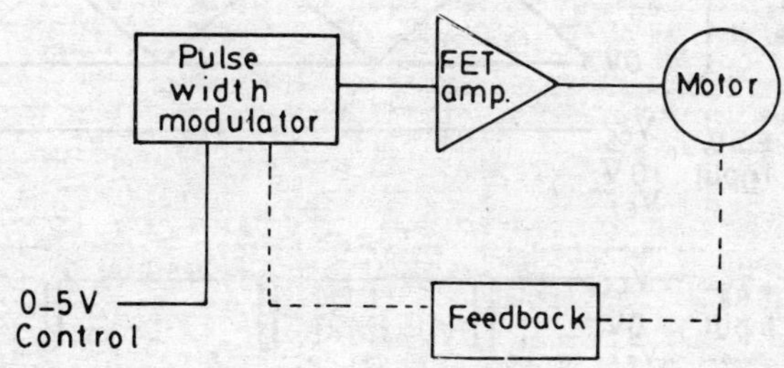


**Fig. 4.51:** Simplified block diagram of the system.

### *(4) Pulse-Width Modulator*

The pulse width modulation circuit can be broken up into two functional blocks, a *triangle ramp generator*, and a *comparator* as shown in Fig. 4.52, Let's examine how then pulse-width modulator functions.

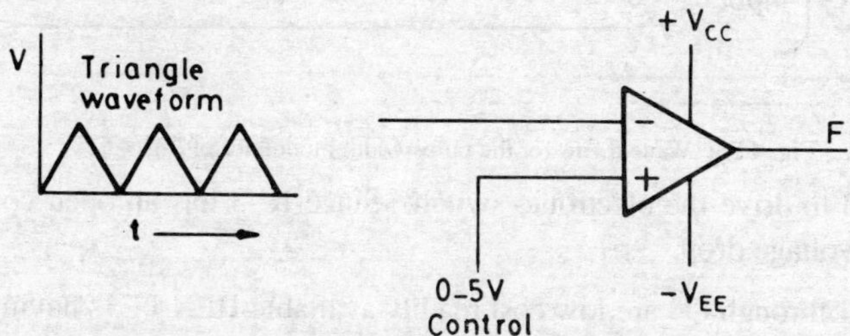


**Fig. 4.52:** Pulse-width modulator.

Using Fig. 4.52 as a guide, if the comparator positive output is made more positive than the negative input, the output F will be high ($+V_{CC}$). If the negative input is made more positive than the positive input, then the output will be low ($-V_{CC}$). With this in mind, Fig. 4.52 shows how pulse-width modulation occurs with a varying positive control voltage.

### *(5) Basic Circuit Operation*

Function generator IC-1 in Fig. 4.54 is used to generate the triangular waveform signal, with the frequency determined by the setting of the potentiometer R26. The triangular waveform output from pin 4 is coupled via C3, to the inverting input of buffer IC-2 with potentiometer R27 determining the zero dc offset for driving comparator IC-3 and R5/R3 determining the gain. The non-inverting input (pin 2) of comparator IC-3 is fed from summer buffer IC-4 that accepts both the 0 to 5V control voltage and the selected current or voltage feedback signal, see Fig. 4.54. The output of IC-3 at pin 7 is the pulse-width

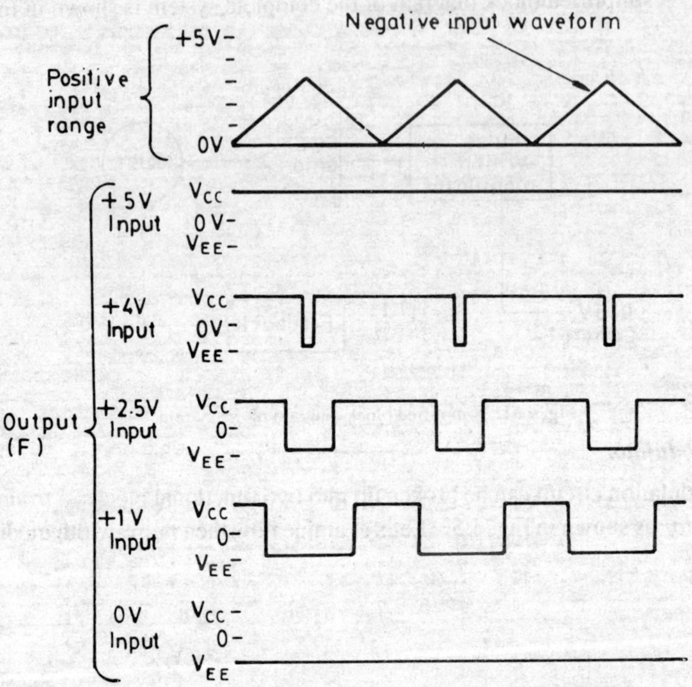


**Fig. 4.53:** Wave forms for the pulse-width modulator of Fig. 4.52.

modulated signal used to drive the electronic switch. Since IC-3 has an open collector circuit, R10 provides the required voltage drop.

Switch elements Q1 through Q4 are low cost readily available HEX FETs having a continuous current rating of 3.5 amperes each and a typical on resistance of 1.3 ohms. Thus, the four in parallel can handle upto 14 amperes and a resultant parallel resistance of less than 0.5 ohms. Their high gate input impedance can be driven directly from the IC-3 output without loading. Since the forward trans-conductance of these devices is a minimum of 1.5 mho, the 12V comparator output is enough to assure complete device saturation. Diode D1 removes any possibly damaging back EMF voltage from the switching circuit. If the load selected requires more or less current, the number of HEX FETs can be increased or decreased accordingly.

Positive voltage and negative current feedback circuits are shown is Fig. 4.55. The voltage feedback circuit increases the voltage across the motor using a differential amplifier. The buffered pulsating motor voltage is switched by R25 and C5 to obtain a dc potential proportional to motor speed. The greater the time constant of the RC network the more damped the system response.

The current feedback also uses a differential amplifier increasing the voltage developed across a motor series resistor R 5. The voltage generated across this resistor is proportional to the motor current. The pulsating output of the differential amplifier is smoothed by R16 and C4 into a dc level proportional to motor current (hence speed). The smoothed dc is level shifted (determined by R31) and can be passed

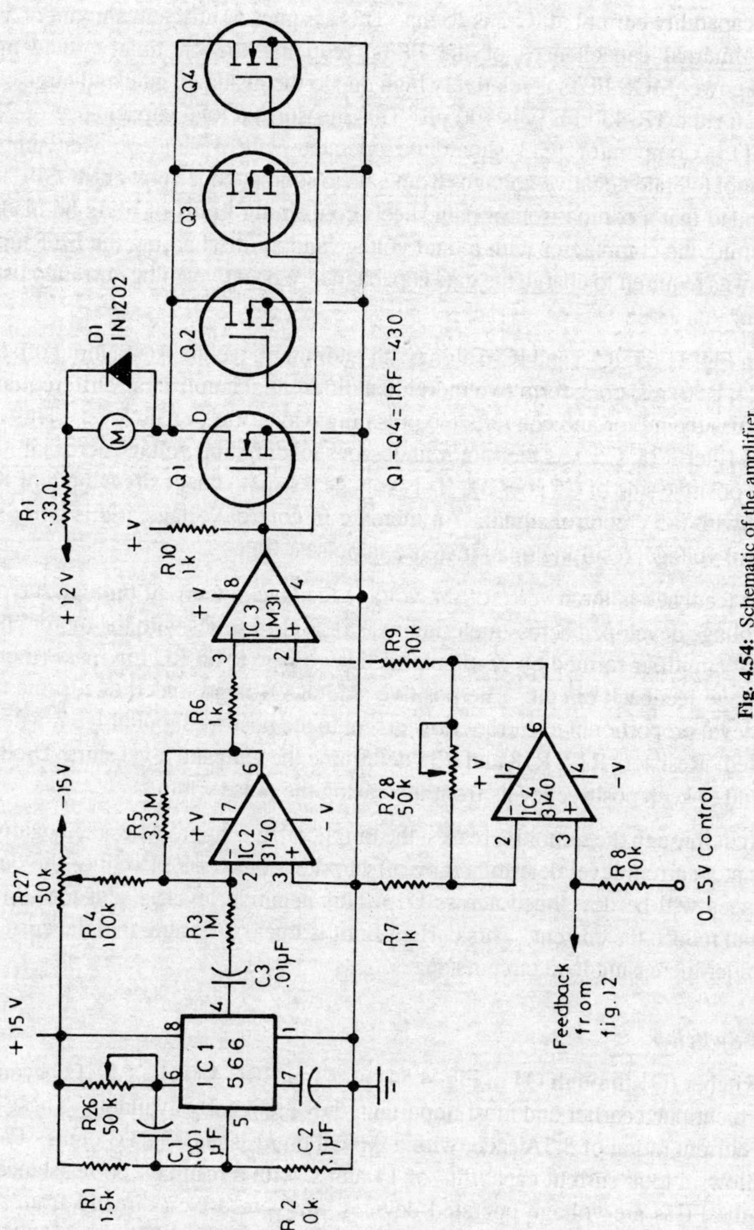


**Fig. 4.54:** Schematic of the amplifier.

to the pulse-width modulator, to be summed with the voltage feedback if required. The feedback circuit can be eliminated in some cases.

The output capability current of IC-3 is 15 ma. This assumes a sufficient amount of available current to charge the internal capacitances of the FETs, reducing the on time. Incidentally, the *input capacitance* to power MOS FETs is relatively high due to the insulated gate and large chip topography. Typical values for the TR-430 FETs is 700 pF. This measures a total capacitance of 2800 pF for four paralleled FETs. Comparator IC-3 should be operated with a single power supply since many MOSFETs cannot tolerate negative voltages from gate to source. Since power MOSFETs are not cheap, it is recommended that a comparison of data sheets concerning limits of FETs be made. There is advantage to running the comparator with a dual voltage supply-discharging the FET input capacitance. Just as current was required to charge the gate capacitance, a negative going current pulse will discharge the capacitance.

As shown in Fig. 11.55, IC-5 and IC-6 alongwith their input, feedback and input offset compensation (R15, R29, R24, R30) resistors, form two individual differential amplifiers. Differential IC-6 measures the voltage across the motor and converts the pulsating waveform to a dc level. This is accomplished via a high-pass filter R25, C5. As the motor undergoes loading, the voltage across it increases and the dc level on the positive side of C5 goes up. This voltage is sent to the positive input of IC-4 (Fig. 11.54) to be summed with 0-5V control signal. An increase in control voltage at this point simulates an increase in control voltage, resulting in an increase in pulse-width.

The current feedback is the reverse of the voltage feedback. Current through the motor is proportional to the voltage developed across metering resistor R11 in series with the motor and power supply. The differential amplifier formed by IC-6 outputs this voltage to an RC low-pass filter identical to the one in the voltage feedback circuit. The positive side of C4 is grounded due to the negative voltage level. The dc level proportional to current flow is sent to the positive input of IC-7, where it is amplified and level shifted. Resistors R17, R18 and R31 determine the gain and level shift. Diode D1 clamps the signal to ground to keep positive levels from increasing the pulse width.

As the current through the motor increases, the output of IC-7 approaches the negative supply, crossing zero volts at a current level determined by offset potentiometer R31. Since the output is clamped, negative voltages will be developed across D1. This negative level is sent to IC-4 to decrease the pulse-width and reduce the current. This current limit is linear, meaning that the current level is maintained even under increasing load circumstances.

### *(6) The FET Switcher*

The power switcher (Q1 through Q4 in Fig. 4.54) uses four IRF-430 HEX FETs because they meet the requirements mentioned earlier and most important, they are readily available. These HEXFETs, have a continuous current rating of 3.5A each, with a typical on resistance of 1.3 ohms. Paralleling, four of these FETs allows a total current capability of 14 amps, with a resultant on resistance of less than 0.5 ohm. Since the FETs are voltage operated devices, they are driven directly from the output of the comparator.

The forward conductance ($G_{fs}$) of the FETs is a minimum of 1.5 mho. Therefore, the +12 V comparator output is enough to assure saturation. The MOSFETs have *internal diode protection*, partially eliminating the need for external protection circuitry. To avoid heat dissipation, all four are mounted on

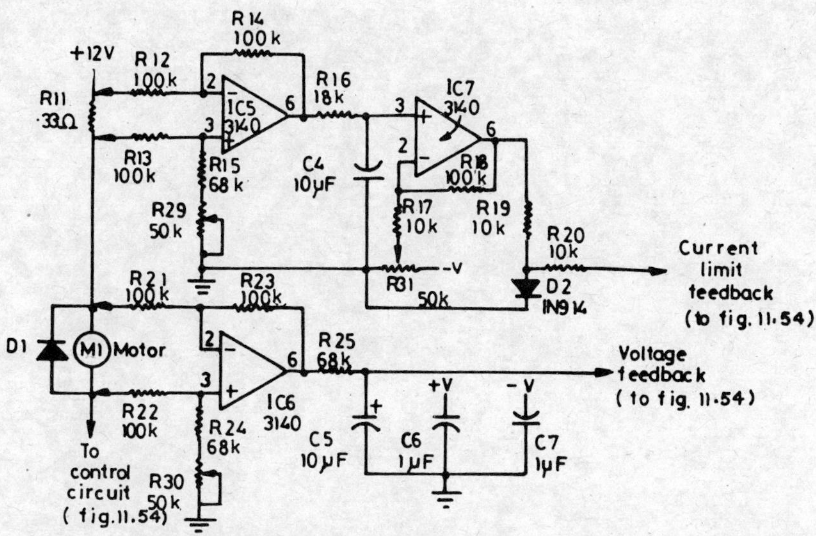


**Fig. 4.55:** Current limit and voltage feedback circuits.

*aluminium heat sinks.* The gates are *tied together* and connected to the output of the PWM. The sources are *grounded* and the drains *tied together* and connected to the negative side of the motor. Wire of sufficient gauge is used to permit heavy current drain. To help eliminate the problem of inductive voltage spikes form the motor, a *free wheeling diode* is used across the armature. Higher current capability can be attained by paralleling more FETs.

The efficiency of the amplifier is greater than 95% at current levels upto 10A. The FETs run extremely cool, even at current levels over the rated 10A. The control voltage can be derived from a potentiometer setup.

# PART-3
# SOLID STATE CLINIC

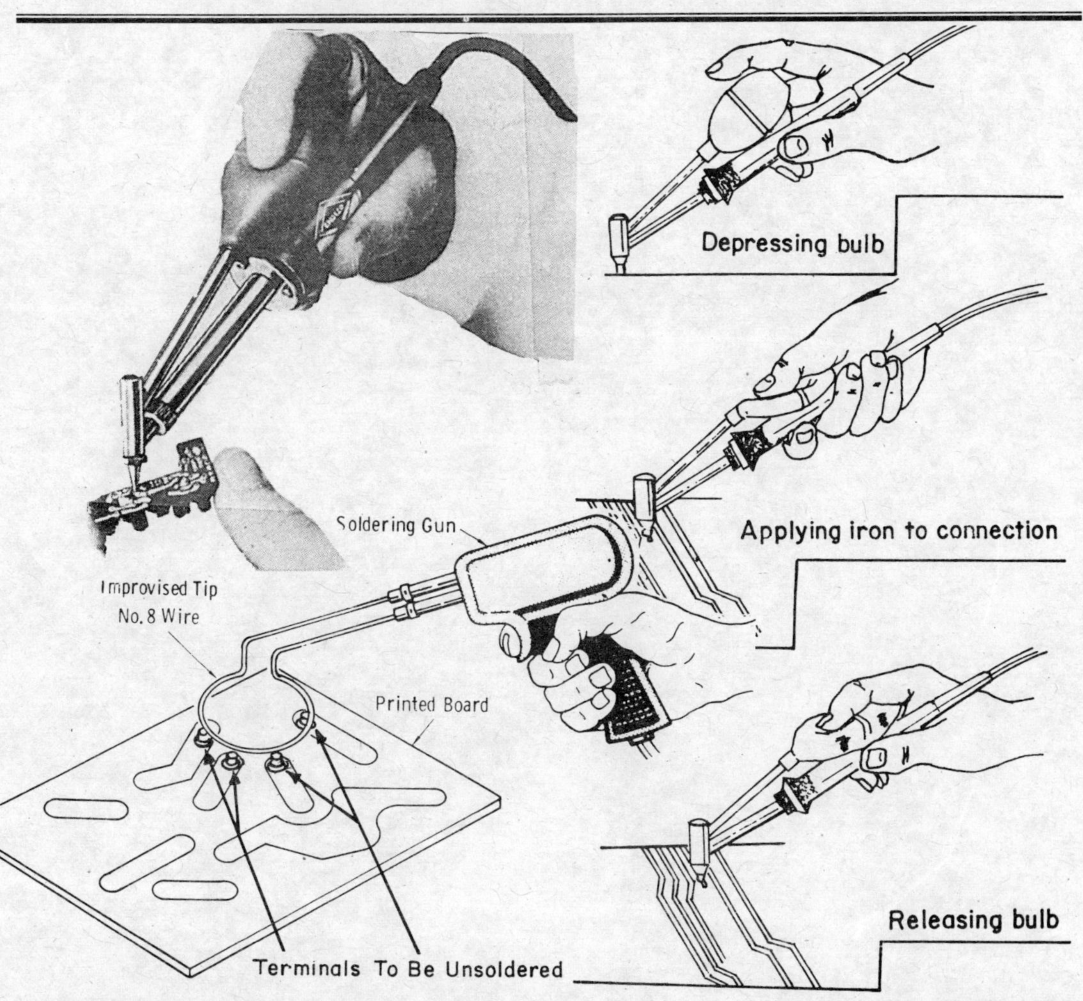


1. PHILOSOPHY OF TROUBLESHOOTING
2. SERVICE AIDS
3. SOLID-STATE CLINIC
4. DO'S AND DONT'S

*Chapter 1*

# THE PHILOSOPHY OF TROUBLE SHOOTING

*Troubleshooting is basically the process of determining why a device, or circuit, does not operate as it should so that it can be remedied within the shortest possible time. To do this satisfactorily, it is necessary to have a broad knowledge of how things work and to be able to recognise the symptoms of a defective device or circuit so as to form a strategy of action for repair. In this chapter we look into the most important aspect of devices and circuits, the search for trouble-which we call troubleshooting.*

## 1.1 The Search For Trouble

Almost three quarters of the servicing time is devoted to *finding the trouble* and the remaining one quarter to *remedying it*. Obviously then, it is in the search for the trouble, which we call *trouble shooting*, that we can make the greatest gains through *logical organisation of thought* and by *efficient execution*.

In troubleshooting, the technician becomes a *detective*, and the defective component or circuit is the *culprit* which must be exposed and eliminated in the shortest possible time. The troubleshooting process involves the following steps:

1. *gathering all possible evidence*;
2. *interpreting the evidence* and putting it together to narrow down the trouble causing possibilities to as small a number as possible; and
3. *making routine tests* that ultimately eliminate each of these possibilities in turn until the actual trouble source alone remains.

The range of time and effort required to accomplish the above three steps is different for different symptoms. At one extreme is the case in which a *particularly significant evidence* immediately tells the technician what is wrong and enables him to remedy it immediately. At the other extreme is the case of an *intermittent condition* which occurs only occasionally. The source of this trouble can be determined only by operating the equipment for long periods of time and thoroughly checking all possibilities each time the intermittent condition occurs. Even though the amount of time in the two extreme cases mentioned above is drastically different, *the logical organisation of thought is the same* with the three main steps listed above necessary in each case.

In the general trouble shooting of devices and circuits there are usually no hard and fast-rules to be followed. The technician must gather all possible evidence and then apply his knowledge of theory, test instruments, tools, backed by experience, to the solution of a particular trouble symptom.

For troubleshooting, the technician sets up a "set of procedures". What he does in a particular case depends on the nature of the trouble symptom and its indication. The efficient technician prepares a mental *check list*. As he goes down the mental check list he *eliminates* many of the items which do not apply to a specific symptom. Thus he makes sure that each possibility has been considered.

While examining an equipment make your own observations. Try to determine by *visual inspection* if any component overheats and then try to *interpret* and find the cause of trouble. It helps in approaching the actual source of trouble more quickly.

The gathered evidence can now be used to *localise* the trouble to one particular section of the equipment. Component tests in that section will then lead to one or more guilty components.

Two well known methods of detective work are *deduction and elimination.* We cannot always deduce that the trouble must be in just one component or in just one location. We can ordinarily place it in one *general section.* Thereafter, by the process of elimination, we eliminate all of the possibilities except, of course, the real trouble cause.

When the trouble is fairly well *localised,* transistors are checked, if necessary, and other *elimination tests* made to determine the actual component(s) at fault so that the necessary remedial action can be taken.

## 1.2 Routine Tests

Routine tests involve visual inspection, smell, hearing and touch, continuity tests, transistor tests, voltage tests and integrated circuit tests. Testing of solid-state devices (transistors and integrated circuits) is explained in Chapter 3 on Solid-State CLINIC.

**1. Visual Inspection:** Unless the probable source of trouble is known, troubleshooting should usually begin with a *thorough visual inspection.* Loose connections, broken wires, poor solder joints, burned out resistors, or even a burned out fuse can very often be located in this way.

Let us take an example of a *dead equipment.* You might adopt a lengthy procedure to come to a conclusion. Perhaps if you had inspected the attachment plug before replacing anything, you would have discovered that a wire had become unfastened from a plug terminal screw. With this corrected, the equipment might be restored to its *normal health.* In this case, a thorough visual inspection would have saved time and possibly some expenses.

In more complicated troubleshooting and repair work, more time and more money can be wasted by procedures which would have been quite unnecessary if a visual inspection had been made first.

**2. Smell, Hearing and Touch:** The senses of smell, hearing and touch can be *valuable aids* in troubleshooting. Overheated resistors, transformers, choke coils and insulation materials can very often be detected by their *peculiar odor.* This can be helpful in locating certain defective components. When the overheating of any component or part in a circuit is noticed, the circuit should be turned off immediately since *it may add to further troubles.*

Overheated transformers and choke coils will sometimes produce a cooking or frying noise that can be easily heard. The source of noise can thus be located.

The sense of *hearing* can be exploited by using a *loudspeaker* as a test instrument. Lack of sound, or distortions of sound, heard through it can indicate specific circuit defect(s) which can be quickly located.

Overheated components can, in many cases, be located by *touch*. This is especially true of transformers, choke coils, electrolytic capacitors, transistors and integrated circuits. If any one of these components *feels excessively* warm there is reason to believe that a defect exists in the component or its associated circuit.

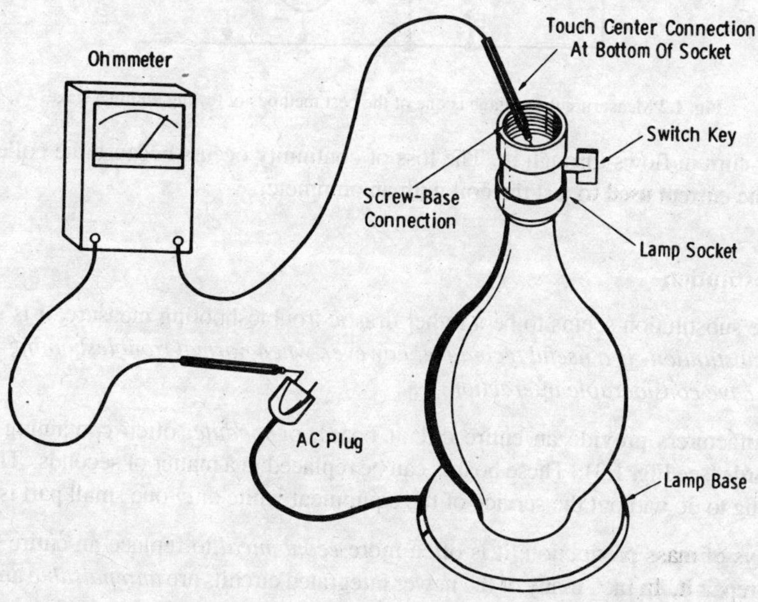


**Fig. 1.1** Ohmmeter used to determine continuity in lamp cord.

**3. Continuity Tests:** Following the preliminary visual inspection and the smell, hearing and-touch tests, troubleshooting involves continuity tests with a suitable ohmmeter. This is illustrated in Fig. 1.1. Such tests should be made when even *open/short circuit conditions* are suspected. *An ohmmeter should never be used in a circuit to which a voltage is applied. Also, the component should be isolated by disconnecting one of its leads before continuity tests are performed.*

**4. Voltage Tests:** Regardless of the method used to find a defective stage, the defective part within the stage must be pinpointed. Measuring the voltage is one of the best methods of *testing within a stage*, because with this method *the circuit is tested with the power applied*. Some defective parts check good when the power is not applied. For example a coil in the collector circuit of a transistor can show continuity when measured with an ohmmeter. However, the continuity of the coil might be lost entirely

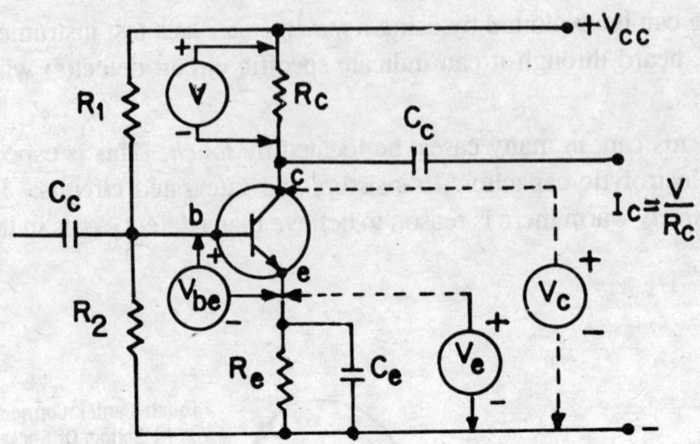


**Fig. 1.2** Measuring the voltage is one of the best methods of testing within a stage.

when collector current flows through it. The loss of continuity occurs because the collector current is several times the current used to test the coil with an ohmmeter.

## 1.3 Stage Substitution

Although stage substitution seems to be a rather drastic troubleshooting measure, it is sometimes justified. *Stage substitution is a useful technique, however, when normal troubleshooting methods fail in circuits which have considerable interaction.*

Some manufacturers provide an entire circuit board, or *module*, often containing several stages mounted on cards (see Fig. 1.3). These boards can be replaced in a matter of seconds. This prevents the customer having to do without the service of the equipment while only one small part is 'in for repair'.

In these days of mass production, it is often more *economical* to replace an entire mass produced circuit than to repair it. In fact, many of the newer integrated circuits are *unrepairable* and, as such, they are *replaced as an entire unit*. Some of these integrated circuits are equivalent to three or more conventional stages with numerous transistors, diodes, resistors, capacitors etc.

## 1.4 Brute Force Technique

One typical example of brute force technique is trying to remove an *intermittent* by soldering every suspected connection on a printed circuit board. Yet another example is deliberately *overloading* a circuit component to determine whether it becomes defective under additional strain. The idea is that *a nondefective part can withstand considerable temporary overloading without damages, but a defective, or almost defective, part will fail when overloaded.* By this means the *intermittent* (or a possible intermittent) is eliminated.

Extreme caution is necessary with any overload test. Brute force technique, as the name suggests, is to be *applied with utmost care*.

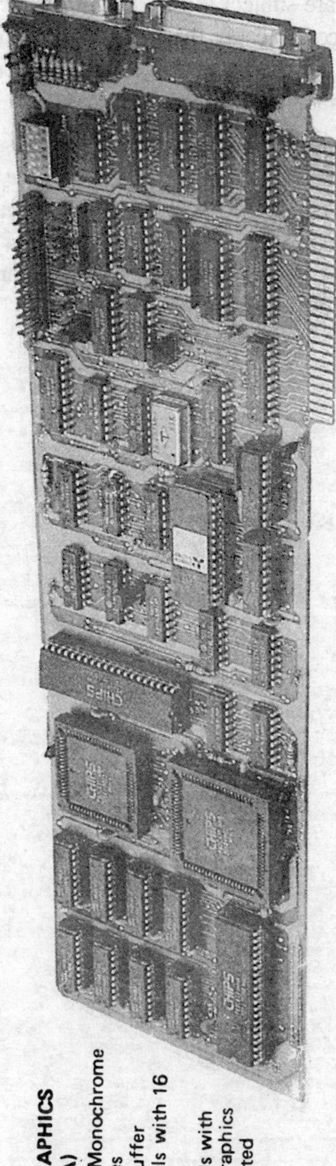


**Fig. 1.3:** Plug in modules for stage substitution.

### 1.5 Duplicating the Trouble

Many defects in electronic circuits are subject to the *'healing' effects of the test procedure*. Either an electrical or a mechanical disturbance can cause the circuit to develop *temporary good health*. Replacing a part that is suspected of being defective can cause enough circuit disturbance to make the trouble disappear temporarily but not permanently.

If you are certain of your diagnosis, and if the trouble does disappear, *return the original part or adjustment to the circuit to determine whether you can duplicate the trouble. Again, replace the part or change the adjustment and note whether the trouble is again removed. If the exact trouble can be duplicated and again removed, you can be reasonably sure that you have found the cause*. This procedure should be used when the trouble is *intermittent*, or when you are not *absolutely sure of your diagnosis* and what you did to eliminate the trouble

# Chapter 2

# SERVICE AIDS

*Service aids are provided by equipment manufacturers in the form of schematics and physical layout diagrams. The specific needs of a service technician are kept in mind while designing these diagrams. In addition to schematics, certain tools and test equipment in general, and some special maintenance tools in particular, are necessary to perform preliminary checks and to carry out extensive repairs on the state-of-art solid-state equipment. These are discussed in this chapter.*

## 2.1 Schematic Diagrams

The greatest aid to servicing is the schematic diagram. It forms the *nucleus* of all service information about a specific make and type of equipment. You can determine the basic function of each part of the equipment and its *trouble causing possibilities* by examining this diagram.

When a given equipment is to be serviced, definite and specific information about that equipment is a practical necessity. Mere possession of a schematic, or service data, is of absolutely no use till you know how best to exploit it as an aid to servicing.

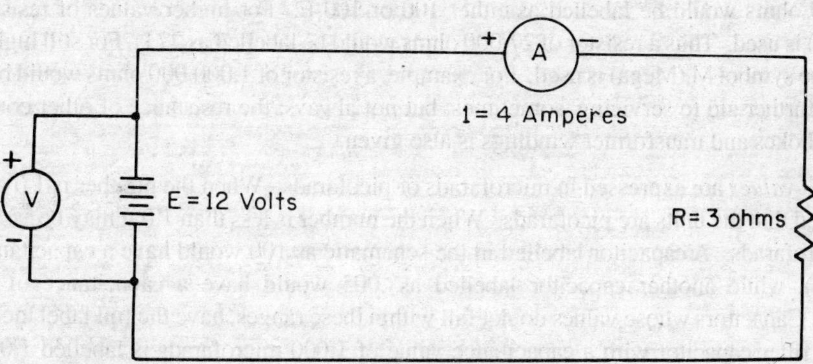


Fig. 2.1: Schematic diagram.

Schematic diagrams are handy tools of the circuit designer, who wished to expresson the paper, *electrical operation of the circuit, isolated completely from physical construction or component layout.*

Schematics supply, at appropriate places in the diagram, vital information necessary for making *observations and checks* to determine the cause of equipment failure.

### *Component Values*

In order to distinguish one component from another, the following symbols are used to label the components:

| | |
|---|---|
| Resistors | $R_1$, $R_2$, $R_3$ . . . . . . . . . . . . and so on. |
| Variable resistors | $VR_1$, $VR_2$, $VR_3$ . . . . . . . . . . . . . . |
| Capacitors | $C_1$, $C_2$, $C_3$ . . . . . . . . . . . . . . . . . |
| Inductors | $L_1$, $L_2$, $L_3$ . . . . . . . . . . . . . . . . . |
| Transformers | $T_1$, $T_2$, $T_3$ or . . . . . . . . . . . . . . . <br> $Tr_1$, $Tr_2$, $Tr_3$ . . . . . . . . . . . . . . . . |
| Diodes | $D_1$, $D_2$, $D_3$ . . . . . . . . . . . . . . . . . |
| Zener Diodes | $ZD_1$, $ZD_2$, $ZD_3$ or $D_1$ $D_2$ $D_3$ . . . . . . . . |
| Transistors | $Q_1$, $Q_2$, $Q_3$ . . . . . . . . . . . . . . . . . |
| Integrated Circuits | $IC_1$, $IC_2$, $IC_3$ . . . . . . . . . . . . . . . . |

The *basic symbol* for the various components is used but the value of the components is also preserved. Resistor values are usually given in ohms, capacitor values in micro farads ($10^{-6}$ Fds) or pico farads ($10^{-12}$ Fds), etc. This information is given twice-once on the schematic diagram and again in the parts list for the equipment.

The *parts list* forms an integral part of the service data. The parts list gives complete information, in the sense that it gives *additional information* such as tolerance and wattage rating of resistors, working voltage of capacitors and so on. This information is useful where a component needs to be replaced.

The values of components also help in locating specific components and identifying them by their values as well as their position in the circuit. *Resistance values* are expressed in ohms upto 1000. A resistor of 100 ohms would be labelled as either 100 or 100 E. For higher values of resistance, the symbol k (kilo) is used. Thus a resistor of 27,000 ohms would be labelled as 27 k. For still higher values of resistance the symbol M (Mega) is used. For example, a resistor of 1,000.000 ohms would be labelled as 1 M. As a further aid to servicing, sometimes, but not always, the resistance of other components, such as filter chokes and transformer windings is also given.

*Capacitance values* are expressed in microfarads or picofarads. When the number is 1.0 or more, it may be assumed that the units are picofarads. When the number is less than 1.0 it may be assumed that the unit is microfarads. A capacitor labelled in the schematic as 100 would have a capacitance of 100 pf (picofarads), while another capacitor labelled as .005 would have a capacitance of .005 MF (microfarads). Capacitors whose values do not fall within these ranges, have the full label including the units. Thus a filter capacitor with a capacitance value of 1,000 microfarads is labelled 1,000 MF or 1,000 μF and not simply 1,000, because it can then be mistaken for a 1,000 pF capacitor.

*Inductance values* are sometimes also indicated, either in milli Henrys ($10^{-3}$H) or in micro Henrys ($10^{-6}$H).

***Test Points (TP)***

Schematic diagrams also give the *voltages and waveforms* at various test points marked TP. The voltages can be marked right on the schematic, or provided in a tabular form-$A \pm 10\%$ variation in these voltages is quite normal, and this will not in any way affect the normal performance of the equipment.

Variations of voltage in excess of ± 10% point to a defect, somewhere in the circuit, which may be due to a change in some component value, or due to its malfunctioning and needs to be rectified.

Many schematic diagrams also include representation of the waveform of the voltage at each of a number of test points in the circuit. This information is provided to give the technician, using an oscilloscope (CRO), an idea of what type of waveform is to be expected at each of these test points. Deviations from normal waveforms can then be noticed and traced. The *peak-to-peak (p-p) value of the voltage* also accompanies the waveform illustration and is used as a *reference* when the amplitude is measured on the oscilloscope screen.

## 2.2 Physical Layout Diagrams

Physical layout diagrams, on the other hand, were developed for the technicians i.e. wire man and service man. In these diagrams *components were represented pictorially, each with its connecting wires and location relative to other components*. With the development of complex electronic equipment, *physical layout diagrams* became too unwieldy for servicing use, and the service technicians turned to the *schematic diagram* as the base to start with further operation. Field experience enabled him to picture out for himself the physical appearance of the components expressed schematically in the diagram.

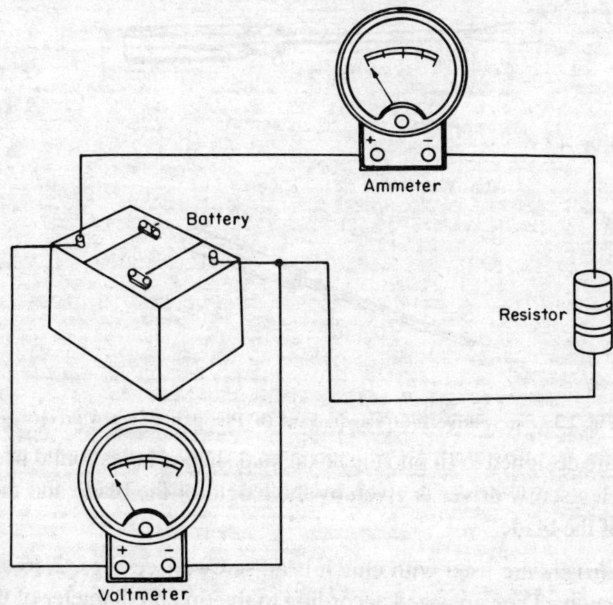


**Fig. 2.2:** Pictorial diagrams

## 2.3 Hand Tools

As in all trades or professions, the recognised tools of the electronics trade are essential to good work. *Without the proper tools, a job can require much more time than is necessary and still may not be done well.* It is almost impossible to list all the special tools required for electronics, but most hardware dealers can supply catalogs that list the available tools. A variety of tools can also be purchased from most of the electronics parts dealers. Gedore, Taparia, Vishwa, Unique,... etc. are just a few of the standard and reliable tools' manufacturers.

### *Screw Drivers*

Screw drivers, Fig. 2.3 are available in various sizes and lengths. The *size* of these screw drivers is usually given by the diameter and the length of the blade. Standard screw drivers are used for routine work. These are often referred to as *electrician's screw drivers*, shown in Fig. 2.3(a). *Philips head screw drivers*, shown in Fig. 2.3(b), are used with cross-slot screw heads and provide a more positive driving action than is possible with standard screw drivers. The size of a Philips head screw driver is most often given by a number such as 1, 2, 3 or 4. The larger the number, the larger the point of the screw driver.

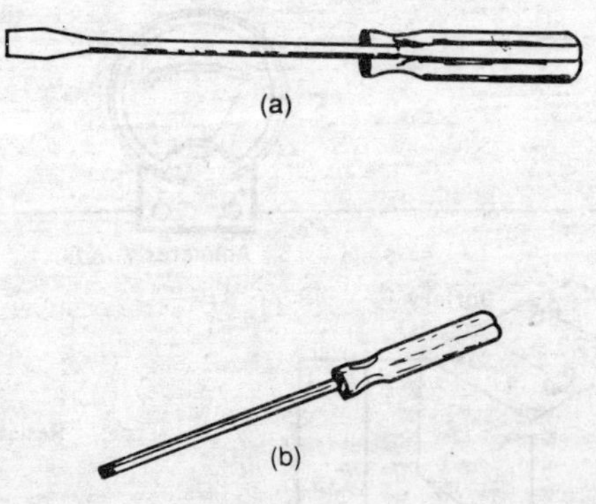


**Fig. 2.3:** Screw drivers: (a) standard and (b) Philips head screw drivers.

*Allen screw drivers* are designed to fit into the hexagonal shaped holes found in some set screws, Fig. 2.4(a). The size of such a screw driver is given by the length of the blade and the distance across the flats, or parallel sides of the blade.

*Clutch-head screw drivers* are used with clutch-head screws (see Figs. 2.46 and 2.6), to provide a very positive driving action. They are sized according to the tip and diameter of the blade.

In electronics, you often have to start a screw in a place where it is impossible or nearly impossible to reach with your fingers or pliers. Two styles of screw-drivers are shown in Fig. 2.5. In the top of Fig.

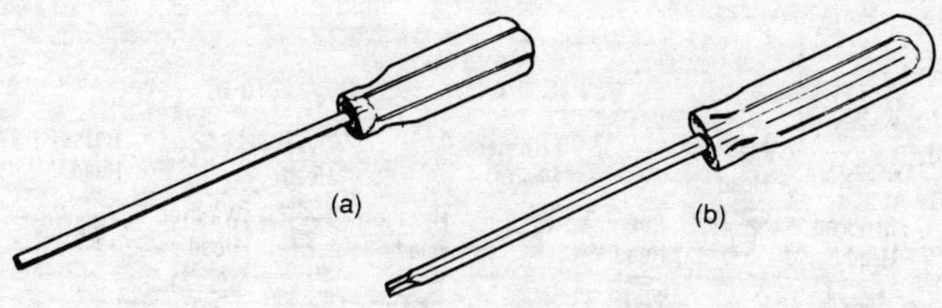


**Fig. 2.4:** Special screw drivers: (a) Allen screw driver and (b) clutch-head screw drivers.

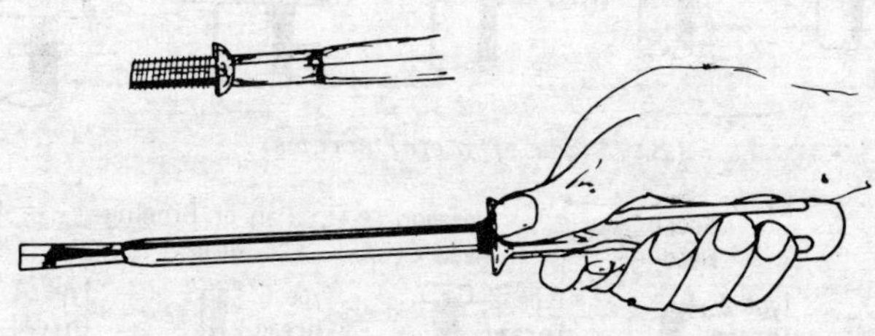

**Fig. 2.5:** Split-blade holding screw driver.

2.5, the screw driver bit is *split*. When the tube is pushed downwards, the bits are forced together causing the tip to widen in the slot and hold the screw. In the bottom of Fig. 2.5, when the handle of the *holding screw driver* is pushed downward, the centre portion of the bit turns.

### *Screws*

Although screws are not hand tools in themselves, they are used with hand tools. The various types of screws used in electronics devices are shown in Fig. 2.6. The *standard metal screw* has a standard thread that requires either a nut or a threaded stud or strap. The most popular sizes used in electronics are numbers 2-56, 4-40, 6-32, 8-32 and sometimes 10-32. The first number indicates the *screw size*. The second number indicates the *number of threads per inch*. A 6-32 screw, for example, is a No. 6 screw with 32 linear threads per inch. The length of the screw depends on its application. Sheet-metal screws, also called *self-tapping screws*, are available in various sizes, lengths and head styles. They are used to fasten brackets, chassis, covers or parts. These screws do not require a nut and serve the same purposes for metal that wood screws serve for wood. Sheet-metal screws are tapered less than wood screws, and they are made from an extremely hard material, therefore, these screws can be forced to cut their way through the metal. To start the metal screw, a small hole is bored through the metal, then the screw is force-fed, tapping its own way. Also available are *drive-type metal screws* that are forced into the metal with a hammer or special pressure tool. The drive screw rotates as it is forced into the metal, threading the metal and holding itself tightly in place.

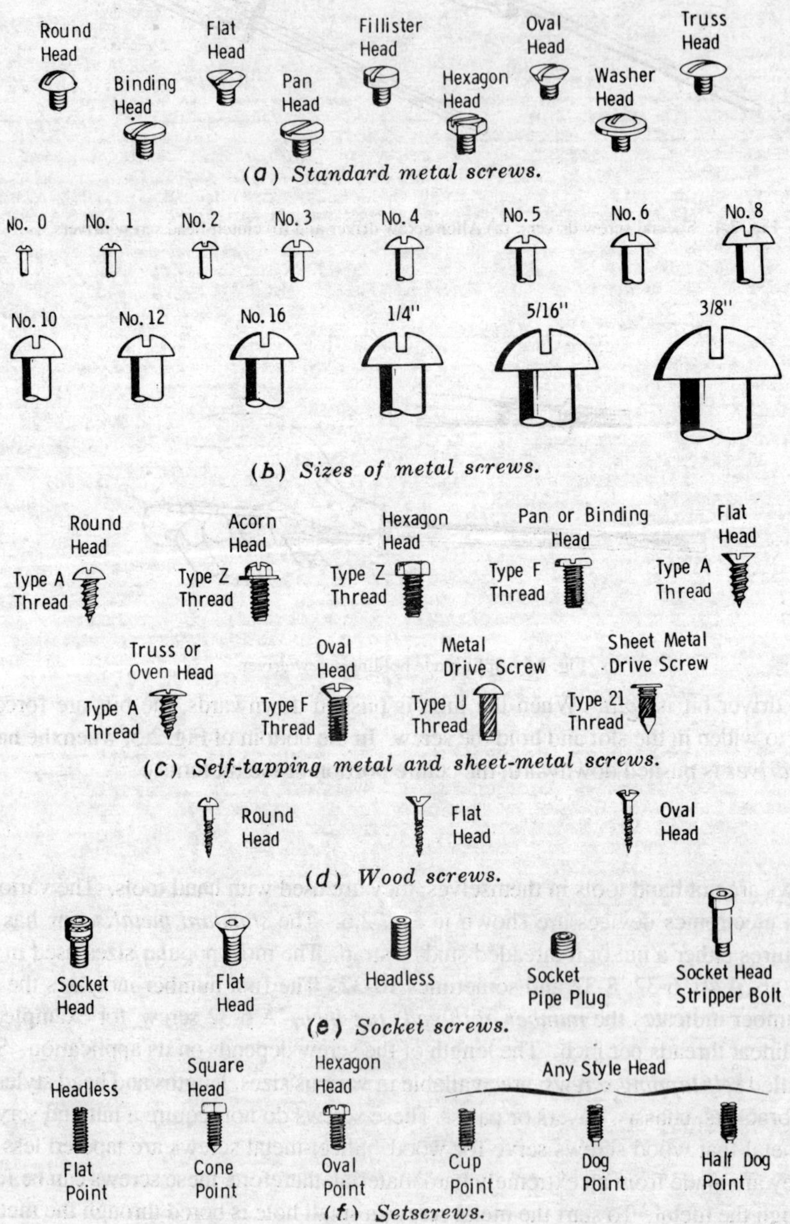


**Fig. 2.6:** Various types of screws used in electronic devices.

*Wood screws* are used in electronics to mount various parts, such as loudspeakers, in wooden cabinets and to mount protective fibre-board covers on the backs of the phonographs, radios and television sets. The slotted round-head, flat-head and oval-head wood screws are used most often in electronics equipment.

*Socket-head screws* are used often in electronics, especially for making adjustments in record players, tape recorders and similar equipment. Usually these screws are either Allen (six-sided), or spliced screws for which a special driver is necessary.

*Set screws* are usually headless; the head of the screw is no longer than the shank of the screw. However, some square-headed and some hexagonal-headed set screws are also used.

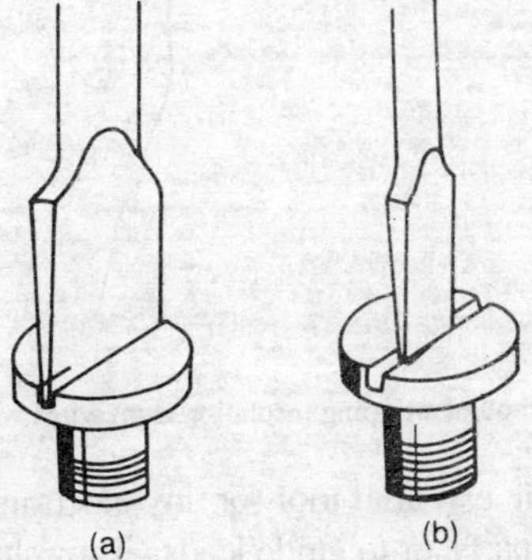


**Fig. 2.7:** The tip of the screw driver should match the head of the screw it is required to drive (a) screw driver tip matched to the head of screw it is driving and (b) screw driver tip mismatched to the head of the screw it is driving.

The correct method of using a screw driver is illustrated in Fig. 2.7(a). The tip of the screwdriver should *match* the head of the screw it is required to drive. In the event of a mismatch, Fig. 2.7(b), either the tip of the screwdriver, or the head of the screw, or both, can possibly get damaged, thus rendering them unfit for normal use.

### Cutters

Anyone in electronic servicing should have at least two or three pairs of *diagonal cutters* (also called *dikes*). Cutters are used to *clip wires* and wire leads from components. They also can be used to *strip insulation* from small hook-up wires, as shown in Fig. 2.8. Grip just hard enough to keep the cutters from slipping, then pull. Never use side cutters or diagonal cutters to cut bolts, heavy wires or thick metals. Don't use them to tighten nuts. With proper care, a pair of cutters can be used for several years.

*Round-nose cutters* are slightly stronger and less likely to become misaligned than sharp-nose cutters, but sharp-nose cutters can be used in closer places. Attempting to cut tough metal ruins the tips of the sharp-nose cutters.

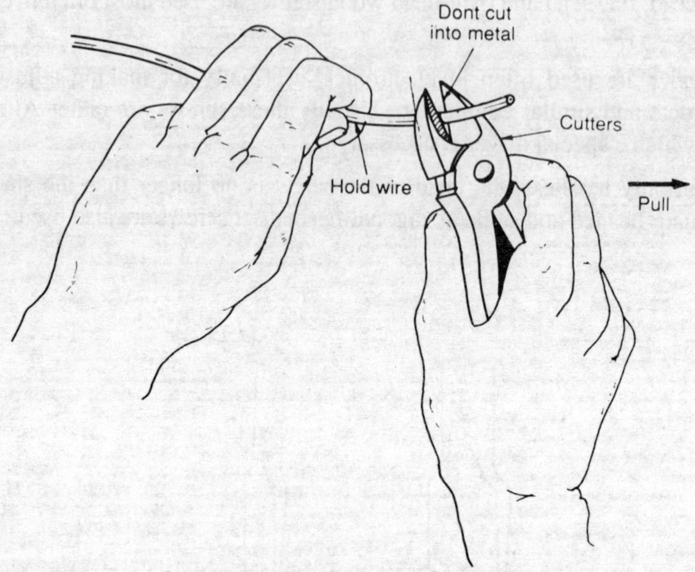


**Fig. 2.8:** Method of stripping insulation from wires with cutters.

*Long-nose pliers* are another basic essential tool for any electronics workbench. Long-nose pliers can be used to shape wires, to insert parts, or to guide leads. They should not be used to tighten nuts or bolts, although they sometimes can be used to start a nut on to a bolt in a tight place.

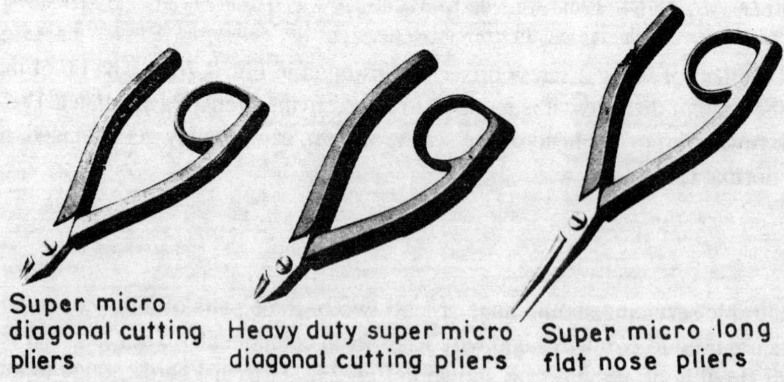


**Fig. 2.9:** Various types of cutters.

Some types of long-nose pliers are provided with *built-in cutters*. However, since the cutter is at the side, rather than at the tip, they cannot replace diagonal cutters when it is necessary to nip a lead with the tip of the cutters. Various types of cutters are shown in Fig. 2.9.

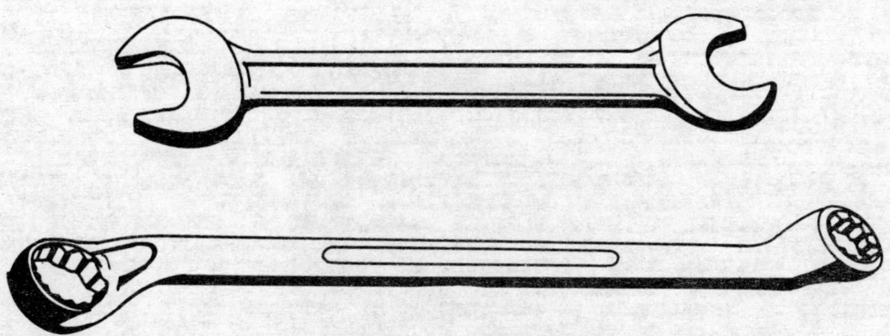

**Fig. 2.10:** Double-ended and ring type nut drivers (spanners).

### *Nut Drivers*

Nut drivers, sometimes called *spin wrenches*, are essential in most electronics work particularly for fixing, assembling and dismantling tv antennas. They are designed for six-sided bolt-heads or nuts and are available in various sizes. Miniaturisation of equipment has caused increased use of smaller sizes. Various lengths of nut drivers are also available.

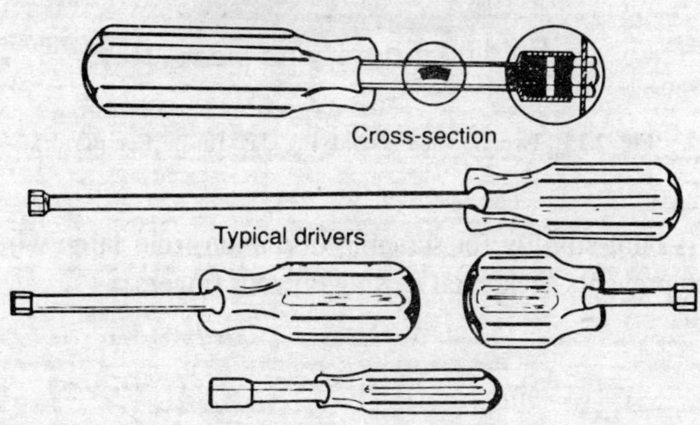


**Fig. 2.11:** Nut drivers with a hollow shaft.

A driver with a long shaft is used when the nut or bolt is recessed and cannot be reached with a *hollow shaft*. Most nut drivers come with a hollow shaft that permits a bolt or volume control shaft to slide into the shaft of the wrench while the wrench moves to reach the nut.

### *Tweezers*

Tweezers and dental-type inspection mirror can save much time when working in cramped space. A good magnifying glass is also frequently needed in tape-recorder and vcr (video cassette recorder) maintenance and repair procedures. These are shown in Fig. 2.12.

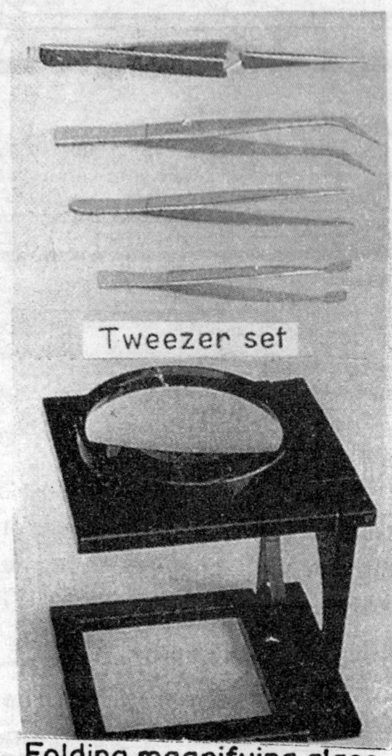


**Fig. 2.12:** Tweezers and magnifying glass (inspection mirror).

### *Electrician's Knife*

An electrician's knife is built sturdily for scraping insulation from large wires, and the screw driver blade locks open, which prevents accidental closing on your fingers.

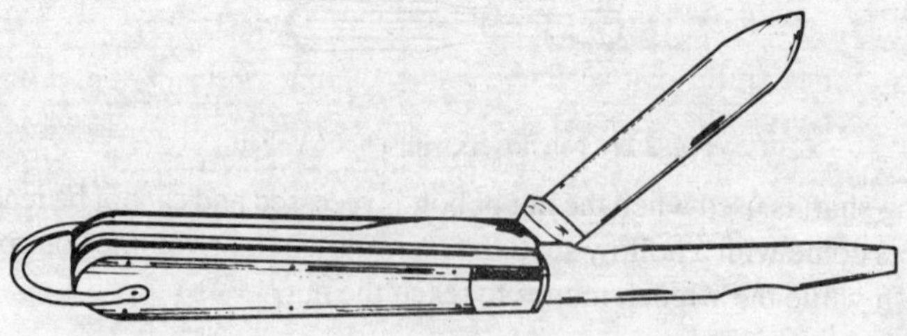

**Fig. 2.13:** Electrician's knife.

### *Seizers*

Tools that are common in the medical field have been known to find their way into the tool boxes of electronic technicians and electrical workers. Seizers, Fig. 2.13, are used by both surgeons and electronic technicians, but, of course, for different purposes. Seizers can be locked onto an electronic part to guide it into place or to hold two wires, as in a vice for soldering.

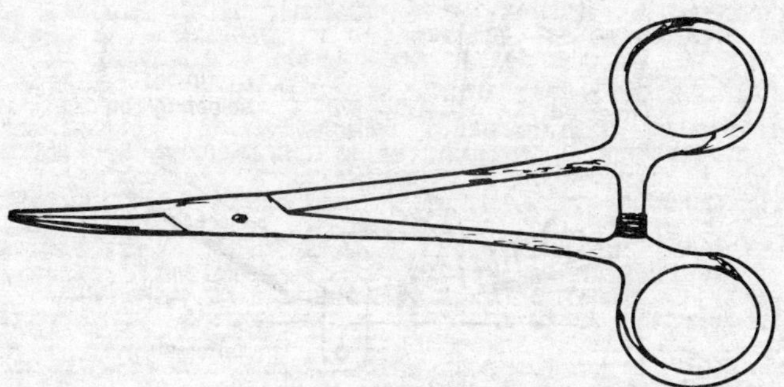

**Fig. 2.14:** Seizers.

Seizers can be locked onto a lead and used as a *heat sink*. A heat sink can be used to carry heat away from a delicate electronic component while its lead is being soldered. In transistor circuits the heat sink is sometimes used to transfer the normal heat away quickly to prevent damage to the electronic parts.

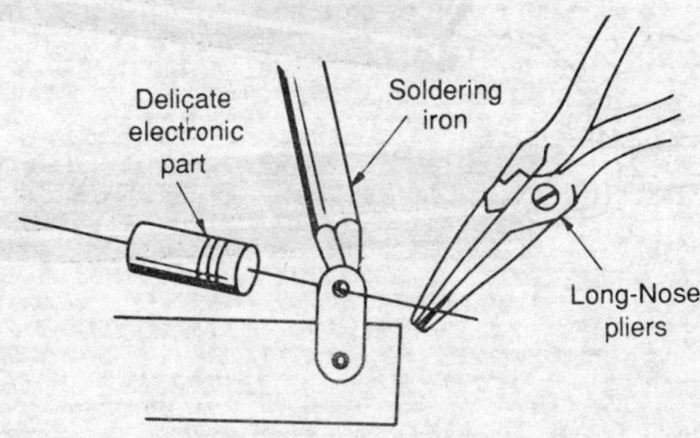


**Fig. 2.15:** Using long, nose pliers to carry the heat away from the part and the terminal.

The use of long nose pliers, Fig. 2.15, and seizers, Fig. 2.16, will carry most of the heat away from the part and terminal. The heat sink then can act as a *baffle* to keep heat from reaching the part through the air.

A specially made heat sink is shown in Fig. 2.17.

## Shears

Shears are made to cut with a *shearing action* rather than with the *pressure action* of the conventional cutters. Shearing action causes the ends of wire leads to cut squarely.

A *square cut* can sometimes be a disadvantage, because the peaked cut made by diagonal or side cutters is easier to insert in a hole in a printed circuit board or terminal strip. The point made by the

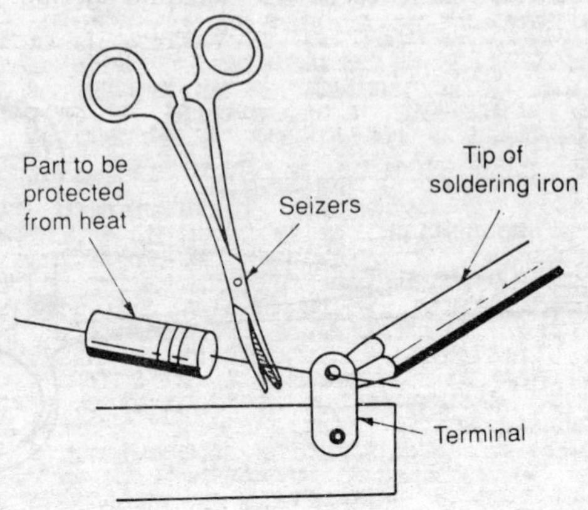


**Fig. 2.16:** Using seizers as a heat sink.

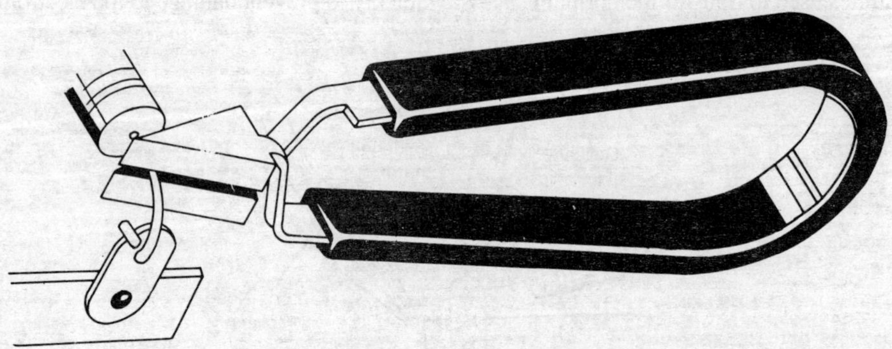

**Fig. 2.17:** A specially made heat sink.

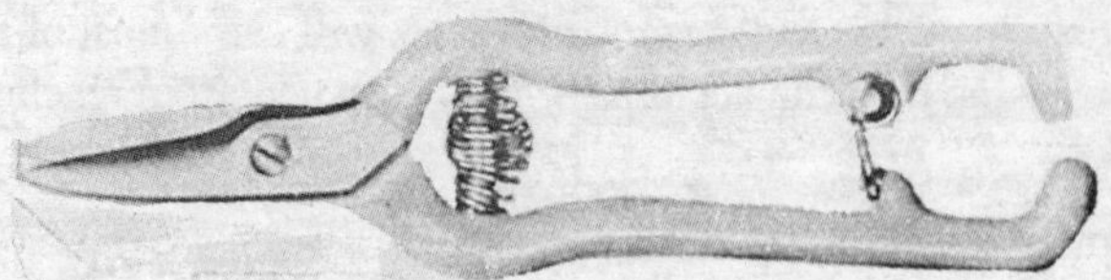

**Fig. 2.18:** Shears.

diagonal cutters, however, results in a slight flattening of the wire. Therefore, the square cut provided by the shears is more desirable if the wire is to be inserted into a hole only slightly larger than the diameter of the wire.

### *Hand Files*

Files are hardened steel tools used for cutting, removing or polishing various substances such as metal and wood. The cutting edge or *teeth*, on the surface are made by diagonal rows of chisel cuts. Files are classified according to (i) size (ii) shape and (iii) cut.

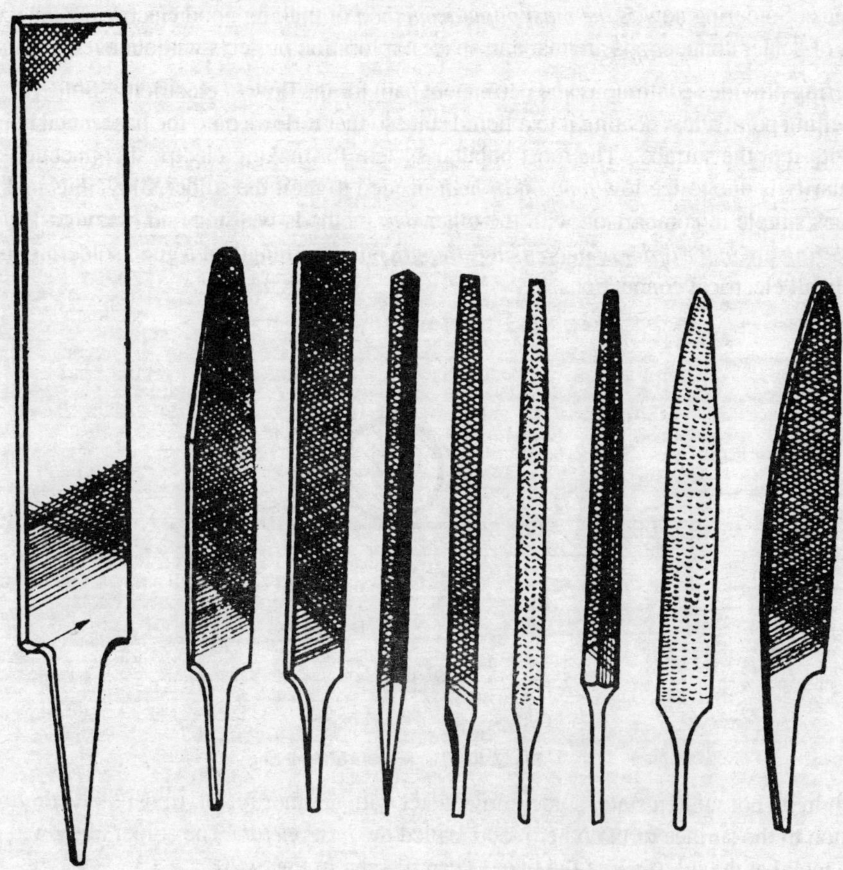

**Fig. 2.19:** An assortment of files.

The *size* of a file depends on its length such as a 15 cm, 20 cm or 25 cm file. The physical appearance of the file dictates its *shape* such as a flat, half-round, round or triangular file. The *cuts* on a file are classified as bastard, second cut, smooth and dead smooth.

For precision work in electronics, smooth and dead smooth files are used. Utmost care should be used while using and preserving files. Once the job is finished, the teeth of the file should be nicely cleaned with a *wire cord* to prevent the accumulation of metal particles.

## 2.4 Soldering

Soldering as an art is not new, it dates back some 4,000 years. Soldering as an exact science is new, it dates back only to about the turn of the century. Alloying precise amount of *tin and lead* made heat joining of the metals of tin cases both economical and practical for using the cans, or containers of foods. The recent rapid increase in the knowledge of solder alloys and fluxes parallels the growth of electronics. Soldering now *is the most popular method* of making good electrical connections. To day millions of solder connections are made in space exploration projects without even a single failure.

Soldering provides continuous and permanent path for the flow of electricity. Soldering entails using a low melting point alloy, heating it to a liquid state so that it flows onto the base metal being joined and alloys with it at the surface. The most popular system for making electrical connections is soldering. Its popularity is due to the low *amount of heat* needed to melt the solder alloy, thus making tools and techniques simple in comparison with the other two methods-welding and brazing. *Welding, brazing and soldering are in the order of decreasing strength* but the strength of a good soldering job is sufficient for nearly all electrical connections.

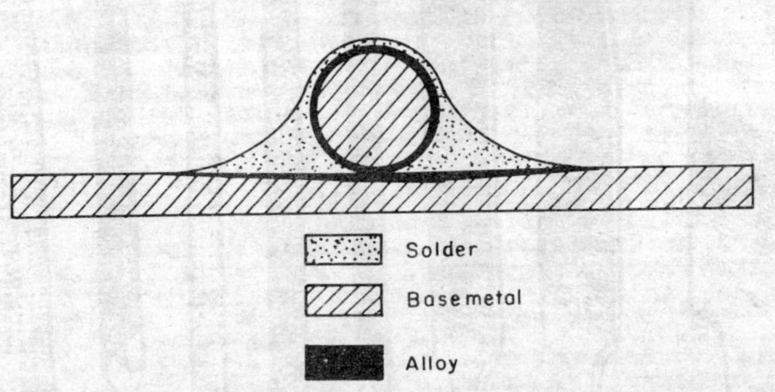


**Fig. 2.20:** The process of soldering.

When in its hot molten state, solder, on contact with another metal, transfers its thermal activity by connection to the surface of the other metal, called the *base metal.* The solder alloy will combine with the base metal at the *interface* of the two, as can be seen in Fig. 2.20.

Molten solder dropped onto a nonmetallic or *unclean surface* will form nearly a round ball, like water on a non-porous surface. Surface tension keeps the solder in the form of a ball, the effect of gravity causes it to flatten. Molten solder dropped onto a *clean surface* of certain metals will develop a capillary action that overcomes the surface tension and this will flow along the surface of the metal, as shown in Fig. 2.21, like a drop of water on a clean paper towel. This is known as *wetting* of the base metal and the secret of good soldering is the ability to properly wet the base metal. Wetting is a metallic solution action between the solder and the base metal.

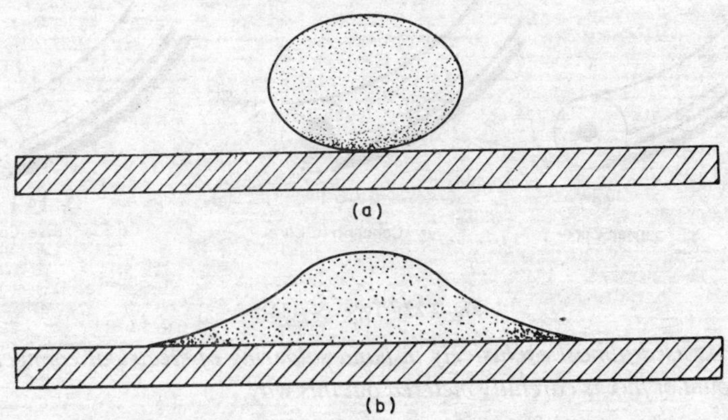


**Fig. 2.21:** Wetting of the base metal-the secret of good soldering.

Wetting of the base metal will not take place unless the base metal is *clean*. 'Clean' here does not mean just clean in appearance, but clean of all invisible surface oxides. *Flux* is used with the solder to remove the oxide and thus permit good solder wetting. Cleanliness of metal surfaces is of prime importance. This is ensured by *pre-cleaning* and the use of *resin-base fluxes*. The tin to lead ratios effect the connection to a degree and the temperature of the heat applied to the connection is of even greater importance.

## 2.5 Solder

Solder is an *alloy* of tin and lead. *Tin* is not attacked by air or water. By itself it has too high a melting temperature to make solder alone. It also is too costly. *Lead* reduces the melting temperature and when alloyed with tin adds to the overall strength.

The proportions of tin and lead in solder are expressed as a *ratio*. The first two digits represent the percentage of tin and the last two digits the percentage of lead. A 60/40 solder, for example, means it is an alloy of 60% tin and 40% lead. Here are some of the tin/lead ratios and their properties.

*60/40* Good general purpose solder.

*50/50* Lower-cost general purpose solder.

*40/60* Lower-cost common solder.

*Rosin-core solders* are available in small packs for the home users. The smaller wire sizes *permit better control* of the amount of solder being fed to small connections such as printed circuit boards. Smaller size wires melt faster because less of the heat is carried away by the rest of the solder. Better control is possible with larges wire sizes on larger connections such as terminals and solder lugs.

Most solders are purchased with a continuous inner core of *flux*. Some are multicored, with as many as five cores of the flux (see Fig. 2.22). Solder is also available without the inner flux core for use with separate liquid flux, if preferred. The flux cores are also different, depending on the activity desired.

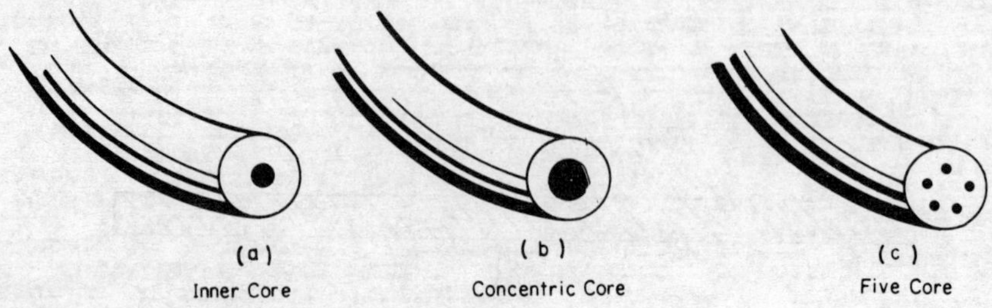


**Fig. 2.22:** Types of solders.

*Flux cored solder is used almost exclusively in hand soldering of electrical connections in electronic industry. The amount of flux is carefully metered out this way.*

With the increasing of printed circuit boards in electronics industry, the modern trend is towards *rapid soldering systems*. These include dip and wave soldering which involve dipping the entire board into a molten bath of solder. The flux then must be applied in advance by brushing on a liquid flux or dipping the board into a flux bath. In dip fluxing no post cleaning will be needed with the use of pure rosin. Resoldering old terminals is sometimes easier if a liquid flux is applied to the terminal first. Excess flux is removed by wiping it with a cloth dampened with alcohol after the connection has cooled.

## 2.6 Soldering Irons and Accessories

The parts diagram of a typical soldering iron is shown in Fig. 2.23. Soldering irons range in size from small irons drawing less than 20 watts to large irons which draw more than 100 watts of power. A 100 watt soldering iron is shown in Fig. 2.23. *The 35 watt soldering iron is used for printed circuit boards.*

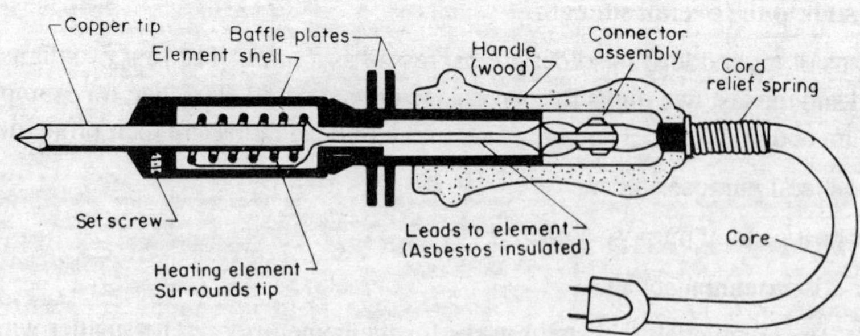


**Fig. 2.23:** Exploded view and parts diagram of a 100 watt soldering iron.

The main parts of an electric soldering iron are a steel shell, copper tip, wooden or plastic handle and line cord. A heating element is mounted inside the steel shell. The hard wooden handle remains fairly cool. A baffle consisting of steel discs is placed between the steel shell and the wooden handle to prevent heat from reaching the wooden handle. *The tip is the business end of the soldering iron.* It is positioned inside the steel shell and heating element and is usually held in place with a set screw. Copper is used for the bit because it conducts heat well and is easy to tin.

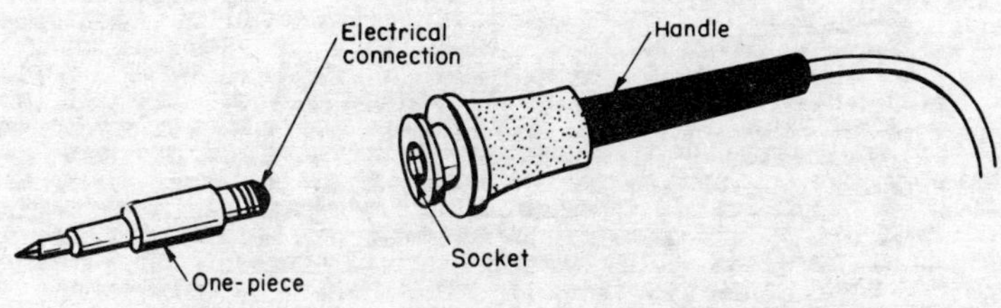


**Fig. 2.24:** A one-piece unit.

In some soldering irons, the tip and the heating element are contained in a one-piece unit that is then screwed into a handle. This is illustrated in Fig. 2.24. *When the tip is no longer serviceable the entire unit is replaced.*

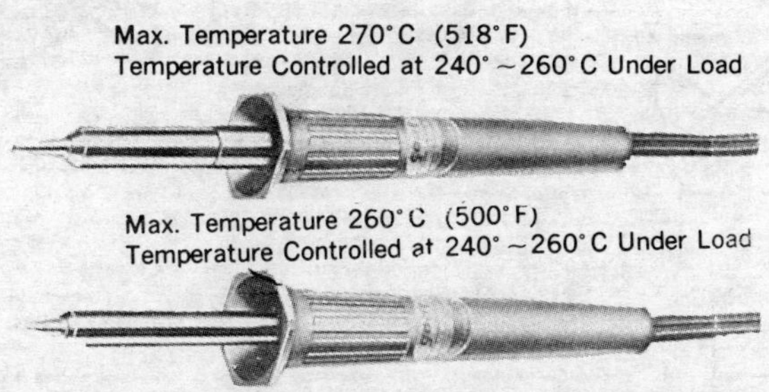


**Fig. 2.25:** Temperature controlled soldering irons.

Temperature controlled, positive-temperature-coefficient (PTC) heater soldering irons, Fig. 2.25, have rapid heat up and recovery, never overheat, have high insulation resistance and are trouble free. These are ideal for printed circuit boards and *precision circuits' soldering work.*

Overheating of a soldering iron can be prevented by merely unplugging the cord of the soldering iron when it is not in use. This method may be unsatisfactory, however, because you must wait for the iron to reheat each time it is needed. A better method is to use *soldering iron holder* which helps to carry the heat away from the tip to the air, Fig. 2.27. *A heated soldering iron should always be placed in its holder or stand when it is not in use.*

Soldering iron tips, Fig. 2.28, are available in several different shapes. They are usually made of copper although some are nickel plated or iron clad to reduce corrosion. *For the most efficient operation of a soldering iron, the replacement tip should always be the size and type recommended for use with a particular iron.*

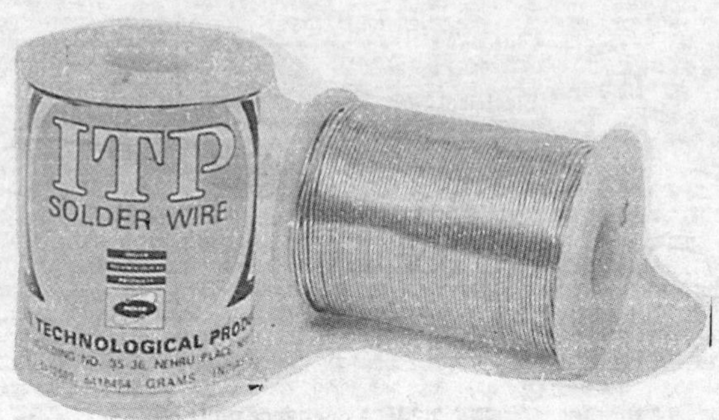


**Fig. 2.26:** Flux-core solder wire.

**Fig. 2.27:** Soldering iron holders.

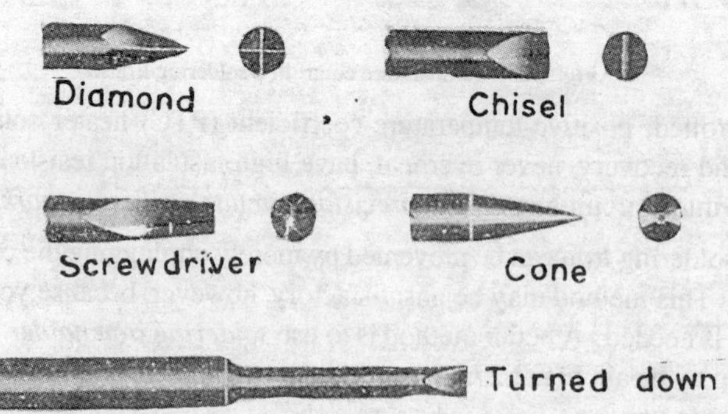


**Fig. 2.28:** Soldering iron tips.

The parts diagram of a soldering gun is shown in Fig. 2.29. The tip of the soldering gun is a part of the shorted turn looped through a special low voltage transformer placed inside the handle. The tip of

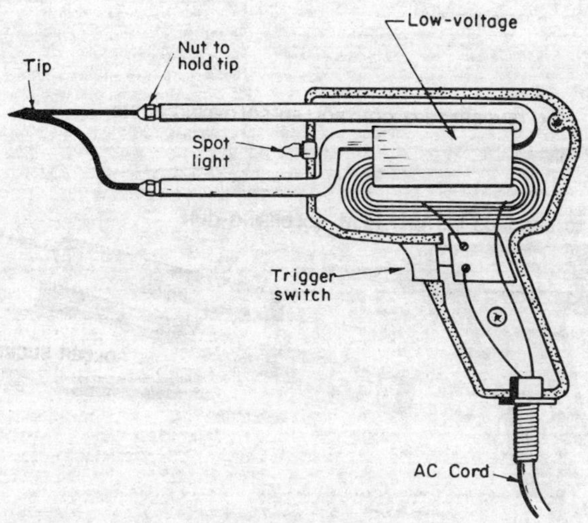


Fig. 2.29: Parts diagram of a soldering gun.

the gun is the *high resistance point* of the shorted turn. Therefore, electron flow creates maximum heat at the tip. The actual voltage across the tip is much lower than 1 volt but the low resistance of the tip and its connections means high current flows to produce nearly instantaneous heating.

Most soldering guns are provided with a *built-in spot light* that focuses on the connection being soldered. These small bulbs receive their current from a separate winding on the handle transformer. If

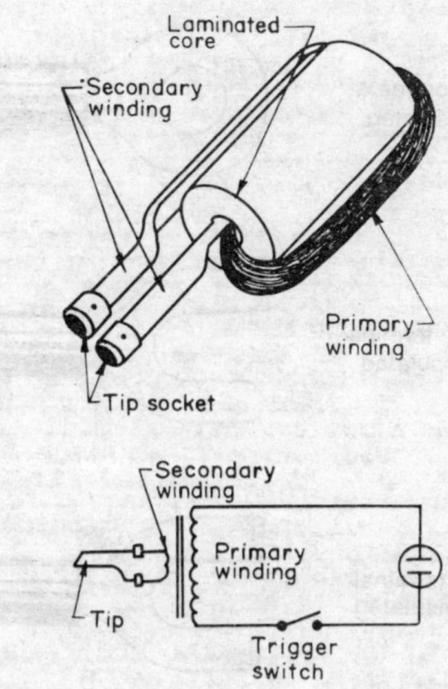


Fig. 2.30: Construction details and schematic diagram of a common soldering gun.

the bulb burns out, there is no effect on the heating of the tip of the gun. The bulbs are usually the types used in small flash lights.

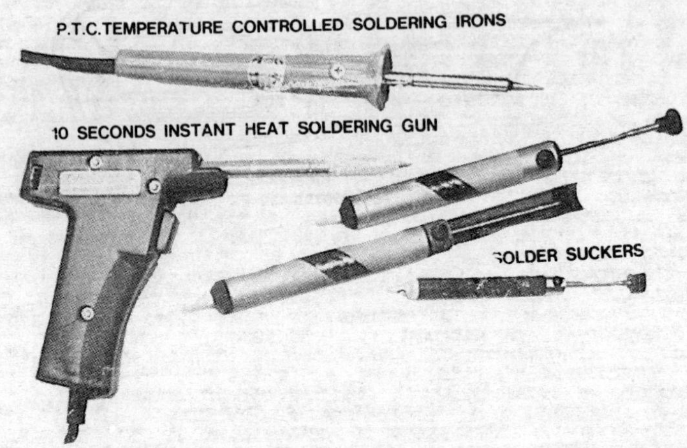


**Fig. 2.31:** Soldering irons and accessories.

## 2.7 Terminal Lugs and Strips

Terminal lugs are used for preparing a cable for screw terminals. Several different types of *terminal lugs* are shown in Fig. 2.32. The method of attaching a solder-type terminal lug to the wire is shown in Fig. 2.33. The insulation is also clamped into the lug.

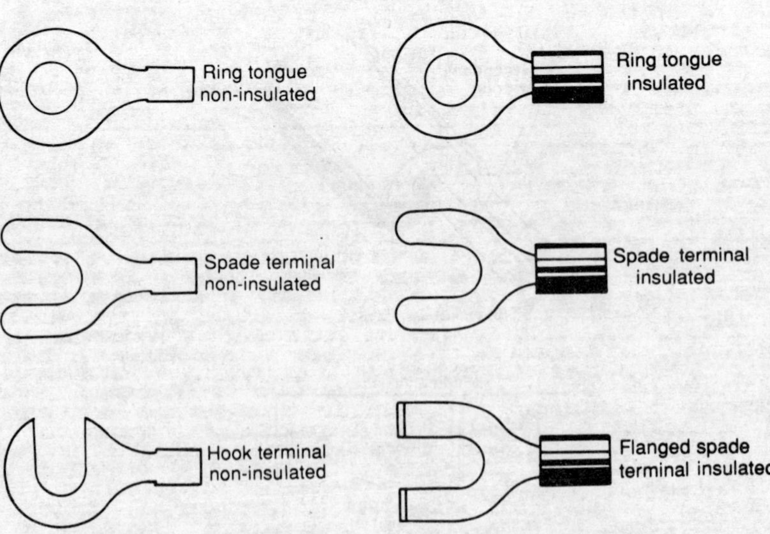


**Fig. 2.32:** Terminal lugs for use on end of wire.

*Lug-type terminal strips* consist of solder lugs and a mounting lug, or lugs, which are usually fastened to a phenolic strip.

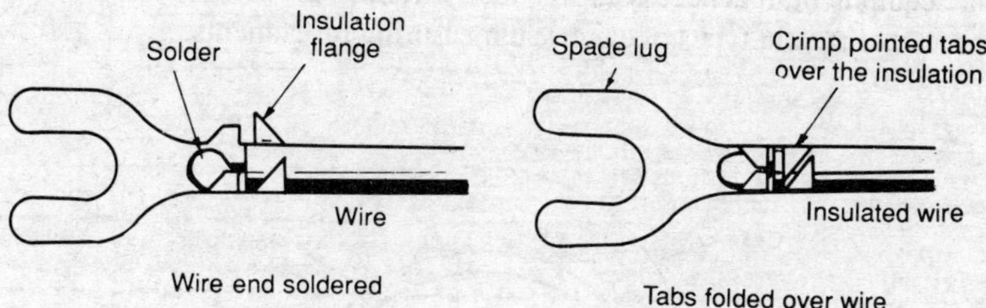


**Fig. 2.33:** Attaching a terminal lug to wire.

*Terminal strips* provide convenient solder terminals for wires and component leads as shown in Fig. 2.34 (a). Whenever possible, connections which require soldering should be made on terminal strips. This is illustrated in Fig. 2.34(b).

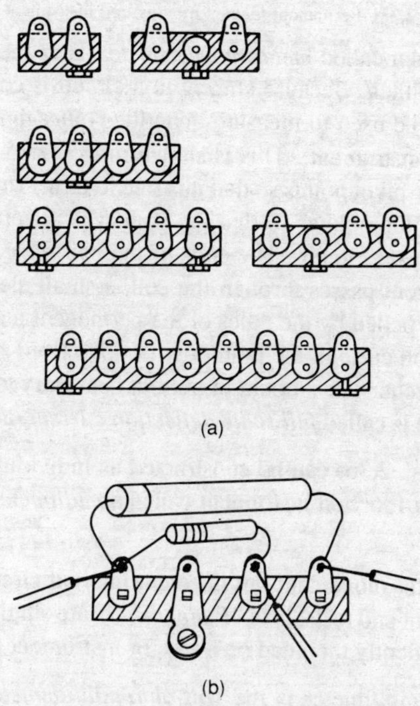


**Fig. 2.34:** Terminal strips.

## 2.8 Multimeters

We cannot see electricity, but we can measure electrical quantities such as voltage, current, power and resistance with instruments, called *meters*. This is necessary if we are to check and test circuits, record consumption of electricity and *verify name plate data*. Also, in order to troubleshoot and service many types of electronic equipment, it is necessary to check various parts of the circuit during normal operation. This is done with different types of electrical measuring instruments.

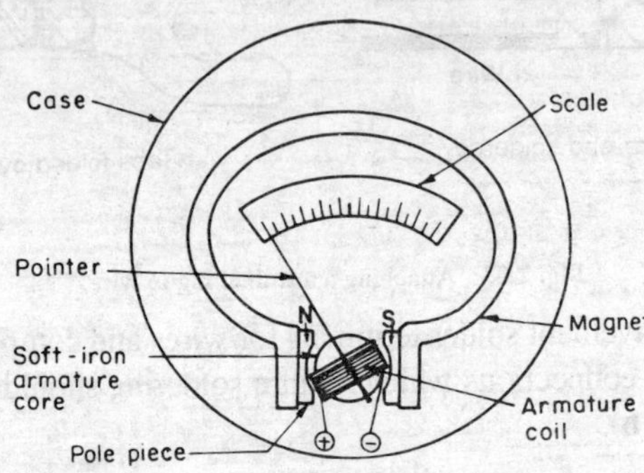


**Fig. 2.35:** Permanent magnet moving coil instrument.

In a typical meter, a mechanism called *meter movement* reacts to the flow of current and rotates a shaft to which is connected a pointer. Because current in a circuit is controlled by voltage and resistance, the same mechanism can be used to measure quantities other than current. Most meters use a permanent magnet moving coil instrument. This is shown in Fig. 2.35. The coil is wound around a frame which is mounted between pivot points so that it is free to turn. The pointer is fastened to the coil assembly. A small spring holds the pointer so that the pointer is at zero on the meter scale when no current is flowing through the meter circuit. The ends of the coil are connected to the stationary terminals of the meter. When current passes through the coil, a small electromagnet is produced. The poles of the electromagnet are repelled by the poles of the permanent magnet. Since the movement of the coil and pointer depends upon current, the pointer will move from zero to a distance that is determined by the strength of the current. The amount of current necessary to move the pointer to the maximum reading on the meter scale is called *full scale deflection current* of the meter.

Voltmeters and ammeters (Fig. 2.36) can be constructed as individual instruments. They can also form part of a multi-range, multi-function instrument called a *multimeter*. The front panel of a typical multimeter is shown in Fig. 2.37.

Multimeters use a single meter movement but different internal circuits to permit measurement of different levels of voltage, current and resistance. *Panel meters* are single-purpose instruments, usually made so that they can be conveniently mounted on a test, or instrument panel.

The most common type of multimeter is the *volt-ohm-milliammeter, or VOM*. This instrument, shown in Fig. 2.38, can be used as a voltmeter, an ammeter or an ohmmeter, by adjusting its *function selector switch* to correspond to the required function. The *range selector switch* can then be adjusted

**Fig. 2.36:** Voltmeters and ammeters.

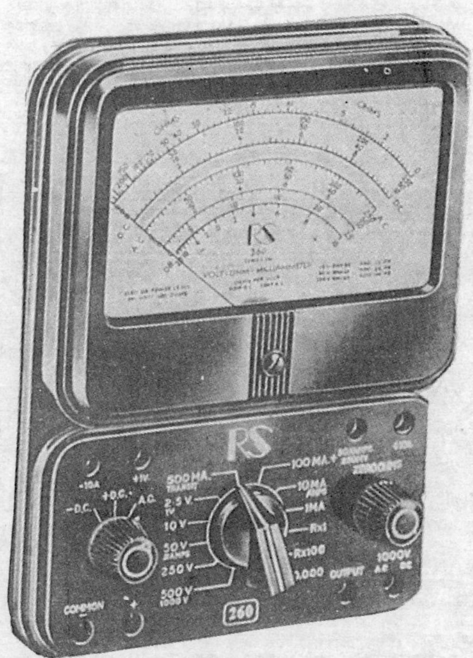


**Fig. 2.37:** A typical multimeter-RS 260 (SIMPSON).

to correspond to the required range for the selected function. Some multimeters are also equipped with a decibel scale which is used when measuring the sound output level of audio amplifiers.

The sensitivity of multimeters is expressed in *ohms per volt*. It is numerically equal to the reciprocal of the full scale deflection current. It is, in fact, *the resistance between the meter terminals with one volt applied across it*. Meters with a sensitivity of 20,000 ohms per volt, or higher, are used for troubleshooting solid-state devices and circuits.

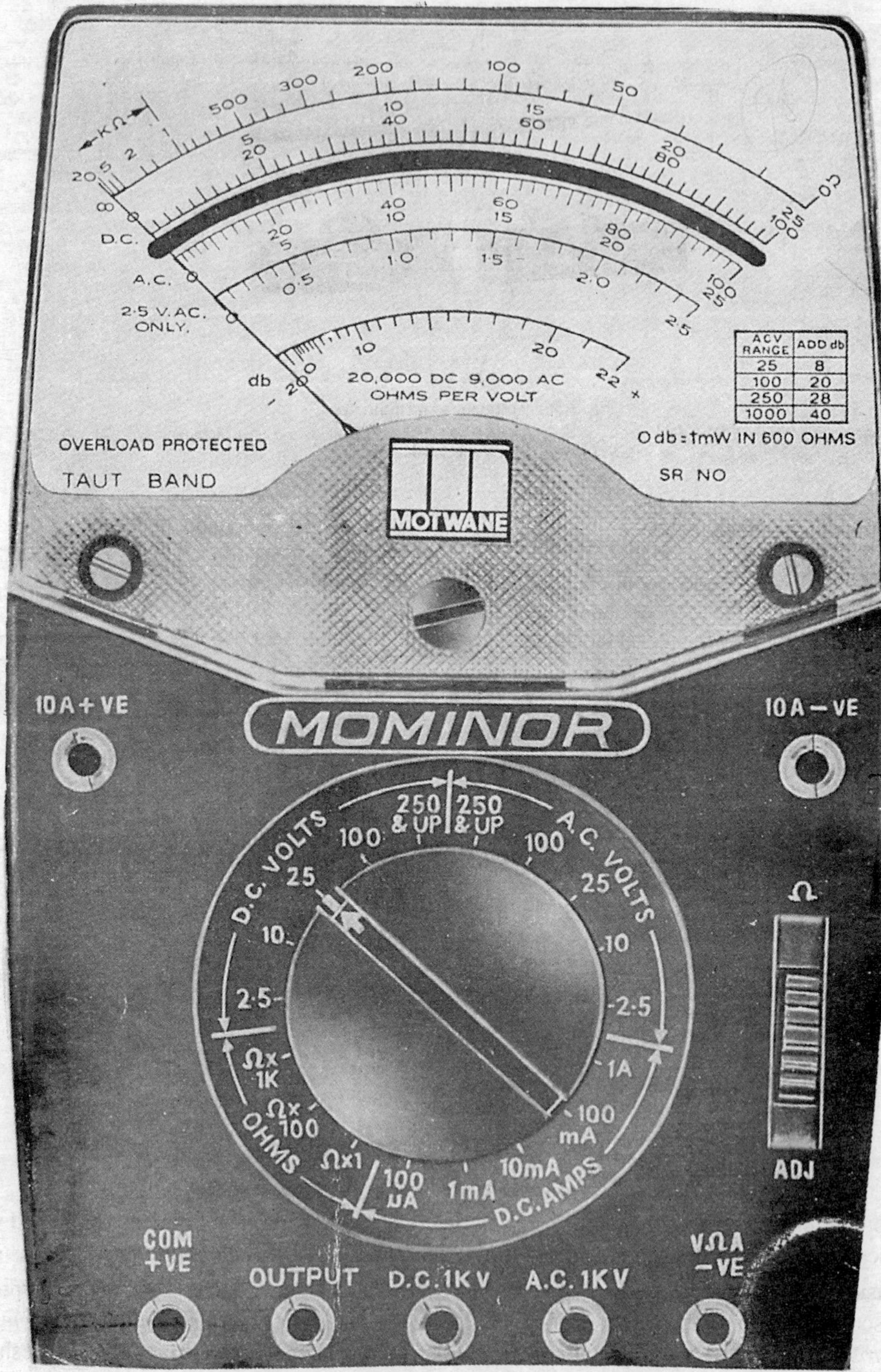


**Fig. 2.38:** MOTWANE Mo-Mnior Multimeter-sensitivity 20,000 ohms per volt.

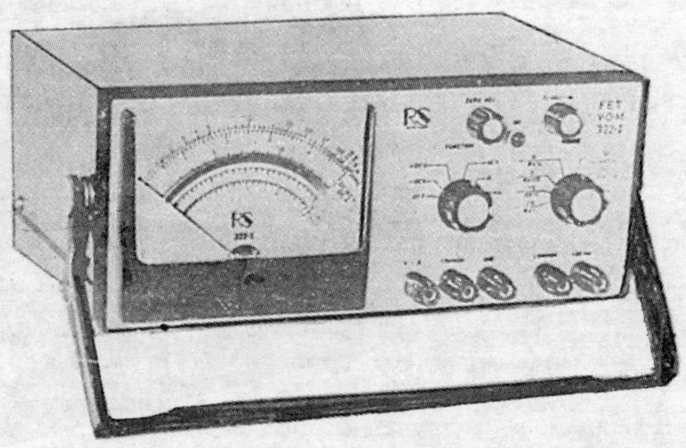

**Fig. 2.39:** The FET volt-ohm milliammeter.

The field-effect transistor volt-ohm milliam meter, FET VOM, shown in Fig. 2.39, uses one or more devices called field-effect transistors which cause the instrument to have an *extremely high input impedance.* As a result the FET VOM is capable of measuring voltages without upsetting the circuit to which it is connected. This is particularly important when measuring voltages in high impedance circuits.

The modern trend is towards the use of digital multimeters, Fig. 2.40, which give a *direct read-out* of the quantity being measured. Almost the same size as calculators, digital multimeters are versatile volt-ohm milliammeters, more accurate and less prone to troubles as compared to conventional VOMs.

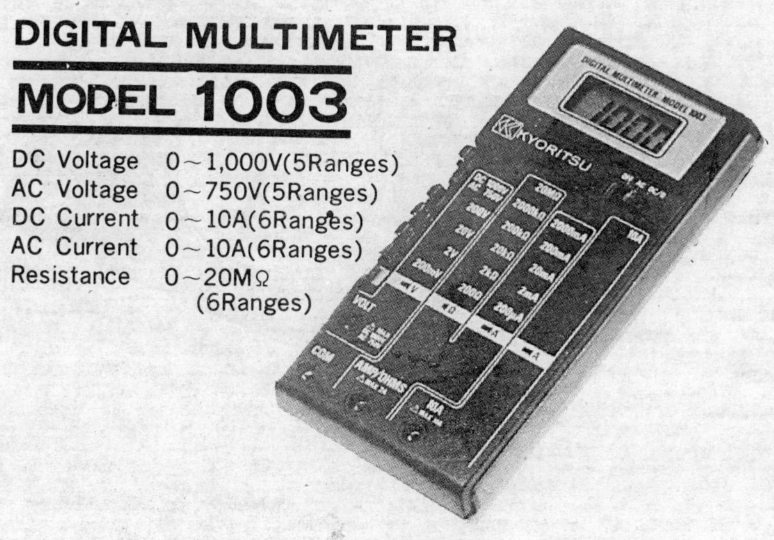


**Fig. 2.40:** Digital multimeter-KYORITSU Model 1003.

**Fig. 2.41:** MOTWANE Digital multimeters DM 452,352 and 350.

Motwane digital multimeters, Fig. 2.41, are ergonomically designed in rugged, fully sealed and attractive ABS casing and provide normal digital multimeter functions of ac/dc voltages and resistance measurement with some additional features.

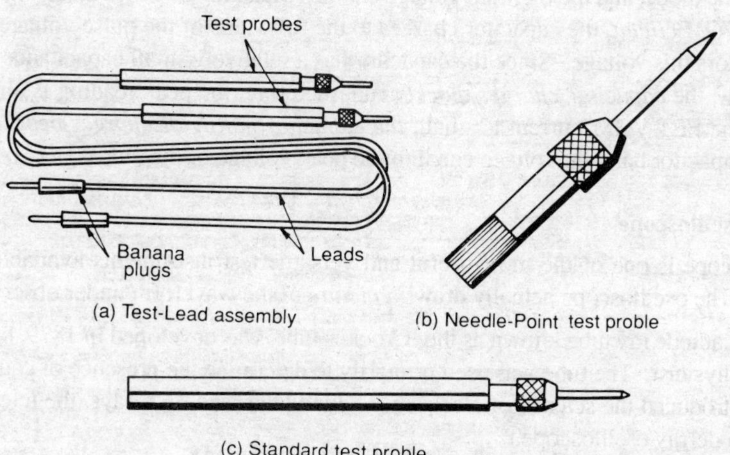


**Fig. 2.42:** Test probes.

Multimeters, whatever the type, are invariably accompanied by a *pair of test probes* which, though separate, form an integral part of the system as a whole. These are illustrated in Fig. 2.42.

## 2.9 Peak-Reading Meters

Most electronic volt meters (EVMs) are provided *with either peak or peak-to-peak* markings on the voltage scales. These scales are not important for sinewave voltages, since the *rms* values on a meter can be multiplied either by 1.414 to obtain the peak voltage, or by 2.828 to obtain the peak-to-peak voltage. This is not true, however, when you measure square waveforms, rectangular waveforms, saw-tooth waveforms, pulse voltages, and non symmetrical wave forms.

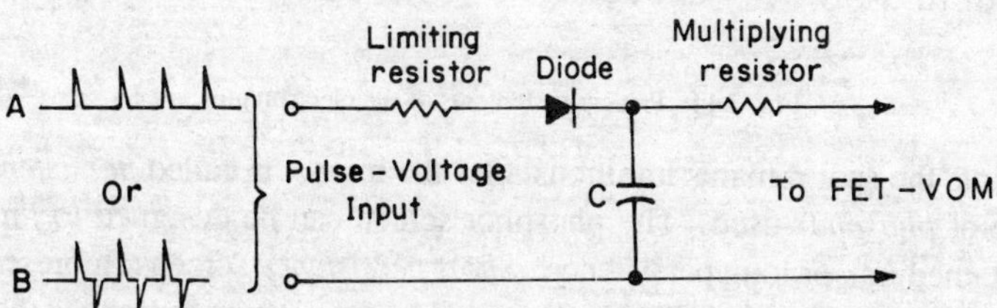


**Fig. 2.43:** Measuring peak voltage.

In pulse circuits, especially, the *peak voltage* is important, because it is the *peak* that normally triggers a transistor or diode, such as those found in computers and television sets, into operation. If the pulses are of short duration and the time interval between pulses is lengthy, the rms of a pulse voltage can be near zero, while the peak voltage can be say 50 volts or more.

A circuit for measuring peak voltages with a meter is illustrated in Fig. 2.43. The pulse voltages are rectified by the diode, and the rectified voltage charges capacitor C. Since *almost no current is required by the FET-VOM circuit*, the capacitor charges to the full value of the pulse voltage and the measuring circuit monitors this voltage. Since the capacitor has a relatively small capacitance and the FET-VOM current is low, the *capacitor charges quickly* and, therefore, the peak reading is almost instantaneous. Also since the FET-VOM current is small, *the capacitor hardly discharges between pulses.* In other words, the capacitor has a dc voltage equal to the peak voltage of the pulses.

## 2.10 The Oscilloscope

The oscilloscope is one of the most useful and versatile test instruments available to the electronics technician. The oscilloscope actually draws a *picture* of the waveform under observation.

The first cathode ray tube known as the 'Crookes tube' was developed in 1879, by William Crookes, an English physicist. The tube was used primarily to determine the presence of cathode rays. Crookes discovery introduced the science of electronics which later lead to X-rays, the television picture tube, and the cathode-ray oscilloscope.

The heart of the oscilloscope is the *cathode-ray tube* which is essentially an electron gun. When powered, the gun *emits* an accelerated stream of electrons that terminates upon impact with a fluorescent screen, producing a *luminous spot*, which can be deflected to *trace* an oscillographic pattern of the waveform (signal) under observation.

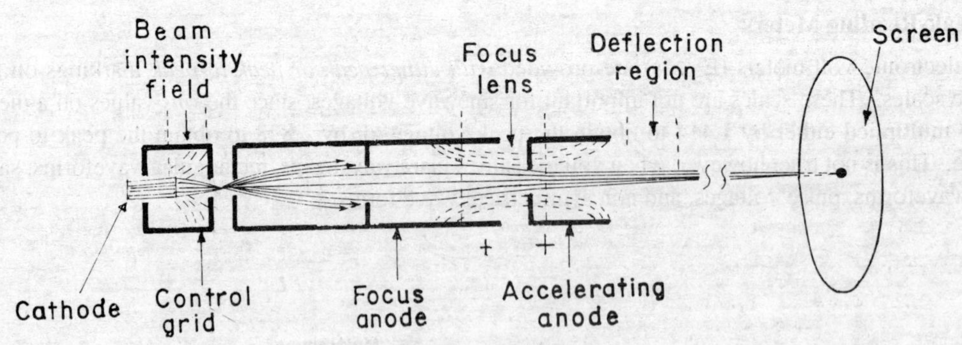


**Fig. 2.44:** Principle elements of the electron gun.

The period of time the spot remains luminous after the impact is called *persistence* and is a characteristic of the type of *phosphor* used. The phosphor screen can be designed to *fluoresce* in a certain colour with a short, medium, or long persistence. *Short persistence* screens fluoresce for about 1/1,000 second, *medium persistence* from 1 to 2 seconds, and *long persistence* for longer than 2 seconds.

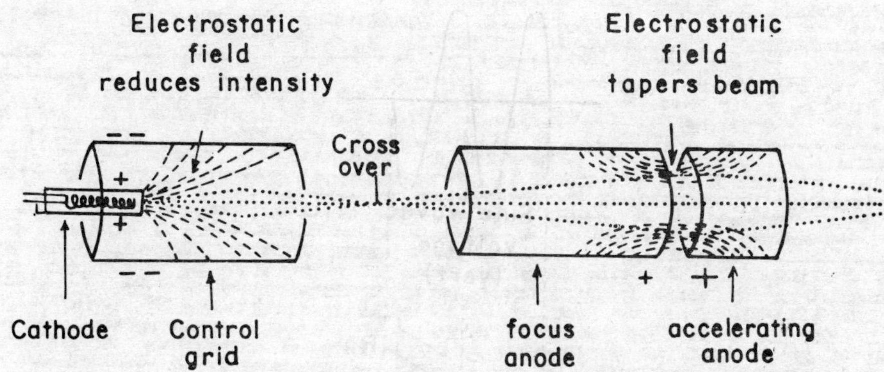


**Fig. 2.45:** Electrostatic fields focus the beam as it travels towards the fluorescent screen.

The effect of the electrostatic field between the cathode and the control grid causes the beam to taper sharply and *cross over*. However, after the first cross over point the beam is subject to *dispersion*. The beam must be focused to produce a *second cross over point* in the vicinity of the fluorescent screen.

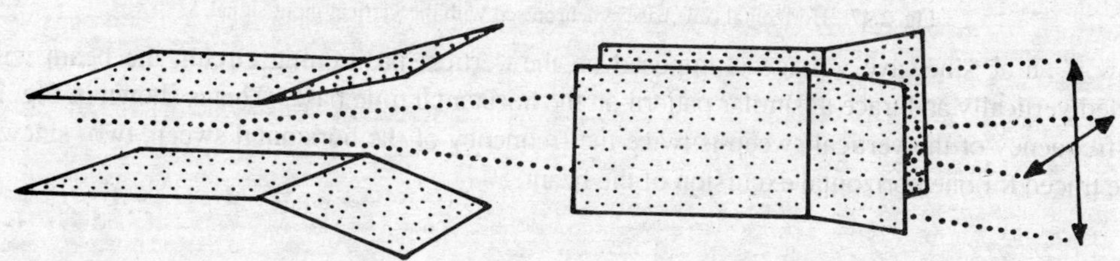

**Fig. 2.46:** Vertical and horizontal deflection plates.

In the absence of deflection voltages the beam strikes the fluorescent screen at its geometric centre. Vertical and horizontal deflection plates move the beam *'up and down' and 'to and fro'*. This is illustrated in Fig. 2.46.

*Deflection sensitivity* of a cathode-ray tube is dependent upon the magnitude of the potential applied to the accelerating anode. The greater this voltage, the less the sensitivity. The vertical deflection plates being located nearer to the cathode show *slightly more* deflection sensitivity than the horizontal deflection plates.

If an ac voltage is applied to the horizontal positioning circuit, the spot will move rapidly across the screen, thereby tracing a straight horizontal line. This horizontal sweep is referred to as the *time base* and is usually produced by a linear sawtooth voltage of known frequency obtained from a sawtooth generator.

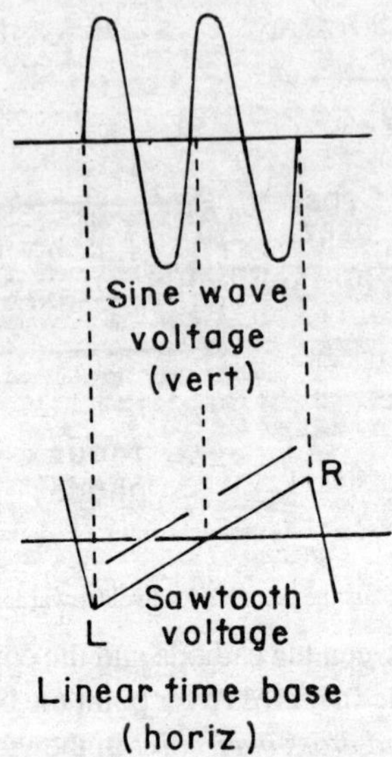


Fig. 2.47: Horizontal time base synchronised with the vertical input signal.

Now if an ac sinewave voltage is imposed on the vertical positioning circuit, the beam will be deflected vertically and trace a similar pattern on the horizontal time base. This is shown in Fig. 2.47. If the frequency of the vertical sweep is twice the frequency of the horizontal sweep, two sinewaves will be traced for one horizontal excursion of the beam.

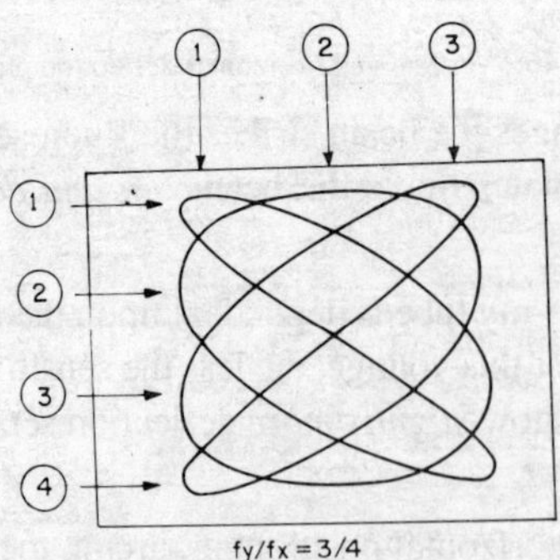


Fig. 2.48: Frequency measurements using Lissajous figures.

Quite often it is necessary to measure an *unknown frequency*. A simple method for determining this information is by the use of *Lissajous figures*, see Fig. 2.48. The ratio $f_y/f_x$ can be used to determine the unknown frequency. When counting the x and y loops, they must be opened and not superimposed upon one another.

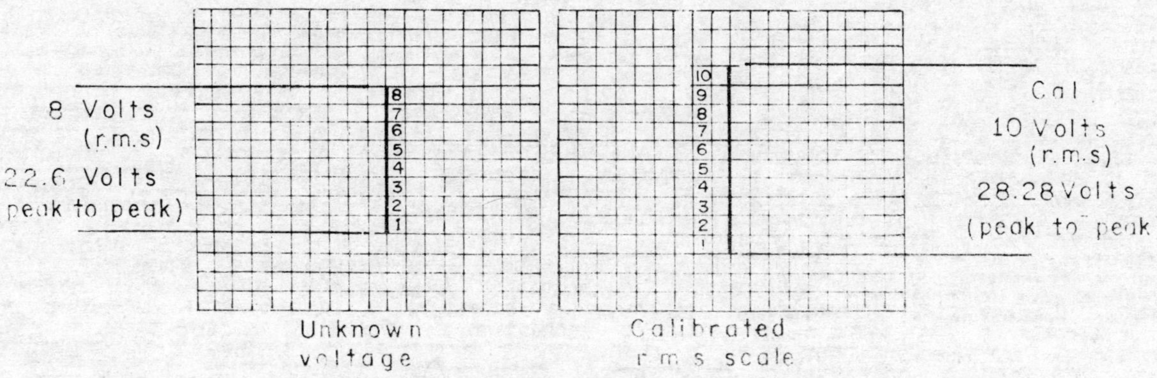


**Fig. 2.49:** Voltage measurement with an oscilloscope.

*Voltage measurement* is another useful application of the oscilloscope, see Fig. 2.49. This is accomplished by *calibrating the vertical trace to a known voltage and comparing it with the vertical trace obtained with the unknown voltage.*

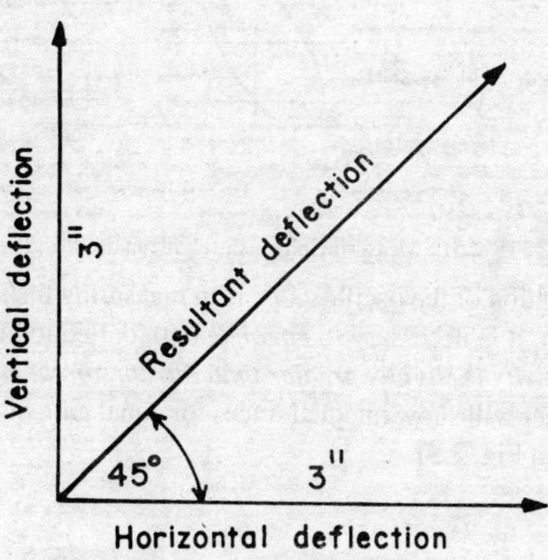


**Fig. 2.50:** Equalising x and y deflection.

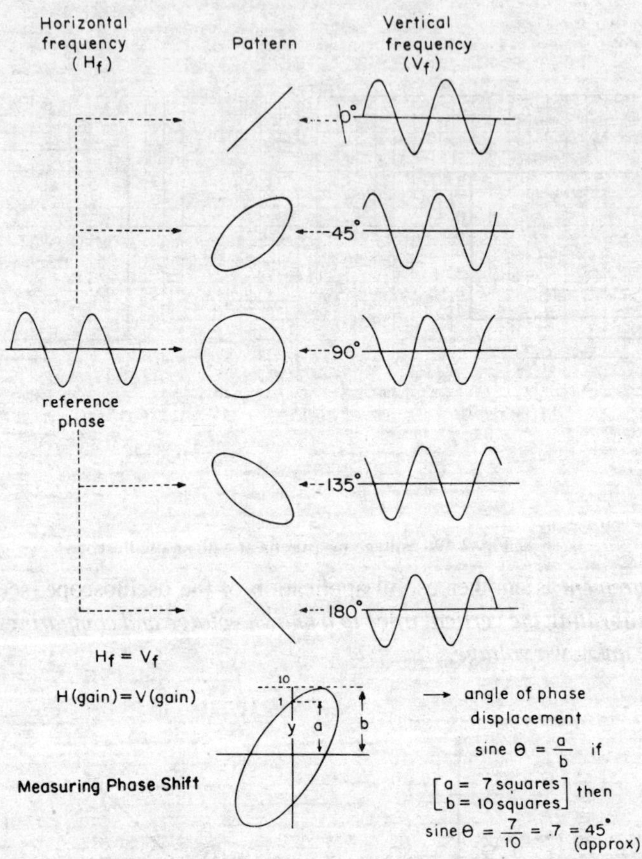


**Fig. 2.51:** Measuring phase shift with an oscilloscope.

Yet another useful application of the oscilloscope is in measuring the *phase shift* of audio amplifiers. When measuring phase shift, it is necessary to *equalise* both deflection traces. This is important since the *vertical deflection sensitivity is slightly greater than the horizontal* and a standard input signal connected to the x and y amplifier will show unequal traces for equal gain of x and y amplifiers. Phase shift measurement is elaborated in Fig. 2.51.

*Chapter 3*

# SOLID-STATE CLINIC

*There are two aspects of interest to every professional, whatever his field - making money and enjoying the work he must do to earn that money. Those of us in electronic servicing will not find it difficult to do both, provided they have an open mind and the desire to learn. The introduction of solid-state devices has made additional study necessary if technicians are to keep abreast of new developments. This chapter presents a simple and logical explanation of how to service solid-state devices and circuits. By following certain definite and straight forward procedures, the technician will be able to save himself much servicing time and in the servicing business, time saved is money earned.*

## 3.1 The Junction Diode

It takes only a small voltage to give the charge carriers in a junction diode *sufficient energy to cross the junction*, and once they start moving, a constant *exchange* of charge carriers takes place. As electrons from the n-region move into the p-region, holes from the p-region cross over to the n-region. Electrons move through the diode and back into the battery. An ammeter placed in the circuit will indicate that the current is flowing.

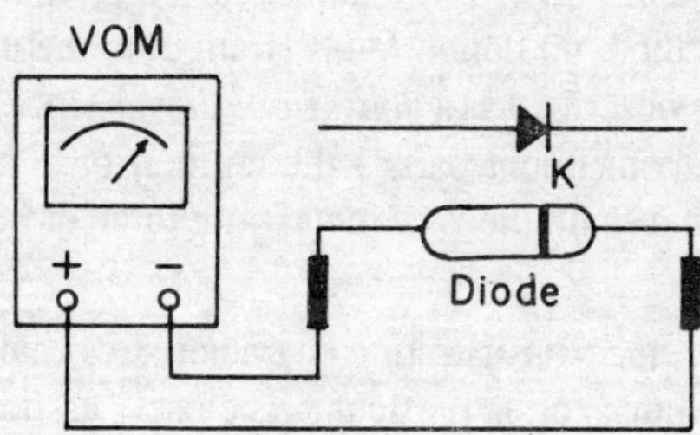


**Fig. 3.1** Forward current-low resistance-in a junction diode

If the same diode is now connected to the battery, but with a *reverse polarity*, electrons from the n-region and holes from the p-region *move away from the junction*, being attracted by the charges at the terminals of the battery. The battery current is now blocked and current flow ceases through the ammeter. However, a very minute *leakage current* does exist because of heat or moisture.

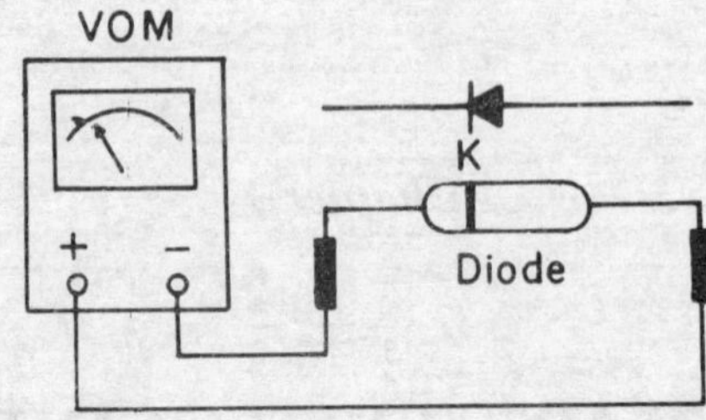


**Fig. 3.2** Reverse leakage current-high resistance-in a junction diode

Instead of a battery, an ohmmeter switched to appropriate ohms range would show the same results but instead of reading current it will show a *low forward resistance* in the first case, and a *high reverse resistance* in the second case. This is so, because the ohmmeter scale, is marked in ohms.

A diode can thus be made a low resistance, or a *relatively* high resistance, depending on the polarity of the ohmmeter battery. This forms the *basis* of all tests on solid-state devices. It is simple, easy to perform and hardly takes any time. For this the technician does not have to know all the complicated theory involved in the working of a junction diode. All that he needs to know is the *polarity of the ohmmeter leads.*

## 3.2 The Transistor

Since the invention of the transistor, thousands of different types of transistors have been introduced over the past two decades. A lot of transistors with quite different numbers are in fact very similar, if not virtually identical, to one another. In addition to the basic category of ordinary transistors, there are many other classes of semiconductors which are related to, or descended from, the original device as first invented. These other solid-state devices, which together serve a wide variety of purposes, include zener diodes, silicon rectifiers, silicon-controlled rectifiers (SCRs or thyristors), silicon unijunction transistors (UJTs), field effect transistors (FETs), varactors and ordinary diodes.

The operation of most of the electronic devices can be better understood by physically breaking down the device into its constituent parts. For example, each component of a vacuum tube can be removed and inspected. This method, no doubt, lends strength to the understanding of vacuum tubes. On this basis, the transistor can be described as a *mysterious device* since no useful information is gained by breaking it down to its three constituent parts - the emitter, base and collector. The observation requires a microscope. Even then a small piece of metal with three or more contacts fastened to it is all that will be observed.

The transistor is *heat sensitive*, since its operating characteristics change as its operating temperature changes. The design of a power transistor, in particular, is closely associated with the amount of power that must be dissipated. The envelope of the power transistor is constructed of a good heat conductor, such as aluminium, copper, and/or combinations of *heat conducting metals*. In a typical power transistor, Fig. 3.3, the case is formed of copper or copper alloy and the collector of the junction is mounted directly to this case. This direct contact between collector and case provides a maximum of *heat transfer* from junction to case.

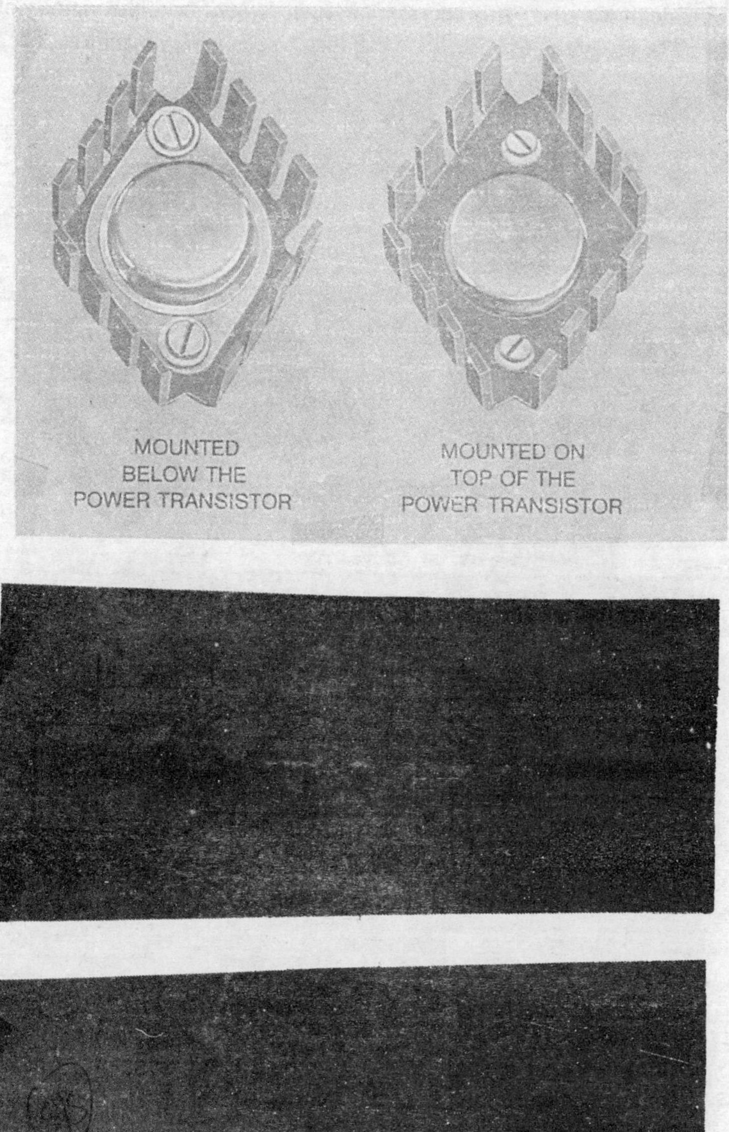


**Fig. 3.3** Heat sink for power transistor TO-3 case -6 W dissipation. For mounting below the power transistor - 4 holes of 4 mm ϕ; for mounting on top of the power transistor - 2 holes of 1 mm ϕ and 1 hole of 23 mm ϕ

One of the parameters encountered in transistor specification is *thermal resistance ($R_T$)*. This is a measure of how well the heat is conducted away from the junction. The unit of thermal resistance is °C/Watt.

The *power rating* of a transistor is directly related to the thermal resistance between the transistor junction and the outside of the case. It is necessary to *remove heat* from the vicinity of the junction as quickly as possible. The temperature of the case can then be *controlled* by external heat sinks.

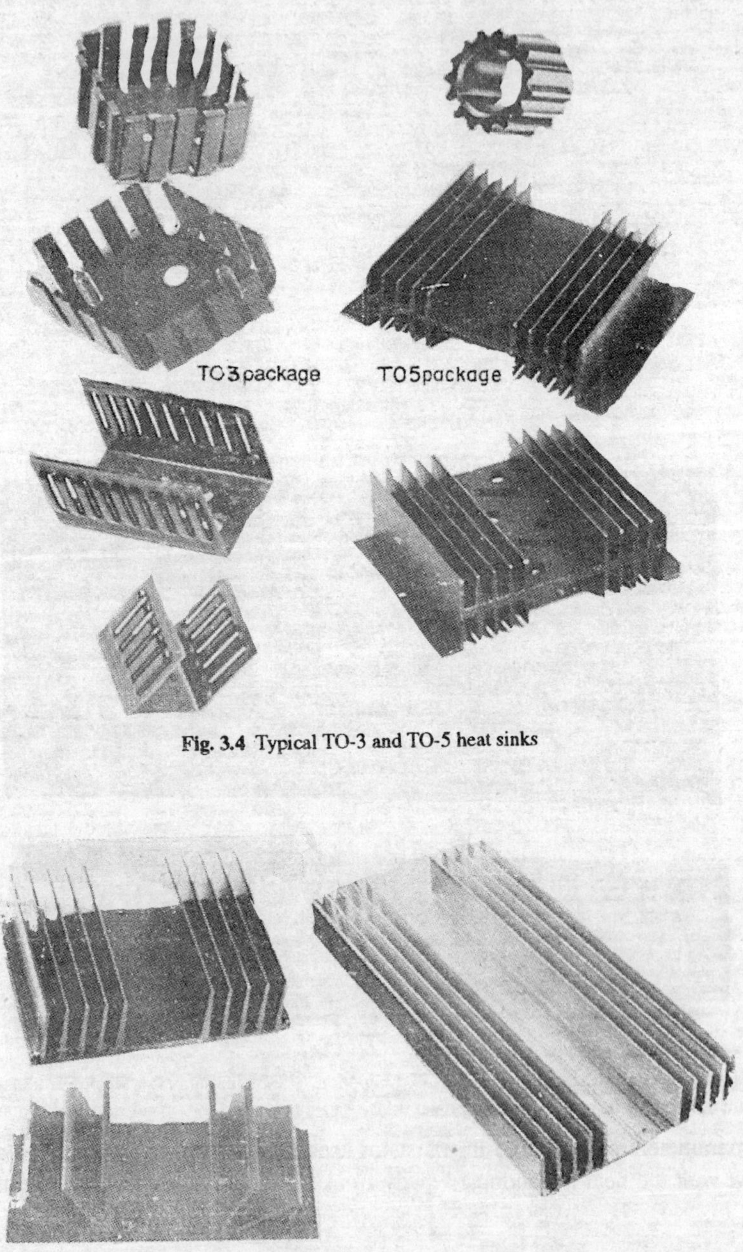


**Fig. 3.4** Typical TO-3 and TO-5 heat sinks

**Fig. 3.5** Heavy duty heat sinks

*The total power that a transistor can safely handle depends upon the efficiency of the heat sink.* When a transistor is operated without a heat sink, the power input must be reduced to protect the transistor. High power output requires that large amounts of heat be dissipated into some form of heat sink. A few of the arrangements for increasing the heat dissipation of power transistors are shown in Figs. 3.3 through Fig. 3.5.

## 3.5 Transistor Pinouts

A few of the transistors have the base, emitter, and collector terminals identified with B, E and C on the case next to the terminals. The case of the *power transistor* has been standardised by most manufacturers. The metal case is the collector terminal and the metal pins are the emitter and base terminals.

General standards have been adopted for the most-often used types. These types fall into three groups and can easily be recognised when they are encountered.

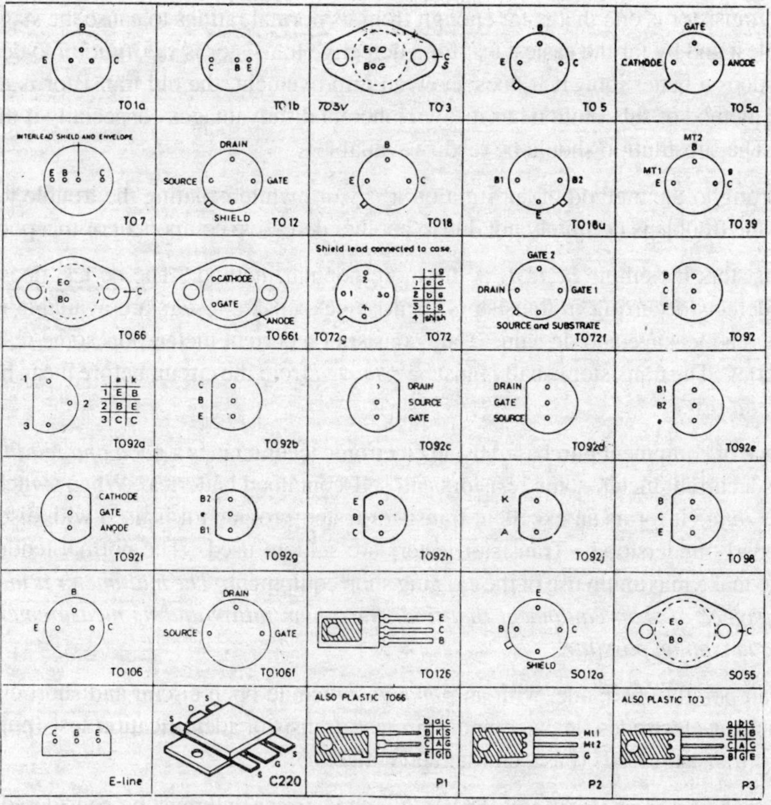


**Fig. 3.6** Transistor pinouts (all viewed from below)

The first group consists of those with a *colored dot* on the side. This dot is usually red, or blue, and identifies the *collector terminal* in the case of pnp transistors and *emitter terminal* in the case of npn transistors. The base terminal is in the centre.

In the second group, the three terminals are *in a line* and there is a *wide spacing* between the collector and base. This is in confirmity with the *physical construction of the transistor* wherein the distance between the collector and base is more than that between base and emitter.

The third group has a *triangular arrangement*. The *base* is in the centre for AF transistors, whereas the *emitter* is in the centre for RF transistors. Transistor pinouts, as viewed from below, are elaborated in Fig. 3.6.

## 3.4 Testing Transistors

There are different types of transistor tests. Transistor tests can be either static or dynamic. *Static tests* on transistors are strictly dc tests, which measure the ability of the transistor to respond to a simple battery current. *Dynamic tests* are made under actual operating conditions, where in ac signals are fed and the results measured.

A defective transistor is one that is far enough from its normal ratings to cause the stage to maltunction. The simplest and by far the easiest test for a defective transistor is *substitution* to determine if the new transistor does a better job. If it does show an improvement, the old transistor is assumed to be defective. The method of substitution has its own inherent disadvantages - especially if the transistor is hard to reach. The substitution should be readily available.

Before resorting to the method of substitution it is worthwhile isolating the trouble to a stage(s) to make sure that the trouble is definitely not due to another defective component or to a poor connection.

Testing transistors by simple *dc tests* is the more popular method. The dc test does catch a high percentage of defects occurring in transistors. Fairly inexpensive testers are available in the market. Most dc testers read leakage and dc gain. They consist of a current meter, plus some resistor and self-contained batteries. The transistor usually must be *removed* from the circuit before it can be tested. This is an *out-of-circuit test*.

The first piece of equipment purchased by an electronic technician is a *good and reliable multimeter*. It also contains a current meter, some resistors and self-contained batteries. When switched to the *appropriate ohms range* it forms an excellent transistor tester, provided it is used with discretion and its limitations properly understood. Transistor testers are seldom used. The normal tendency amongst technicians is to make maximum use of the existing shop equipment. *The multimeter is universally used and is the only piece of shop equipment that finds maximum utility and its measurement potential is exploited to the maximum possible*.

Tests that can possibly be made, with an *ohmmeter* include open circuit and short circuit tests between the transistor electrodes, leakage and gain test, transistor identification test (pnp or npn) and identification of transistor leads (base, emitter and collector).

On the lower ohms ranges (RX1 and RX10), it is the current that must be considered. The current on the (RX1) range, for example, is in excess of 100 milliamperes. These ranges should not be used for checking RF and IF transistors.

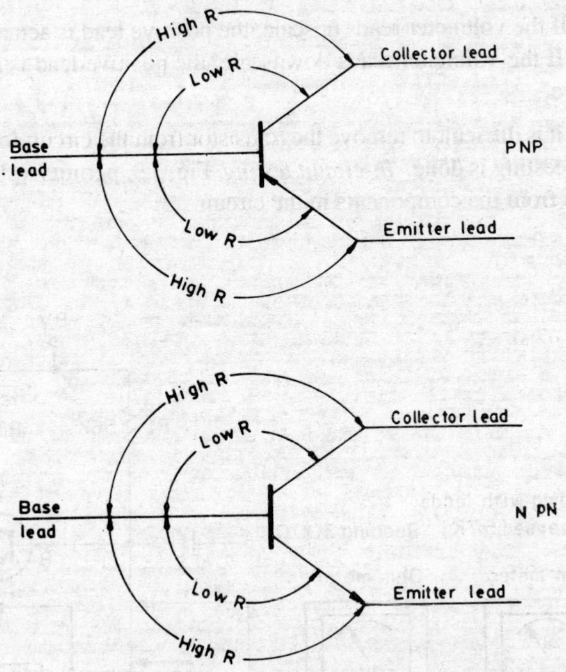


**Fig. 3.7** **Transistor resistor checks (out-of-circuit check)**

On the higher ohms ranges (RX1000 and RX10,000), it is the *battery voltage* that must be considered. Voltages above its maximum voltage rating could immediately damage the transistor. Transistor resistance check is illustrated in Fig. 3.7. The safest ranges for use in transistor testing are the (RX1000 or RX100) range for IF and RF transistors and (RX100 or in some cases RX10 range) for power transistors.

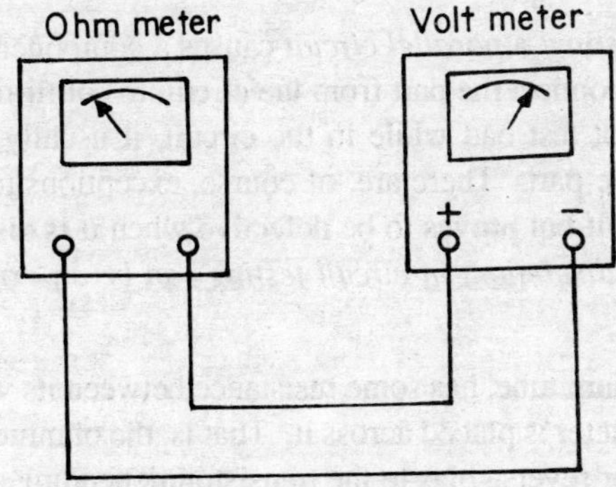


**Fig. 3.8** Checking the ohmmeter lead polarity

Before performing the above tests, *the ohmmeter lead polarity should first be confirmed.* This, in fact, should be done at the time of purchasing the meter. Connect the ohmmeter to a voltmeter, as illustrated in Fig. 3.8. If the voltmeter reads upscale, the positive lead is actually positive and the common lead is negative. If the voltmeter reads downscale, the positive lead is actually negative and the common lead is positive.

In certain instances, it is difficult to remove the transistor from the circuit for testing. In these instances in-circuit transistor testing is done. *In-circuit testing,* Fig. 3.9, permits testing the transistor without disconnecting the leads from the components in the circuit.

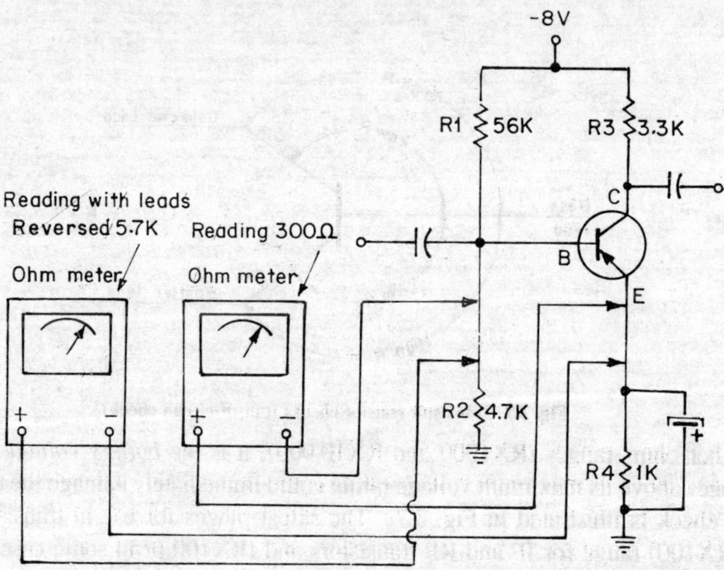


**Fig. 3.9** In-circuit testing of transistors

Quite often, with in-circuit testing, a *parallel circuit* causes a component to *appear to be defective*; then it becomes necessary to disconnect the part from the circuit to confirm if it is *really defective*. On the other hand, if a part does not test bad while in the circuit, it usually is not faulty and it can be eliminated as a possible defective part. There are, of course, exceptions to this rule. A defective part sometimes tests good in the circuit but proves to be defective when it is disconnected from the circuit. Nevertheless, *if used with some discretion, in-circuit testing can be an extremely valuable diagnostic tool.*

The transistor, unlike the vacuum tube, has some resistance between its various elements. Therefore it is *circuit-active* when an ohmmeter is placed across it. That is, the ohmmeter, with its in-built voltage supply, provides either forward or reverse bias to the transistor, depending on the polarities of the transistor and meter leads.

Therefore, *in-circuit tests must be made in two directions* with the test leads connected in one polarity and then in the reverse polarity. The *difference* in the two readings indicates that the transistor provides *diode action* and is neither shorted nor open between the base and emitter. The same diode action should be obtained between the base and collector. To the ohmmeter, a transistor appears similar to the circuit in Fig. 3.10 - *two diodes facing each other or back-to-back*. Also, there seems to be a *third diode* in series with a resistor between the emitter and the collector.

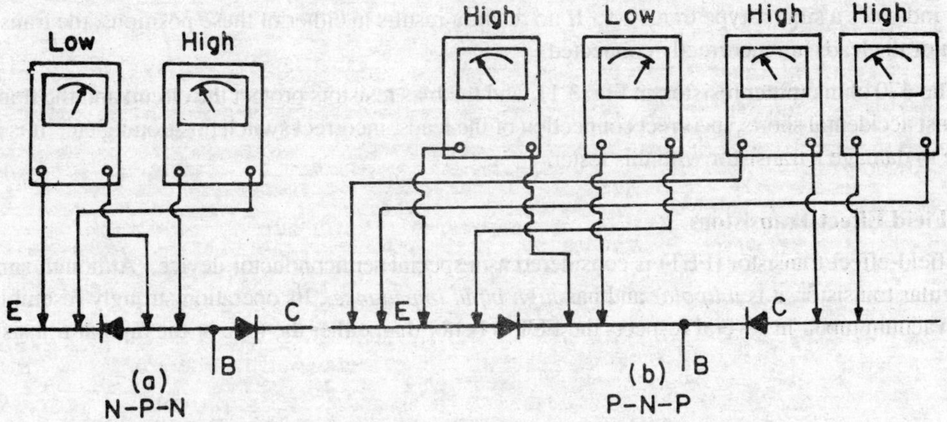


**Fig. 3.10** Equivalent circuit of a transistor when checked with an ohmmeter

The circuit resistance is usually high compared with the low, or forward resistance of the transistor. Therefore, an in-circuit ohmmeter test for the presence of transistor diode action is a good preliminary test for quickly finding out an open or shorted transistor. For example, if you observe that the resistance measurement between two transistor elements is high in both directions, an *open transistor* is suspected, and if it is low in both directions, a *shorted transistor* is suspected. This test is not valid in circuits in which the external resistance between transistor elements is extremely low.

A simple circuit of a transistor tester is shown in Fig. 3.11. This tester can be used to identify germanium and silicon transistors, to indicate shorts, opens, dc gain and excessive leakage, and to identify a pnp or an npn transistor.

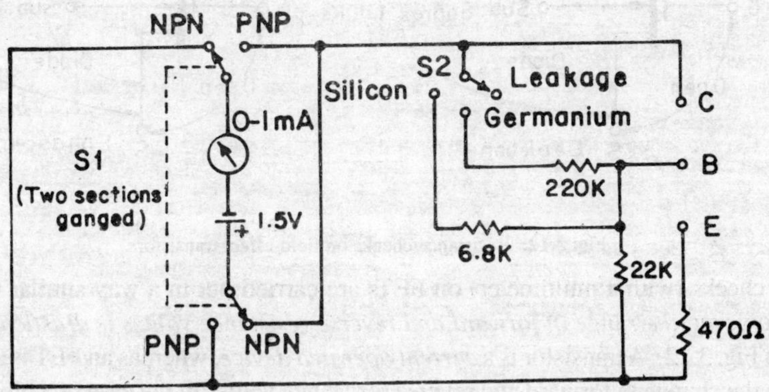

In using the tester, merely connect the leads to the transistor, and move switch S1 to pnp or npn position. Little, or preferably no, meter reading should be registered in the *LEAKAGE* position of the function switch S2. Then move switch to *GERMANIUM*, a reading on the meter indicates germanium type transistor. If there is no reading on the meter, move the switch to *SILICON*, a reading on the meter then indicates a silicon-type transistor. If no reading results in either of these positions, the transistor is *open* or the leads are incorrectly connected.

The 470 ohm emitter resistor, in Fig. 3.11, and the bias resistors protect the circuit and the transistors against accidental shorts, incorrect connection of the leads, incorrect switch positioning etc. It is impossible to damage a transistor with this tester.

## 3.5 Field Effect Transistors

The field-effect transistor (FET) is considered as a special semiconductor device. Although similar to a regular transistor, it is *unipolar* and has *high input impedance*. Its operation strongly resembles that of a vacuum tube. In several respects the FET is better than either the tube or the bipolar transistor.

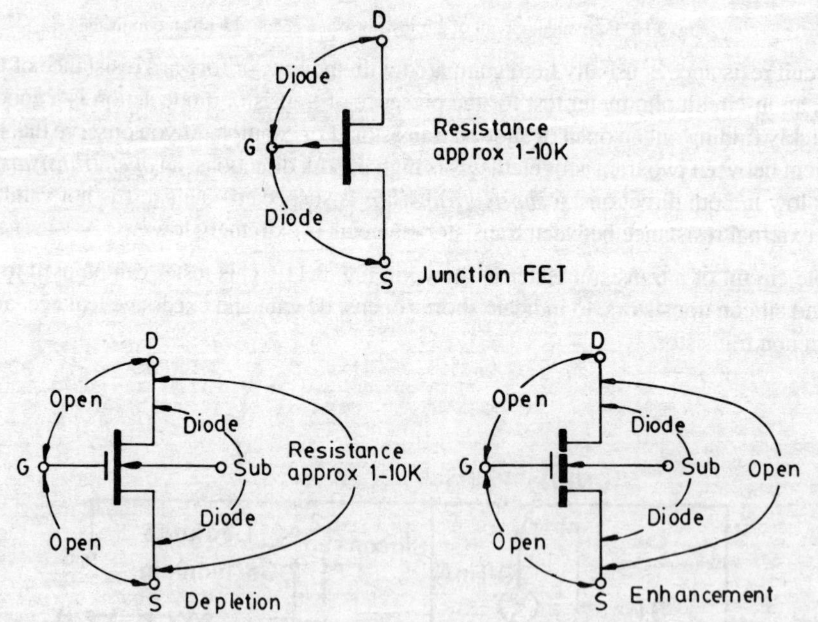


**Fig. 3.12:** Resistance checks on field-effect transistors.

Resistance checks (with a multimeter) on FETs are carried out in a way similar to that for bipolar transistors. However, *the range of forward and reverse resistance values is drastically different*. This is illustrated in Fig. 3.12. A transistor is a *current operated device*, whereas an FET is a voltage *operated device*. Hence the change in forward and reverse resistance values in the case of a bipolar transistor and that in the case of a unipolar FET is quite obvious.

All the correct procedures for handling bipolar transistors are equally valid for the FET. Certain special precautions, however, apply especially to the FET, since the FET is more easily damaged than is the bipolar transistor. The FET is also more susceptible to improper operation. These precautions are discussed in Chapter 4, 'Do's and Don'ts'.

### 3.6 Integrated Circuits

Integrated circuits are electronic circuits that consist of many individual circuit elements, such as transistors, diodes, resistors, capacitors, inductors and other *active and passive semiconductor devices* formed on a single chip of semiconducting material and mounted on a single piece of substrate material. The components in an IC are so integrated as to form a *functioning electronic circuit.*

While testing ICs, the first step is to *isolate* the fault to an IC stage by *signal checks* by feeding the required signal at the input pin and measuring at the output pin. The required signal can be obtained from a signal generator. The output signal may be checked by an oscilloscope.

If an oscilloscope is not available, the signal supposed to be present at the output pin, may be taken from a signal generator and injected at the output pin to make sure if the stage(s) *beyond* the IC are working. If the signal is passing from the output pin of the IC, but does not pass from the input pin of the IC, the IC stage is obviously defective. The fault might be in the IC or in the external components associated with the IC.

An easy method of testing ICs is by testing the voltage on IC pins. Standard voltages at different IC pins are normally given in the schematic of the equipment you are working on. This information is also supplied by IC manufactures in the form of application information whenever a new IC is inducted into service. Data manuals (priced publications) of different IC manufacturers are commercially marketed and these include comprehensive information about the various applications in which a particular IC can be put to service. *The voltages actually measured on different IC pins are compared with the standard voltages on its corresponding pins and discrepancies, if any, noted. Next, the components external to the suspected IC pin* are checked. If these components are found OK, the IC is removed from the circuit.

To confirm that the IC is really the culprit , it is given a *further resistance check*. The forward and reverse resistance between the different IC pins and the common or ground pin of the IC are measured and compared with the resistance check carried out on another good IC of the same make and type. Discrepancy between the two resistance checks further confirms that the IC under observation is the real culprit. It is then replaced. *The results of the resistance check are tabulated and preserved for future reference.*

ICs are either *directly soldered* into the circuit, or *fixed in IC sockets*, Fig. 3.13, which are soldered into the circuit. In the first case removal of the IC from the circuit demands a careful *desoldering operation.* A desoldering pump is most suitable for the job. In case a desoldering pump is not readily available, a small brush can be used for removing the melted solder from the IC, during desoldering operation, to make it easier and convenient to remove the IC. Both the methods work equally well.

If the IC is fixed in a socket, removal of the IC from its socket demands a bit of care, patience and, of course, some common sense. Lift the IC from its base *equally from both the sides* with the help of a poker, or the tip of a small screw driver, or a tweezer and then remove it straight off.

**Fig. 3.13:** IC Sockets-solder-tail design to allow through-hole board patterns.

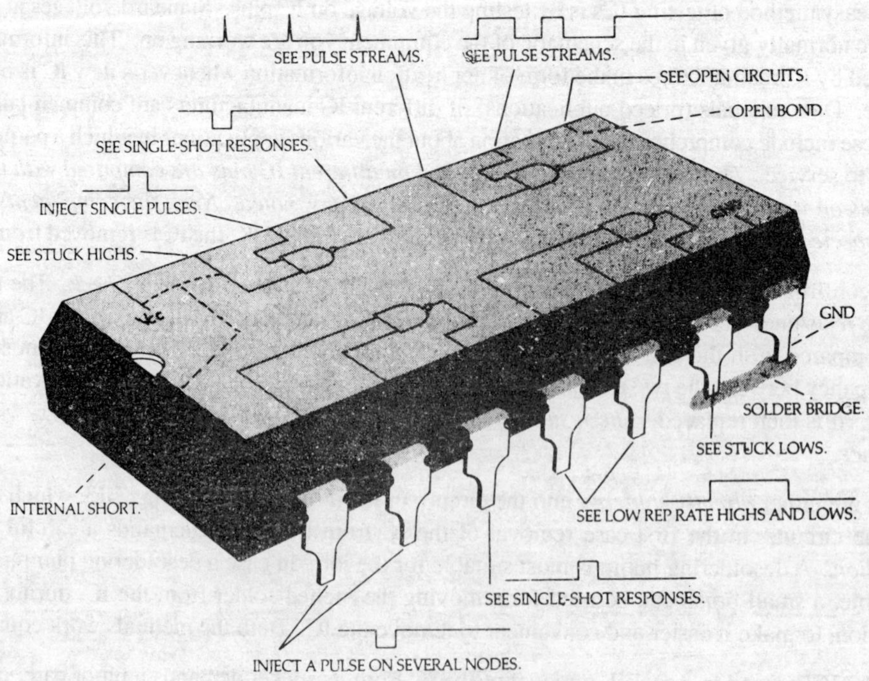


**Fig. 3.14:** Hewlett-Packard IC trouble shooters.

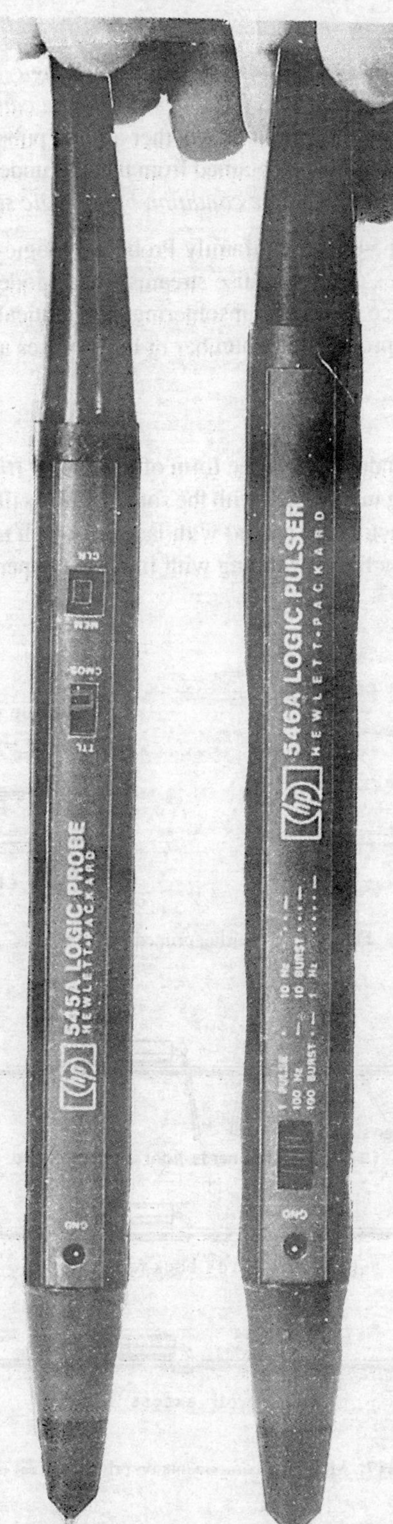


**Fig. 3.15:** The Hewlett-Packard Model 545A/546A Logic Probe/Pulser.

### 3.7 Logic Probes

Logic probes are IC troubleshooters that *indicate the state of logic levels* in TTL, DTL, or any other system with thresholds between 0.7 and 2.15 volts. *A strobe input can be used to detect the coincidence of logic signals at two points.* An indication of whether a logic pulse has, or has not, occurred can be obtained in a 'store' mode. Power may be obtained from the unit under test or any other suitable supply. *Two bright lights in the probe tip indicate the condition of the logic signal.*

The Hewlett-Packard Model 545 A Multifamily Probe sees logic highs and lows, bad levels, open circuits, captures one-shot pulses and sees pulse streams. The Model 546 A Pulser injects a pulse, or pulses, into a circuit without trace cutting or unsoldering, automatically drives low nodes high, or high nodes low, creates the required programmed number of logic pulses and has multi family capability.

### 3.8 Printed Circuits

If a *break* in a printed circuit conductor is in the form of a *hair line (thin) crack* across the conductor, it can usually be repaired by filling in the crack with the solder. This is illustrated in Fig. 3.16(a). On some printed circuit boards the conductors are coated with lacquer which must be removed before the crack is soldered. This can be done by lightly rubbing with fine sand paper or by light scraping with a knife blade.

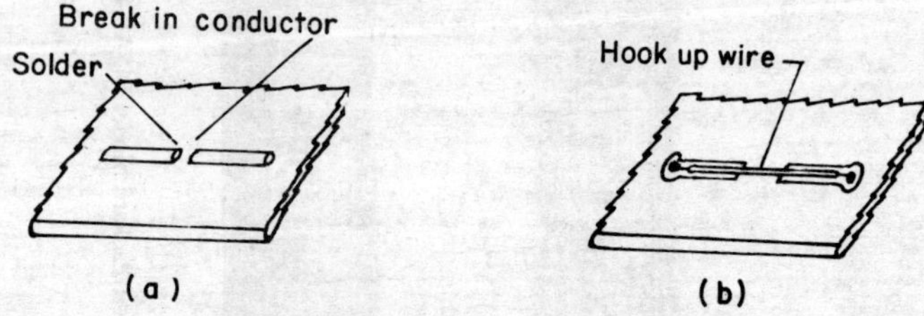


**Fig. 3.16:** Repairing printed circuit boards.

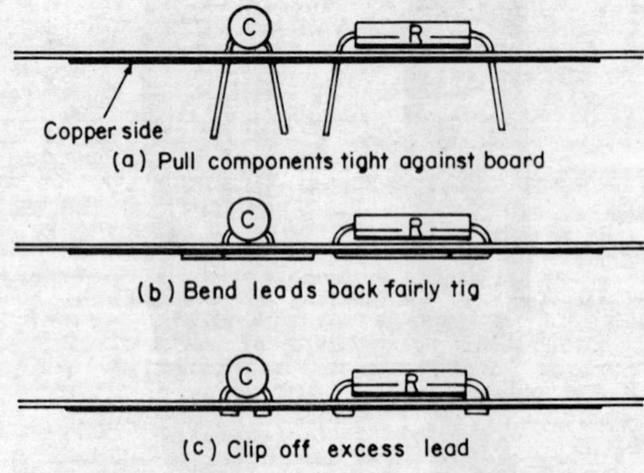


**Fig. 3.17:** Mounting components on printed circuit boards.

A *broken* printed circuit can be repaired by *bridging over the break* with hook-up wire, which is connected to the conductor terminal points as shown in Fig. 3.16(b). This method of repairing is recommended when a piece of a conductor has been destroyed, or badly damaged.

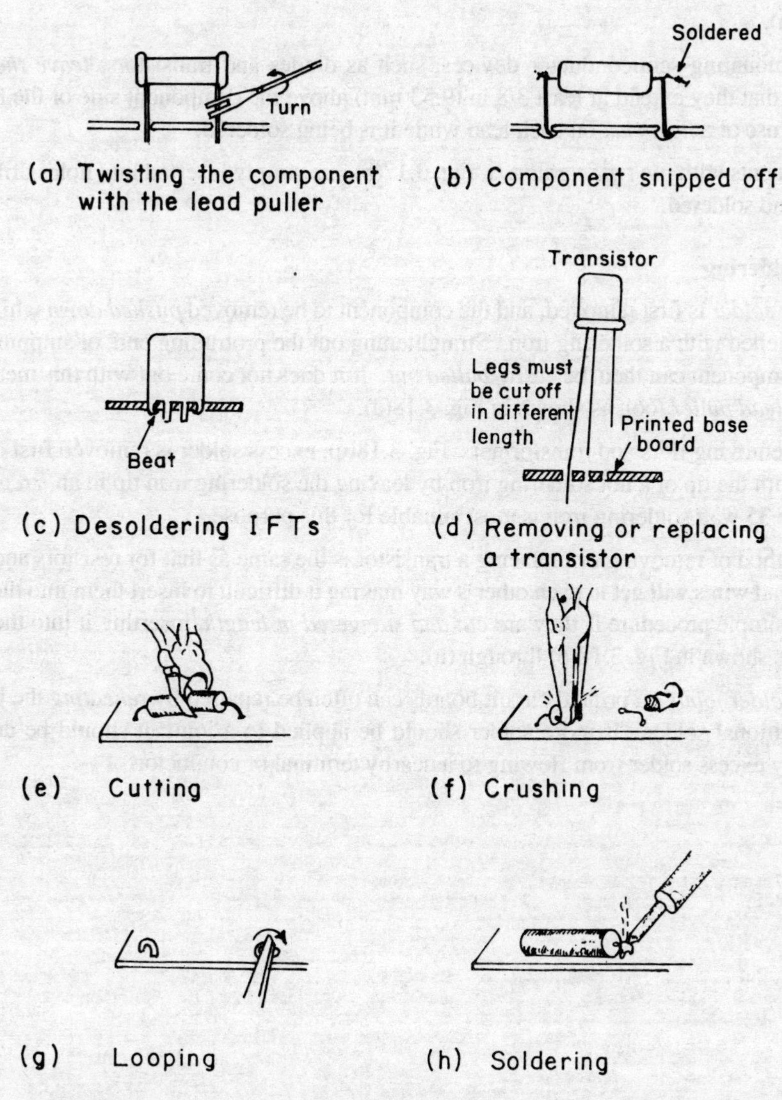


**Fig. 3.18:** Desoldering.

## 3.9 Mounting of Components on Printed Circuit Boards

When mounting small components, first insert their *full length leads* through the copper mounting holes. The components should be mounted in such a way that any *information* written on them can be easily read. This would be of help when a component needs to be replaced. *Cut the leads* so that they extend approximately 1/8 in (3.18 mm) beyond the wiring side of the board. *Bend the leads* at right angles so that a larger surface of each end is in contact with the terminal point surface to which it is to be soldered, Fig. 3.17(a).

When mounting semiconductor devices, such as diodes and transistors, *leave their leads longer enough* so that they extend at least 3/8 in (9.53 mm) above the component side of the board. This will enable the use of a *heat sink* on each lead while it is being soldered.

Components with *pin type terminals*, Fig. 3.17(b), are simply inserted into holes drilled in the board for them and soldered.

## 3.10 Desoldering

The *excess solder* is first removed, and the component to be removed *pushed down* while the remaining solder is melted with a soldering iron. Straightening out the protruding end, or snipping off the calked end, the component can then be easily *pulled out*. If it does not come out with this method it is twisted out with a *lead puller* tool, as shown in Fig. 3.18(a).

While removing IFTs and transformers, Fig. 3.18(c), excess solder is removed first and the bent end pried up with the tip of a hot soldering iron by leaving the soldering iron tip in an *irregular manner*. A pencil type 35 watt soldering iron is most suitable for this purpose.

The method of removing or replacing a transistor is the same as that for resistors and capacitors, but the three lead wires will get in each other's way making it difficult to insert them into their holes. It will become a simple procedure if they are *cut and staggered in length*, inserting it into the holes from the long one as shown in Fig. 3.18(e) through (h).

*Loose solder joints* on printed circuit boards can often be repaired by *reheating* the joint without the use of additional solder. If more solder should be applied to a joint, it should be done carefully to prevent any excess solder from flowing to a nearby terminal or conductor.

*Chapter 4*

# DO'S AND DON'TS

---

*There is a code of ethics in every discipline-it applies equally well to electronics servicing. In this, chapter we discuss, what the technician should liberally do-the DO'S, and what the technician should religiously avoid-the DON'TS. We learn a lot from others experience also. There are some tricks of the trade which have been passed on to us. These are also discussed. After sustained practice, they become a matter of routine, save a lot of valuable time and avoidable headache for the technician. The end result is optimum performance-the devices breath and the equipment ticks-i.e. the devices and circuits work the way they should.*

---

## 4.1 Essentials of Good Soldering

1. Use the *correct type of iron* for the job in hand, the correct type of soldering materials and ancillary tools.
2. *A miniature soldering iron*, with its low thermal capacity, must be used for all repairs carried out on miniaturised equipment.
3. The electric iron must be *operated at its rated voltage.*
4. A *thermal shunt* should be used to avoid transfer of heat from the soldering iron to nearby components.
5. *Keep* the soldering bit *in closest contact with* the heating element.
6. *Check* the lead of the soldering iron *for intermittent breaks periodically.*
7. Keep a bright, clean, *well trimmed face* on the bit, preferably tinning only one face.
8. A *new surface* of the bit must be *prepared* whenever the old *surface* becomes *pitted.*
9. The bit should be discarded when it is *reduced to half its original length.*
10. *Do not use a file on a plated copper bit.*
11. *Tin the bit* by melting rosin-cored solder on the working face of the iron *as it heats up.*
12. The *distribution of solder* may be assisted by rubbing the hot bit on a tin plate wiping pad. Remove excess solder from the tinned face.
13. Do not allow the iron to reach *full heat* before trying to tin.

14. Switch off the iron whenever a lengthy break occurs in soldering operations. If this is not practical, use a *heatsink to reduce the bit temperature*.
15. *Rub the working face* of the bit with a *circular motion* on a tin plate wiping pad-a piece of tin plate on which a little rosin-cored solder has been melted.
16. Use an alloy of *60 parts tin to 40 parts lead* for soldering in electronic equipments.
17. *Disturb original wiring runs as little as possible.*
18. Keep your work place *clean* and the tools and equipment *laid out neatly* on the work bench.
19. The *power lead* for the soldering iron should be *long enough for easy manipulation* of the iron but not so long that it may be tripped over.

## 4.2 Precautions In Using Multimeters

1. *Never connect an ohm meter to a circuit in which voltage is applied.* Always remove the line cord/batteries when testing equipment with a multimeter set to read ohms.
2. Use the *lower resistance side* of the ohmmeter scale for more accuracy.
3. Use the *lowest range* (RX1) when checking for *continuity and short circuit.*
4. Use the *highest range* (RX 10,000) when checking for *open circuit.*
5. Always *zero-adjust the equipment* before measuring resistance. This should also be done when switching over different ohms ranges.
6. *Never touch the ohmmeter leads when testing a high resistance* (RX 100 scale and above). Your body resistance, coming in parallel with the meter, will lower the actual reading.
7. *Rubbing the test probes back and forth* on the wire with which you want to make contact usually *results in a better contact.*
8. If an unknown voltage/current is to be measured, *start with the highest range and go on reducing one range at a time till a respectable reading can be observed.* When switching over ranges switch off the circuit/equipment under observation.

## 4.3 Transistor Handling Instructions

Careless handling, installation and operation of semiconductors can cause damage to them or can be responsible for their poor performance. Here•are some guidelines when working with semiconductors.

1. Avoid *rough handling*-dropping, hammering or forcibly squeezing-of semiconductors.
2. Always *instal* the semiconductor *last* in a circuit.
3. While dismantling a circuit, always *remove* the semiconductor *first.*
4. For soldering/desoldering operations *use a good quality 15 W soldering iron.*
5. Use *good quality solder* along with the flux.
6. *Avoid soldering/desoldering while the equipment is on.*
7. *Use a three pin plug. Earth the tip of the soldering iron.*
8. From time to time *check the soldering iron for leakage.*

9. Thoroughly clean the printed circuit board (PCB) with a *cleansing solution* after removing any semiconductor from the circuit.
10. The terminals on a PCB should be *thoroughly tinned* before mounting a semiconductor.
11. Use a *suitable heatsink* when soldering semiconductors into the circuit.
12. While inserting an IC into a socket *straighten its terminals* before hand.
13. Employ the *exact values of components* specified in the circuit diagram.
14. *The alignment of the circuit should not be disturbed.*
15. When replacing fuses always *maintain their original rating.*
16. Provide *adequate sleeving* for the leads, and *sufficient spacing* between the leads of semiconductors.
17. *Observe supply polarity.* Make sure that supply polarity is not reversed.
18. *Supply voltages should be as per specifications of the circuit.*
19. *Protect* the semiconductors from *excessive heat*, whether internal or external.
20. When a *heat sink* is specified, it must be used and should be according to specification.
21. *Protect semiconductors and their circuits from external magnetic fields.*

**4.4 FET Handling Instructions**

Here are some general instructions for handling and testing of field effect transistors.

1. *Avoid needlessly dropping an FET, or otherwise subjecting it to severe mechanical shock.*
2. *Don't operate an FET near hot components.*
3. *Avoid exceeding maximum voltage and current levels, even temporarily.*
4. Follow *standard transistor procedure* when installing an FET.
5. *Instal* an FET *last* in the circuit. Complete all other wiring before hand.
6. A MOSFET requires special handling. *Don't remove its protective, short circuiting device until all the MOSFET leads have been connected.*
7. *Ground your soldering iron to a good earth* connection.
8. *Never allow the gate of MOSFET to float.* It can easily pick up a static charge that will permanently damage the MOSFET. Connect a resistor of several megohms between gate and ground if necessary, to simulate an open circuit.
9. All FETs are susceptible to stray pick-up. *Keep any possible signal emitting component away from the FET.*
10. Keep all FET circuit input signal leads *as short as possible.*
11. When a MOSFET has an *external substrate terminal*, connect this terminal to the source terminal, unless otherwise instructed.
12. *Observe the correct polarity of each electrode* voltage.

## Handling & Operating Considerations-COS/MOS ICs

BEL is having in its range a number of COS/MOS Integrated Circuits ranging from simple gates to programmable counters. Some of the important practices for handling and operating MOS integrated circuits that will guard against device damage and assure optimum preformance are presented below.

### Handling Considerations

The *input protection networks* incorporated in all BEL COS/MOS ICs are effective in a wide variety of device handling situations. However to be totally safe and to avoid damage to the devices during handling, testing or actual operation, the following should be adhered to:

1. *The leads of devices should be in contact with a conductive material, except when being tested or in actual operation*, to avoid build-up of static charge.
2. Soldering iron tips, metal parts of fixtures and tools, and other handling facilities should be *grounded.*
3. *Devices should not be inserted into or removed from circuits with the power ON* because transient voltages may cause permanent damage.
4. *Signals should not be applied to the inputs while the device power supply is OFF.*
5. All *unused input leads* must be connected to either $V_{ss}$ (ground) or $V_{DD}$ (device supply) whichever is appropriate for the logic circuit involved.
6. Dry weather (RH%) tends to multiply the accumulation of static charges on any surface. Conversely higher humidity levels tend to reduce the magnitude of the static voltage generated. In a *low humidity environment*, the handling precautions listed above have an added importance and should be adhered to without exceptions.

After COS/MOS devices have been mounted on circuit boards, proper *handling precautions* should still be observed. Until these sub-assembly are inserted into a complete system, the board is no more than an extension of the leads of the device mounted on the board. It is a good practice to *put conductive clips or conductive tape on the circuit board terminals to prevent static charges from being transmitted to the devices through the board wiring.*

All device leads can withstand exposure to temperatures as high as 65°C for as long as 10 seconds, and as close as $\frac{1}{16} \pm \frac{1}{32}$ inch from the body of the device.

### Operating Considerations

When operating near the maximum supply voltage range of 15 volts, *care should be taken to avoid or suppress power supply turn ON and turn OFF transients, supply ripple and ground noise*; any of the above conditions must not cause ($V_{DD}$-$V_{SS}$ ) to exceed the absolute maximum ratings. Another good power supply practice is to *use a zener protection diode suitably in parallel with power bus.*

All unused input leads must be connected to either $V_{SS}$ or $V_{DD}$ depending on the logic circuit involved. A *floating input* not only can result in faulty logic operation, but also can cause the maximum power dissipation to be exceeded and thus damage the device.

Signals shall not be applied to the inputs while the device power supply is OFF unless the input current is limited to a steady state value of typically less than 10mA. *Whenever the possibility of exceeding 10mA of input current exists, a resistor of about 10kΩ in series with the input is recommended.* Particular attention should be given to long input signal lines where high inductance can increase the likelihood of large signal pick-up in noisy environments.

All BEL COS/MOS devices have a *dc fan-out of 50.* The input capacitance for most of our types is 5 pF. Noise immunity is typically 30% of the supply voltage for the range from +3 to +15 volts for all COS/MOS devices.

# INDEX